METAMATERIALS

METAMATERIALS
Physics and Engineering Explorations

Edited by
NADER ENGHETA
RICHARD W. ZIOLKOWSKI

IEEE PRESS

A JOHN WILEY & SONS, INC., PUBLICATION

Copyright © 2006 by the Institute of Electrical and Electronics Engineers, Inc. All rights reserved.

Published by John Wiley & Sons, Inc.
Published simultaneously in Canada.

For general information on our other products and services or for technical support, please contact our Customer Care Department within the United States at (800) 762-2974, outside the United States at (317) 572-3993 or fax (317) 572-4002.

Wiley also publishes its books in a variety of electronic formats. Some content that appears in print may not be available in electronic formats. For more information about Wiley products, visit our web site at www.wiley.com.

Library of Congress Cataloging-in-Publication Data is available.

ISBN-13 978-0-471-76102-0
ISBN-10 0-471-76102-8

Printed in the United States of America.

10 9 8 7 6 5 4 3 2 1

To the pursuit of knowledge,
the thrill of discovery, and
the hope for their use
for the betterment of humankind

CONTENTS

CHAPTER 3   *WAVEGUIDE EXPERIMENTS TO CHARACTERIZE PROPERTIES OF SNG AND DNG METAMATERIALS*   87

             Silvio Hrabar

CHAPTER 4   *REFRACTION EXPERIMENTS IN WAVEGUIDE ENVIRONMENTS*   113

             Tomasz M. Grzegorczyk, Jin Au Kong, and Ran Lixin

SECTION II

TWO-DIMENSIONAL PLANAR NEGATIVE-INDEX STRUCTURES   141

PART II

ELECTROMAGNETIC BANDGAP (EBG) METAMATERIALS **211**

SECTION I

THREE-DIMENSIONAL VOLUMETRIC EBG MEDIA **213**

CHAPTER 8   **HISTORICAL PERSPECTIVE AND REVIEW OF FUNDAMENTAL PRINCIPLES IN MODELING THREE-DIMENSIONAL PERIODIC STRUCTURES WITH EMPHASIS ON VOLUMETRIC EBGs** **215**

Maria Kafesaki and Costas M. Soukoulis

CHAPTER 9   **FABRICATION, EXPERIMENTATION, AND APPLICATIONS OF EBG STRUCTURES** **239**

Peter de Maagt and Peter Huggard

SECTION II

TWO-DIMENSIONAL PLANAR EBG STRUCTURES **285**

CHAPTER 11   *REVIEW OF THEORY, FABRICATION, AND APPLICATIONS OF HIGH-IMPEDANCE GROUND PLANES* **287**

Dan Sievenpiper

PREFACE

While we were organizing, coordinating, and guest editing the October 2003 special issue of the *IEEE Transactions on Antennas and Propagation* on the topic of metamaterials, we began toying with the idea of editing a book on this topic with contributions from experts who are active in this area of research. The senior acquisitions editor of the IEEE Press, Cathy Faduska, was also interested in a book project on this timely topic, and during the 2002 IEEE Antennas and Propagation Society International Symposium in San Antonio, Texas, she suggested to us, and encouraged us, to begin this project. And finally, with a longer time dilation than expected, we have completed this book.

The amount of research in this metamaterials area has grown extremely quickly in this time frame. We have tried to capture, through the selected authors, both some interesting physics and engineering explorations in this area. Why this two-pronged approach? We note that physics asks how nature works, and engineering asks how the works of nature can be used. Thus, we wanted to include some of the metamaterial fundamentals and how they are already being applied.

What is a "metamaterial"? In recent years, there has been a growing interest in the fabricated structures and composite materials that either mimic known material responses or qualitatively have new, physically realizable response functions that do not occur or may not be readily available in nature. The unconventional response functions of these *metamaterials* are often generated by artificially fabricated inclusions or inhomogeneities embedded in a host medium or connected to or embedded on a host surface. Exotic properties for such metamaterials have been predicted; many experiments have confirmed our basic understanding of many of them. The underlying interest in metamaterials is the potential to have the ability to engineer the electromagnetic and optical properties of materials for a variety of applications. The impact of metamaterials may be enormous: If one can tailor and manipulate the wave properties, significant decreases in the size and weight of components, devices, and systems along with enhancements in their performance appear to be realizable.

The pursuit of artificial materials for electromagnetic applications is not new; this activity has a long history which dates back to Jadagis Chunder Bose in 1898 when he worked and experimented on the constructed twisted elements that exhibit properties nowadays known as chiral characteristics. In the early part of the twentieth century, Karl Ferdinand Lindman studied wave interaction with collections of metallic helices as artificial chiral media. Artificial dielectrics were explored, for example, in the 1950s and 1960s for lightweight microwave antenna lenses. Artificial chiral materials were investigated extensively in the 1980s and 1990s for microwave radar absorbers and other applications. The developments of electromagnetic bandgap (EBG) structured materials and single-negative (SNG) and double-negative (DNG) materials and their fascinating properties have driven the recent explosive interest in metamaterials.

We have divided this book into two major classes of metamaterials: the SNG and DNG metamaterials and the EBG structured metamaterials. The SNG and DNG metamaterials involve inclusions and interinclusion distances that are much smaller than a wavelength and, as a consequence, such media can be described by homogenization and effective media concepts. On the other hand, the EBG metamaterials involve distances that are on the order of half a wavelength or more and are described by the Bragg reflection and other periodic media concepts. We have furthered subdivided each of these classes into their three-dimensional (3D volumetric) and two-dimensional (2D planar or surface) realizations. Examples of these types of metamaterials are presented, and their known and anticipated properties are reviewed in this book.

This book begins with DNG metamaterial concepts, simulations, and experiments in Chapters 1 to 6. In Chapter 1 we present a brief recapitulation of the history of artificial materials and metamaterials and their exotic properties, including negative indices of refraction, negative angles of refraction, and focusing using planar slabs. This is followed in Chapter 2 with theoretical and numerical studies of SNG and DNG metamaterials and their particular applications to waveguiding environments and to antennas and is presented by us and our students, Andrea Alù and Aycan Erentok. Next in Chapter 3 Silvio Hrabar describes several waveguide experiments that have been used to characterize the properties of SNG and DNG metamaterials. Tomasz Grzegorczyk, Jin Au Kong, and Ran Lixin present in Chapter 4 their several experiments in waveguide environments to demonstrate the negative refraction properties of DNG metamaterials. In Chapter 5 George Eleftheriades discusses the realization of planar metamaterials and their demonstration of many of the exotic properties of DNG metamaterials, including evanescent wave growth and subwavelength focusing. The use of a planar metamaterial to realize resonance cone antennas is shown by Keith Balmain and Andrea Lüttgen in Chapter 6. Christophe Caloz and Tatsuo Itoh describe in Chapter 7 a variety of microwave coupler and resonator applications of negative-refractive-index planar structures. The book is then transitioned into a review of EBG metamaterial concepts, simulations, and experiments in Chapters 8 to 14. Maria Kafesaki and Costas Soukoulis provide a historical perspective and a review of the fundamental principles in modeling 3D periodic structures with an emphasis on volumetric EBGs in Chapter 8. Peter de Maagt and Peter Huggard describe in Chapter 9 the fabrication, experimentation, and applications of EBG structures. In Chapter 10 Boris Gralak, Stefan Enoch, Gérard Tayeb present their work on superprism effects and EBG antenna applications. Dan Sievenpiper provides in Chapter 11 a review of the theory, fabrication, and applications of high-impedance ground planes. In Chapter 12 Yahya Rahmat-Samii and Fan Yang discuss their development of complex artificial ground planes for antenna engineering. Stefano Maci and Alessio Cucini address frequency-selective EBG surfaces in Chapter 13. Finally, John McVay,

Nader Engheta, and Ahmad Hoorfar describe in Chapter 14 their application of space-filling curves to realize high-impedance ground planes.

In all chapters, the authors have presented recent research advances associated with a diverse set of metamaterials. As noted, the chapters include a combination of theoretical, numerical, and experimental contributions to the understanding of the behavior of metamaterials and to their potential applications in components, devices, and systems. We sincerely hope that the work presented provide the newcomer to metamaterial research with the ability to come up to speed with a basic understanding of metamaterials and their potential for a variety of applications. For the advanced metamaterial researcher, the material reviews the state-of-the-art as viewed by many seasoned veterans in this area. In both cases, the extensive reference lists should provide ample additional reading materials for further considerations.

We would like to thank Cathy Faduska, Anne Reifsnyder, Developmental Editor, and Lisa Van Horn, Production Editor, at IEEE Press for their efforts in interfacing between us, IEEE Press, and John Wiley & Sons. When problems arose, they provided excellent support. Most of all, we would like to thank all the contributing authors for their time and wonderful efforts. We believe that the outcome is an impressive resource for future efforts.

We sincerely hope that the materials presented here will stimulate discussions and new avenues of research in this very exciting research area of metamaterials. We note

> *Science never solves a problem without creating ten more.*
> *George Bernard Shaw (1856–1950)*

Have fun reading!

NADER ENGHETA
RICHARD W. ZIOLKOWSKI

Philadelphia, Pennsylvania
Tucson, Arizona
May 2006

CONTRIBUTORS

ANDREA ALÙ
Department of Electrical and Systems Engineering
University of Pennsylvania
Philadelphia, Pennsylvania

KEITH G. BALMAIN
Edward S. Rogers Sr. Department of Electrical and Computer Engineering
University of Toronto
Toronto, Ontario, Canada

CHRISTOPHE CALOZ
École Polytechnique de Montréal
Montreal, Quebec, Canada

ALESSIO CUCINI
Department of Information Engineering
University of Siena
Siena, Italy

PETER DE MAAGT
European Space Agency, Antenna and Submillimetre Wave Section, Electromagnetics and Space Environments Division, TEC-EE
AG Noordwijk, The Netherlands

GEORGE V. ELEFTHERIADES
Edward S. Rogers Sr. Department of Electrical and Computer Engineering
University of Toronto
Toronto, Ontario, Canada

NADER ENGHETA
Department of Electrical and Systems Engineering
University of Pennsylvania
Philadelphia, Pennsylvania

STEFAN ENOCH
Faculte de St Jerome
Institut Fresnel
Marseille, France

AYCAN ERENTOK
Department of Electrical and Computer Engineering
University of Arizona
Tucson, Arizona

BORIS GRALAK
Faculte de St Jerome
Institut Fresnel
Marseille, France

TOMASZ M. GRZEGORCZYK
Massachusetts Institute of Technology
Center for Theory and Applications, Research Laboratory of Electronics,
Cambridge, Massachusetts
and
Electromagnetics Academy at Zheijiang University
Zheijiang University
Hangzhou, China

PETER HUGGARD
Space and Science Technology Department
CCLRC Rutherford Appleton Laboratory
Chilton, Didcot, Oxfordshire, United Kingdom

AHMAD HOORFAR
Department of Electrical and Computer Engineering
Villanova University
Villanova, Pennsylvania

SILVIO HRABAR
Faculty of Electrical Engineering and Computing
Department of Radiocommunications and Microwave Engineering
University of Zagreb
Zagreb, Croatia

TATSUO ITOH
Department of Electrical Engineering
University of California Los Angeles (UCLA)
Los Angeles, California

MARIA KAFESAKI
Institute of Electronic Structure and Laser (IESL)
Foundation for Research and Technology Hellas (FORTH)
Heraklion, Crete, Greece
and
Department of Materials Science and Technology, University of Crete, Greece

JIN AU KONG
Massachusetts Institute of Technology
Center for Theory and Applications, Research Laboratory of Electronics
Cambridge, Massachusetts
and
Electromagnetics Academy at Zheijiang University, Zheijiang University,
Hangzhou, China

RAN LIXIN
Electromagnetics Academy at Zheijiang University
Zheijiang University, Hangzhou, China

ANDREA A. E. LÜTTGEN
Edward S. Rogers Sr. Department of Electrical and Computer Engineering
University of Toronto
Toronto, Ontario, Canada

STEFANO MACI
Department of Information Engineering
University of Siena
Siena, Italy

JOHN MCVAY
Department of Electrical and Computer Engineering
Villanova University
Villanova, Pennsylvania

YAHYA RAHMAT-SAMII
Department of Electrical Engineering
University of California Los Angeles (UCLA)
Los Angeles, California

DAN SIEVENPIPER
HRL Laboratories LLC
Malibu, California

COSTAS M. SOUKOULIS
Ames Laboratory–USDOE and Department of Physics and Astronomy
Iowa State University
Ames, Iowa

GÉRARD TAYEB
Faculte de St Jerome
Institut Fresnel
Marseille, France

FAN YANG
Department of Electrical Engineering
University of Mississippi
University, Mississippi

RICHARD W. ZIOLKOWSKI
Department of Electrical and Computer Engineering
University of Arizona
Tucson, Arizona

PART I

DOUBLE-NEGATIVE (DNG) METAMATERIALS

THREE-DIMENSIONAL VOLUMETRIC DNG METAMATERIALS

CHAPTER 1

INTRODUCTION, HISTORY, AND SELECTED TOPICS IN FUNDAMENTAL THEORIES OF METAMATERIALS

Richard W. Ziolkowski and Nader Engheta

1.1 INTRODUCTION

To the best of our knowledge, the first attempt to explore the concept of "artificial" materials appears to trace back to the late part of the nineteenth century when in 1898 Jagadis Chunder Bose conducted the first microwave experiment on twisted structures—geometries that were essentially artificial chiral elements by today's terminology [1]. In 1914, Lindman worked on "artificial" chiral media by embedding many randomly oriented small wire helices in a host medium [2]. In 1948, Kock [3] made lightweight microwave lenses by arranging conducting spheres, disks, and strips periodically and effectively tailoring the effective refractive index of the artificial media. Since then, artificial complex materials have been the subject of research for many investigators worldwide. In recent years new concepts in synthesis and novel fabrication techniques have allowed the construction of structures and composite materials that mimic known material responses or that qualitatively have new, physically realizable response functions that do not occur or may not be readily available in nature. These metamaterials can in principle be synthesized by embedding various constituents/inclusions with novel geometric shapes and forms in some host media (Fig. 1.1). Various types of electromagnetic composite media, such as double-negative (DNG) materials, chiral materials, omega media, wire media, bianisotropic media, linear and nonlinear media, and local and nonlocal media, to name a few, have been studied by various research groups worldwide.

As is well known, in particulate composite media, electromagnetic waves interact with the inclusions, inducing electric and magnetic moments, which in turn affect the macroscopic effective permittivity and permeability of the bulk composite "medium." Since metamaterials can be synthesized by embedding artificially fabricated inclusions in a specified host medium or on a host surface, this provides the designer with a large collection of independent parameters (or

Metamaterials: Physics and Engineering Explorations, Edited by N. Engheta and R. W. Ziolkowski

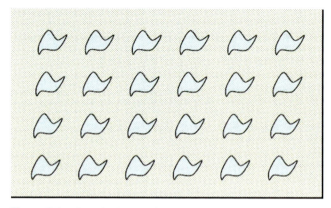

Figure 1.1 Generic sketch of a volumetric metamaterial synthesized by embedding various inclusions in a host medium.

degrees of freedom)—such as the properties of the host materials; the size, shape, and composition of the inclusions; and the density, arrangement, and alignment of these inclusions—to work with in order to *engineer* a metamaterial with specific electromagnetic response functions not found in each of the individual constituents. All of these design parameters can play a key role in the final outcome of the synthesis process. Among these, the geometry (or shape) of the inclusions is one that can provide a variety of new possibilities for metamaterials processing.

Recently, the idea of complex materials in which both the permittivity and the permeability possess negative real values at certain frequencies has received considerable attention. In 1967, Veselago theoretically investigated plane-wave propagation in a material whose permittivity and permeability were assumed to be simultaneously negative [4]. His theoretical study showed that for a monochromatic uniform plane wave in such a medium the direction of the Poynting vector is antiparallel to the direction of the phase velocity, contrary to the case of plane-wave propagation in conventional simple media. In recent years, Smith, Schultz, and their group constructed such a composite medium for the microwave regime and demonstrated experimentally the presence of anomalous refraction in this medium [5,6].

For metamaterials with negative permittivity and permeability, several names and terminologies have been suggested, such as "left-handed" media [4–10]; media with negative refractive index [4–7,9]; "backward-wave media" (BW media) [11]; and "double-negative (DNG)" metamaterials [12], to name a few. Many research groups all over the world are now studying various aspects of this class of metamaterials, and several ideas and suggestions for future applications of these materials have been proposed.

It is well known that the response of a system to the presence of an electromagnetic field is determined to a large extent by the properties of the materials involved. We describe these properties by defining the macroscopic parameters permittivity ε and permeability μ of these materials. This allows for the classification of a medium as follows. A medium with both permittivity

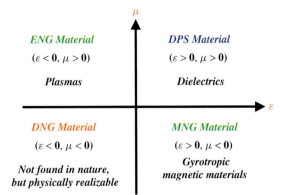

Figure 1.2 Material classifications.

and permeability greater than zero ($\varepsilon > 0$, $\mu > 0$) will be designated a double-positive (DPS) medium. Most naturally occurring media (e.g., dielectrics) fall under this designation. A medium with permittivity less than zero and permeability greater than zero ($\varepsilon < 0$, $\mu > 0$) will be designated an epsilon-negative (ENG) medium. In certain frequency regimes many plasmas exhibit this characteristic. For example, noble metals (e.g., silver, gold) behave in this manner in the infrared (IR) and visible frequency domains. A medium with the permittivity greater than zero and permeability less than zero ($\varepsilon > 0$, $\mu < 0$) will be designated a mu-negative (MNG) medium. In certain frequency regimes some gyrotropic materials exhibit this characteristic. Artificial materials have been constructed that also have DPS, ENG, and MNG properties. A medium with both the permittivity and permeability less than zero ($\varepsilon < 0$, $\mu < 0$) will be designated a DNG medium. To date, this class of materials has only been demonstrated with artificial constructs. This medium classification can be graphically illustrated as shown in Figure 1.2.

While one often describes a material by some constant (frequency-independent) value of the permittivity and permeability, in reality all material properties are frequency dependent. There are several material models that have been constructed to describe the frequency response of materials. Because the magnetic field of an electromagnetic wave is smaller than its electric field by the wave impedance of the medium in which it is propagating, one generally focuses attention on how the electron motion in the presence of the nucleus and, hence, the basic dipole moment of this system are changed by the electric field. Understanding this behavior leads to a model of the electric susceptibility of the medium and, hence, its permittivity. On the other hand, there are many media for which the magnetic field response is dominant. One can generally describe the magnetic response of a material in a fashion completely dual to that of the electric field using the magnetic susceptibility and, hence, its permeability. While the magnetic dipoles physically arise from moments associated with current loops, they can be described mathematically by magnetic charge and current analogs of the electric cases.

One of the most well-known material models is the *Lorentz* model. It is derived by a description of the electron motion in terms of a driven, damped harmonic oscillator. To simplify the discussion, we will assume that the charges are allowed to move in the same direction as the electric field. The Lorentz model then describes the temporal response of a component of the polarization field of the medium to the same component of the electric field as

$$\frac{d^2}{dt^2}P_i + \Gamma_L \frac{d}{dt}P_i + \omega_0^2 P_i = \varepsilon_0 \chi_L E_i \tag{1.1}$$

The first term on the left accounts for the acceleration of the charges, the second accounts for the damping mechanisms of the system with damping coefficient Γ_L, and the third accounts for the restoring forces with the characteristic frequency $f_0 = \omega_0/2\pi$. The driving term exhibits a coupling coefficient χ_L. The response in the frequency domain, assuming the engineering $\exp(+j\omega t)$ time dependence, is given by the expression

$$P_i(\omega) = \frac{\chi_L}{-\omega^2 + j\Gamma_L\omega + \omega_0^2}\varepsilon_0 E_i(\omega) \tag{1.2}$$

With small losses $\Gamma_L/\omega_0 \ll 1$ the response is clearly resonant at the natural frequency f_0. The polarization and electric fields are related to the electric susceptibility as

$$\chi_{e,\text{Lorentz}}(\omega) = \frac{P_i(\omega)}{\varepsilon_0 E_i(\omega)} = \frac{\chi_L}{-\omega^2 + j\Gamma_L\omega + \omega_0^2} \tag{1.3}$$

The permittivity is then obtained immediately as $\varepsilon_{\text{Lorentz}}(\omega) = \varepsilon_0[1 + \chi_{e,\text{Lorentz}}(\omega)]$.

There are several well-known special cases of the Lorentz model. When the acceleration term is small in comparison to the others, one obtains the *Debye* model:

$$\Gamma_d \frac{d}{dt}P_i + \omega_0^2 P_i = \varepsilon_0 \chi_d E_i \qquad \chi_{e,\text{Debye}}(\omega) = \frac{\chi_d}{j\Gamma_d\omega + \omega_0^2} \tag{1.4}$$

When the restoring force is negligible, one obtains the *Drude* model:

$$\frac{d^2}{dt^2}P_i + \Gamma_D \frac{d}{dt}P_i = \varepsilon_0 \chi_D E_i \qquad \chi_{e,\text{Drude}}(\omega) = \frac{\chi_D}{-\omega^2 + j\Gamma_D\omega} \tag{1.5}$$

where the coupling coefficient is generally represented by the plasma frequency $\chi_D = \omega_p^2$. In all of these models, the high-frequency limit reduces the permittivity to that of free space.

Assuming that the coupling coefficient is positive, then only the Lorentz and the Drude models can produce negative permittivities. Because the Lorentz model is resonant, the real part of the susceptibility and, hence, that of the permittivity become negative in a narrow frequency region immediately above the resonance. On the other hand, the Drude model can yield a negative real part of the permittivity over a wide spectral range, that is, for $\omega < \sqrt{\omega_p^2 - \Gamma_D^2}$.

Similar magnetic response models follow immediately. The corresponding magnetization field components M_i and the magnetic susceptibility χ_m equations are obtained from the polarization and electric susceptibility expressions with the replacements $E_i \rightarrow H_i$, $P_i/\varepsilon_0 \rightarrow M_i$. The permeability is given as $\mu(\omega) = \mu_0[1 + \chi_m(\omega)]$.

Metamaterials have necessitated the introduction of generalizations of these models. For instance, the most general second-order model that has been introduced for metamaterial studies is the two-time-derivative Lorentz metamaterial (2TDLM) model [13–15]:

$$\frac{d^2}{dt^2}P_i + \Gamma_L \frac{d}{dt}P_i + \omega_0^2 P_i = \varepsilon_0 \chi_\alpha \omega_p^2 E_i + \varepsilon_0 \chi_\beta \omega_p \frac{d}{dt}E_i + \varepsilon_0 \chi_\gamma \frac{d^2}{dt^2}E_i$$

$$\chi_{e,\text{2TDLM}}(\omega) = \frac{\chi_\alpha \omega_p^2 + j\chi_\beta \omega_p \omega - \chi_\gamma \omega^2}{-\omega^2 + j\Gamma_L \omega + \omega_0^2} \tag{1.6}$$

This 2TDLM model incorporates all the standard Lorentz model behaviors including the resonance behavior at ω_0 but allows for additional driving mechanisms that are important when considering time-varying phenomena. It satisfies a generalized Kramers–Krönig relation and is causal if $\chi_\gamma > -1$. It has the limiting behaviors $\lim_{\omega \to 0} \chi_{e,\text{2TDLM}}(\omega) \to \chi_\alpha$ and $\lim_{\omega \to \infty} \chi_{e,\text{2TDLM}}(\omega) \to \chi_\gamma$. The high-frequency behavior has the peculiar property that if $-1 < \chi_\gamma < 0$, then $0 < \lim_{\omega \to \infty} \varepsilon(\omega) < 1$, which leads to the interesting but still controversial trans-vacuum-speed (TVS) effect [16, 17].

1.2   WAVE PARAMETERS IN DNG MEDIA

One must exercise some care with the definitions of the electromagnetic properties in a DNG medium. Ziolkowski and Heyman thoroughly analyzed this concept mathematically and have shown that in DNG media the refractive index can be negative [12]. In particular, in a DNG medium where $\varepsilon < 0$ and $\mu < 0$, one should write for small losses:

$$\sqrt{\varepsilon} = \sqrt{\varepsilon_r \varepsilon_0 - j\varepsilon''} \approx -j\left(|\varepsilon_r \varepsilon_0|^{1/2} + j\frac{\varepsilon''}{2|\varepsilon_r \varepsilon_0|^{1/2}}\right)$$

$$\sqrt{\mu} = \sqrt{\mu_r \mu_0 - j\mu''} \approx -j\left(|\mu_r \mu_0|^{1/2} + j\frac{\mu''}{2|\mu_r \mu_0|^{1/2}}\right) \tag{1.7}$$

accounting for the branch-cut choices. This leads to the following expressions for the wavenumber and the wave impedance, respectively:

$$k = \omega\sqrt{\varepsilon}\sqrt{\mu} \approx -\frac{\omega}{c}|\varepsilon_r|^{1/2}|\mu_r|^{1/2}\left[1 + j\frac{1}{2}\left(\frac{\varepsilon''}{|\varepsilon_r|\varepsilon_0} + \frac{\mu''}{|\mu_r|\mu_0}\right)\right]$$

$$\eta = \frac{\sqrt{\mu}}{\sqrt{\varepsilon}} \approx \eta_0 \frac{|\mu_r|^{1/2}}{|\varepsilon_r|^{1/2}}\left[1 + j\frac{1}{2}\left(\frac{\mu''}{|\mu_r|\mu_0} - \frac{\varepsilon''}{|\varepsilon_r|\varepsilon_0}\right)\right] \tag{1.8}$$

where the speed of light $c = 1/\sqrt{\varepsilon_0\mu_0}$ and the free-space wave impedance $\eta_0 = \sqrt{\mu_0/\varepsilon_0}$. One sees that the index of refraction

$$n = \frac{kc}{\omega} = \sqrt{\frac{\varepsilon}{\varepsilon_0}}\sqrt{\frac{\mu}{\mu_0}} = -\left[\left(|\varepsilon_r||\mu_r| - \frac{\varepsilon''}{\varepsilon_0}\frac{\mu''}{\mu_0}\right) + j\left(\frac{\varepsilon''|\mu_r|}{\varepsilon_0} + \frac{\mu''|\varepsilon_r|}{\mu_0}\right)\right]^{1/2}$$

$$\approx -|\varepsilon_r|^{1/2}|\mu_r|^{1/2}\left[1 + j\frac{1}{2}\left(\frac{\varepsilon''}{|\varepsilon_r|\varepsilon_0} + \frac{\mu''}{|\mu_r|\mu_0}\right)\right] \tag{1.9}$$

has a negative real part. Its imaginary part is also negative corresponding to the passive nature of the DNG medium.

The index of refraction of a DNG metamaterial has been shown theoretically to be negative by several groups (e.g., [8, 12, 18]), and several experimental studies have been reported confirming this negative-index-of-refraction (NIR) property and applications derived from it, such as phase compensation and electrically small resonators [19], negative angles of refraction (e.g., [6, 19–24]), subwavelength waveguides with lateral dimension below diffraction limits [25–30], enhanced focusing (see [7, 31]), backward-wave antennas [32], Čerenkov radiation [33], photon tunneling [34, 35], and enhanced electrically small antennas [36]. These studies rely heavily on the concept that a continuous-wave (CW) excitation of a DNG medium leads to a NIR and, hence, to negative or compensated phase terms.

1.3   FDTD SIMULATIONS OF DNG MEDIA

In this chapter and in Chapter 2, we present several finite-difference time-domain (FDTD) simulation results for wave interactions with DNG media, in addition to analytical descriptions. Consequently, we briefly discuss some of the features of the FDTD simulator specific to the DNG structures. It should be emphasized that the use of this purely numerical simulation approach does not involve any choices in defining derived quantities to explain the wave physics, for example, no wave vector directions or wave speeds are stipulated a priori. In this manner, it has provided a useful approach to studying the wave physics associated with DNG metamaterials.

As in [12,22,23,37], lossy Drude polarization and magnetization models are used to simulate the DNG medium; specifically the permittivity and permeability are described in the frequency domain as

$$\varepsilon(\omega) = \varepsilon_0 \left(1 - \frac{\omega_{pe}^2}{\omega(\omega - j\Gamma_e)} \right)$$

$$\mu(\omega) = \mu_0 \left(1 - \frac{\omega_{pm}^2}{\omega(\omega - j\Gamma_m)} \right)$$

(1.10)

where ω_{pe}, ω_{pm} and Γ_e, Γ_m denote the corresponding plasma and damping frequencies, respectively. These models are implemented into the FDTD scheme by introducing the associated electric and magnetic current densities and the equations that govern their temporal behavior.

$$J_{i,\text{Drude}} = \frac{d}{dt} P_i$$

$$\frac{d}{dt} J_{i,\text{Drude}} + \Gamma_e J_{i,\text{Drude}} = \varepsilon_0 \omega_p^2 E_i$$

$$K_{i,\text{Drude}} = \frac{d}{dt} M_i$$

$$\frac{d}{dt} K_{i,\text{Drude}} + \Gamma_m K_{i,\text{Drude}} = \omega_p^2 H_i$$

(1.11)

The choices of the space and time locations of the discretized electric and magnetic currents, as well as the polarization and magnetization fields, are made self-consistently following the conventional FDTD method [38]. The simulation space is truncated with a metamaterial-based absorbing boundary condition [15,39]. The FDTD cell size in all of the cases presented here was $\lambda_0/100$ to minimize the impact of any numerical dispersion on the results.

Although in some of the analytical and numerical studies, as well as experiments, considered by other groups (e.g., [5, 6, 18, 40–43]) the Lorentz model and its derivatives have been used, here the Drude model is preferred for the FDTD simulations for both the permeability and permittivity functions because it provides a much wider bandwidth over which the negative values of the permittivity and permeability can be obtained. This choice is only for numerical convenience and it does not alter any conclusions derived from such simulations; that is, the negative refraction is observed in either choice. However, choosing the Drude model for the FDTD simulation also implies that the overall simulation time can be significantly shorter, particularly for low-loss media. In other words, the FDTD simulation will take longer to reach a steady state in the corresponding Lorentz model because the resonance region where the permittivity and permeability acquire their negative values would be very narrow in this model.

1.4   CAUSALITY IN DNG MEDIA

As for the causality of signal propagation in a DNG medium, we note that if one totally ignores the temporal dispersion in a DNG medium and considers carefully the ramifications of a homogeneous, *nondispersive* DNG medium and the resulting NIR, one will immediately encounter a causality paradox in the time domain, that is, a nondispersive DNG medium is noncausal. However, a resolution of this issue was uncovered in [44] by taking the dispersion into account in a time-domain study of wave propagation in DNG media. The causality of waves propagating in a *dispersive* DNG metamaterial was investigated both analytically and numerically using the one-dimensional (1D) electromagnetic plane-wave radiation from a current sheet source in a dispersive DNG medium. A lossy Drude model of the DNG medium was used, and the solution was generated numerically with the FDTD method. The basic 1D geometry is shown in Figure 1.3. The signal direction of propagation (D.O.P.) is from left to right.

A causal result would show that the signal arrives at point 1 before point 2; an NIR result would then show that the peaks of the signals received at point 2 occur before those at point 1 once steady state is reached. The FDTD predicted results, shown in Figure 1.4, confirm this behavior.

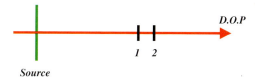

Figure 1.3   One-dimensional FDTD simulation region.

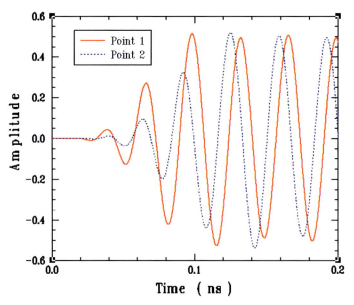

Figure 1.4 Time-domain electric fields predicted by the FDTD simulator at points 1 and 2 shown in Figure 1.3. From [44]. Copyright © 2003 by the American Physical Society.

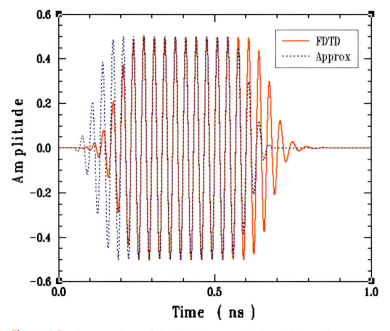

Figure 1.5 A comparison of the FDTD-generated time history of a wave propagating in a dispersive Drude medium and an approximate solution consisting of a causal propagating envelope and the expected NIR sinusoidal signal shows very good agreement in the steady-state region away from the leading and trailing edges where dispersion plays a significant role. From [44]. Copyright © 2003 by the American Physical Society.

The analogous problem in a nondispersive DNG medium was also considered, and it was shown that the solution to this problem is not causal, in agreement with similar observations given in [18]. An approximate solution was constructed that combined a causal envelope with a sinusoid which has the nondispersive NIR properties; it compared favorably with the FDTD results for the dispersive DNG case, as shown in Figure 1.5. It was thus demonstrated that causal results do indeed require the presence of dispersion in DNG media and that the dispersion is responsible for a dynamic reshaping of the pulse to maintain causality. The CW portions of a modulated pulse (i.e., excluding its leading and trailing edges) do obey all of the NIR effects expected from a time-harmonic analysis in a bandlimited "nondispersive" DNG medium. Therefore, one can conclude that the CW analyses of DNG media are credible as long as very narrow bandwidth pulse trains are considered for any practical realizations. This has been the case in all of the experimental results reported to date of which we are aware. Moreover, time delays for the realization of the NIR effects are inherent in the processes dictated by the dispersive nature of the physics governing these media.

1.5   SCATTERING FROM A DNG SLAB

The reflection and transmission coefficients associated with a normally incident plane wave that scatters from a DNG slab embedded in a medium have been derived. The geometry is shown in Figure 1.6. The slab has an infinite extent in the transverse directions; it has a thickness d in the direction of propagation of the incident plane wave. Let the medium before and after the slab be characterized by ε_1, μ_1 and the slab be characterized by ε_2, μ_2. For a normally incident plane wave, the reflection and transmission coefficients for the slab are

$$R = \frac{\eta_2 - \eta_1}{\eta_2 + \eta_1} \frac{1 - e^{-j2k_2 d}}{1 - [(\eta_2 - \eta_1)/(\eta_2 + \eta_1)]^2 e^{-j2k_2 d}}$$

$$T = \frac{4\eta_2\eta_1}{(\eta_2 + \eta_1)^2} \frac{e^{-jk_2 d}}{1 - [(\eta_2 - \eta_1)/(\eta_2 + \eta_1)]^2 e^{-j2k_2 d}}$$

(1.12)

where the wavenumber $k_i = \omega\sqrt{\varepsilon_i}\sqrt{\mu_i}$ and wave impedance $\eta_i = \sqrt{\mu_i}/\sqrt{\varepsilon_i}$ for $i = 1, 2$. For the case of normal incidence, if we consider a matched DNG medium, one would have $\eta_2 = \eta_1$ so that $R = 0$ and $T = e^{-jk_2 d} = e^{+j|k_2|d}$. The

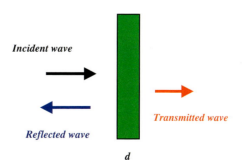

Figure 1.6   Plane-wave scattering from slab of thickness d.

medium would thus add a positive phase to the wave traversing the slab, whereas in a DPS medium the wave would experience a negative phase variation. This means that a matched DNG slab could be used to compensate for phase changes incurred by passage of a plane wave through a DPS slab; that is, one can force $k_{DPS}d_{DPS} + k_{DNG}d_{DNG} = 0$. This phase compensation, to be discussed later in this chapter, is an interesting feature of DNG metamaterials that can lead to exciting potential applications.

When the plane wave is obliquely incident, Eq. (1.12) is straightforwardly modified by introducing the transverse impedance and longitudinal wavenumber components. If, in addition, the incident wave is evanescent, that is, when the transverse component of the wave vector of the incident wave is greater than the wavenumber of the medium ($k_t^2 > \omega^2\mu_1\varepsilon_1$ and $k_t^2 > \omega^2\mu_2\varepsilon_2$), the transverse wave impedance in each medium (with a no-loss assumption) becomes purely imaginary, that is, $\eta_{1,\text{transverse}} = jX_{1,\text{transverse}}$ and $\eta_{2,\text{transverse}} = jX_{2,\text{transverse}}$, and the longitudinal component of the wave vector in each medium also becomes purely imaginary, that is, $k_{1,\text{longitudinal}} = j\alpha_1$ and $k_{2,\text{longitudinal}} = j\alpha_2$ [45,46]. (The proper choice of sign for α_1 and α_2 will be discussed shortly.) However, it can be shown that this transverse wave impedance in the DPS and DNG media have opposite signs; that is, if one has a capacitive reactance, the other will have an inductive reactance so that $\text{sgn}[X_{1,\text{transverse}}] = -\text{sgn}[X_{2,\text{transverse}}]$, where $\text{sgn}(x) = +1 (-1)$ for $x > 0 (x < 0)$ [45, 46]. When we choose the so-called matched condition for which $\mu_2 = -\mu_1$ and $\varepsilon_2 = -\varepsilon_1$, one can demonstrate that $X_{1,\text{transverse}} = -X_{2,\text{transverse}}$. Inserting these features into the generalized form of Eq. (1.12), one would observe that again $R = 0$, but now $T = e^{-jk_{2,\text{longitudinal}}d} = e^{\alpha_2 d}$. What is the proper sign for α_2? A detailed discussion, including a variety of physical insights, on this issue can be found in, for example, [45, 46]. Here we briefly review this point. It is known that at the interface between a DPS and a DNG medium, the tangential components of the electric and magnetic fields should be continuous according to Maxwell equations. However, since the permittivity and the permeability of these two media each has opposite signs, the normal spatial derivatives (normal with respect to the interface) of these tangential components are discontinuous at this boundary [19]. In other words, if the tangential component of the electric field (or the magnetic field) is decreasing as the observation point gets closer to this interface from the DPS side, the same tangential component should be increasing as the observer is receding from the interface in the DNG side. Moreover, one should remember that according to Eq. (1.12) the overall reflection coefficient for the "incident" evanescent wave in this case is $R = 0$. Therefore, as this evanescent wave reaches the first interface of the matched DNG slab from the DPS region, it is decaying, that is, $\alpha_1 < 0$, and no "reflected" evanescent wave will be present in this DPS medium. However, as we move just past the first interface into the DNG region, the tangential components of the field in the vicinity of the interface inside the DNG region should "grow" in order to satisfy the discontinuity condition of the normal spatial derivative mentioned above. (Note that if the evanescent wave decayed inside this matched DNG slab, the tangential components of the field at the DPS–DNG interface would have similar slopes, inconsistent with the boundary condition mentioned above.) Therefore, in the transmission coefficient expression $T = e^{\alpha_2 d}$, one should have $\alpha_2 > 0$. As a result, such a matched DNG slab can compensate the decay of the evanescent wave in the DPS region through the growth of the evanescent wave inside the DNG slab. This issue was originally pointed out by Pendry in [7] and is the basis behind the idea of subwavelength focusing and "perfect" lensing [7]. We emphasize that this growth of an evanescent wave inside the DNG slab does not violate any physical law, since each of these *evanescent* waves carries no real power, and indeed this scenario represents the presence of an interface resonance at the boundary between the DPS and DNG regions [45,46]. Furthermore, this phenomenon can also be described and justified using distributed circuit elements, which provide further insight into related features associated with this problem [45,46].

An interesting question arises here: If one gets a growing evanescent wave inside the slab, as justified above for the case of a finite-thickness matched DNG slab, what should one see for a semi-infinite matched DNG medium when an "incident" evanescent wave is approaching this interface? In other words, when we have a single interface between matched semi-infinite DPS and semi-infinite DNG media, what will happen for an 'incident' evanescent wave? This is a markedly different problem. In this case, we only deal with one interface, and the reflection and transmission coefficients for such an interface can be easily expressed as

$$R_{\text{DPS-DNG}} = \frac{\eta_{2,\text{transverse}} - \eta_{1,\text{transverse}}}{\eta_{2,\text{transverse}} + \eta_{1,\text{transverse}}}$$

and

$$T_{\text{DPS-DNG}} = \frac{2\eta_{2,\text{transverse}}}{\eta_{2,\text{transverse}} + \eta_{1,\text{transverse}}}$$

For the matched condition, as discussed above, we have $\eta_{1,\text{transverse}} = jX_{1,\text{transverse}}$ and $\eta_{2,\text{transverse}} = jX_{2,\text{transverse}}$ with $X_{1,\text{transverse}} = -X_{2,\text{transverse}}$. Therefore, one gets $R_{\text{DPS-DNG}} = \infty$ and $T_{\text{DPS-DNG}} = \infty$, which implies that there is an interface resonance at this boundary. This is indeed another indication that such an interface can indeed support a surface plasmon wave, which is an important factor in understanding the behavior of this interface [45,46]. Similarly, the Fresnel "reflection" and "transmission" coefficients for an incident evanescent wave for this configuration become infinite (the circuit analog of this phenomenon has also been studied [45,46]). This is analogous to exciting a resonant structure (such as an $L-C$ circuit) at its resonant frequency, which also leads to infinite fields in the structure when there are no losses present. Consequently, when there is a source in front of the interface between two semi-infinite matched DNG and DPS media, a resonant surface wave may be excited along the interface, resulting in an infinitely large field value. However, the fields on both sides of this interface, albeit infinitely large, *decay* exponentially as they move away from it; that is, the field distribution represents a surface wave propagating along the interface. This explains and justifies the presence of $R_{\text{DPS-DNG}} = \infty$ and $T_{\text{DPS-DNG}} = \infty$. In summary, for a single matched DPS–DNG interface, one finds an evanescent wave in each medium whose amplitude is infinite at the interface but decays exponentially as the wave recedes away from it. On the other hand, for a matched DNG slab the presence of two interfaces allows a resonant interaction that produces a net exponential growth of the evanescent wave components inside the slab despite their exponential decay outside of it.

1.6   BACKWARD WAVES

Consider the source problems shown in Figure 1.7. A current sheet of the form

$$\mathbf{J}_s = I_0 e^{-jk_{0x}x}\delta(z)\hat{x} \tag{1.13}$$

is located on the interface between two semi-infinite media. In one case both regions are DPS media, and in the other one is a DPS medium and the other is a DNG medium. The wavenumbers in each medium satisfy the dispersion relation

$$k_{i,x}^2 + k_{i,z}^2 = k_i^2 = \omega^2 \varepsilon_i \mu_i \tag{1.14}$$

where region 1 labels $z > 0$ and region 2 labels $z < 0$. Boundary conditions require the wave numbers tangential to the interface be the same in each medium, that is, $k_{1,x} = k_{2,x} = k_{0x}$; thus for propagating waves they also require the propagation constants normal to the interface be given as

$$k_z^{\text{DPS}} = +\sqrt{\omega^2 \varepsilon_{\text{DPS}} \mu_{\text{DPS}} - k_{0x}^2} \qquad k_z^{\text{DNG}} = -\sqrt{\omega^2 \varepsilon_{\text{DNG}} \mu_{\text{DNG}} - k_{0x}^2} \tag{1.15}$$

The wave vectors in each region are thus given by the expressions

$$\mathbf{k}_1 = k_{0x}\hat{x} + k_{1z}\hat{z} \qquad \mathbf{k}_{2,\text{DPS}} = k_{0x}\hat{x} - k_{2z}\hat{z} \qquad \mathbf{k}_{2,\text{DNG}} = k_{0x}\hat{x} + |k_{2z}|\hat{z} \tag{1.16}$$

Similarly, the Poynting's vector in each region is determined to be

$$\langle \mathbf{S}_1 \rangle (x, y, z, \omega) = \frac{1}{2\omega\varepsilon_1} \left| \frac{k_{2z}}{\varepsilon_2} \frac{I_0}{k_{1z}/\varepsilon_1 + k_{2z}/\varepsilon_2} \right|^2 (k_{0x}\hat{x} + k_{1z}\hat{z}) \tag{1.17}$$

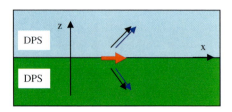

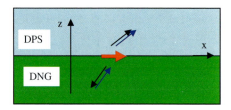

Figure 1.7   Current sheet located at interface between two semi-infinite half spaces. Blue arrows represent the wave vectors, black arrows the Poynting vectors.

$$\langle \mathbf{S}_{2,\text{DPS}} \rangle (x, y, z, \omega) = \frac{1}{2\omega\varepsilon_2} \left| \frac{k_{1z}}{\varepsilon_1} \frac{I_0}{k_{1z}/\varepsilon_1 + k_{2z}/\varepsilon_2} \right|^2 (k_{0x}\hat{x} - k_{2z}\hat{z})$$

$$\langle \mathbf{S}_{2,\text{DNG}} \rangle (x, y, z, \omega) = \frac{1}{2\omega|\varepsilon_2|} \left| \frac{k_{1z}}{\varepsilon_1} \frac{I_0}{k_{1z}/\varepsilon_1 + |k_{2z}|/|\varepsilon_2|} \right|^2 (-k_{0x}\hat{x} - |k_{2z}|\hat{z})$$

$$(1.18)$$

Thus one finds that the Poynting vector and wave vector directions are the same when both regions are DPS media and the generated waves are in the forward direction, that is, in the direction of positive phase advance along the source. In contrast, the Poynting vector in the DNG medium is pointed causally away from the source and is opposite to the wave vector direction, which is toward the source. Moreover, the flow of power of the wave generated in the DNG medium is opposite to the positive phase direction of the source. The backward-wave nature of the wave generated in the DNG medium is thus established. The details of this problem can be found in [47].

1.7 NEGATIVE REFRACTION

The phenomenon of negative refraction is studied by considering the scattering of a wave that is obliquely incident on a DPS–DNG interface as shown in Figure 1.8. Enforcing the electromagnetic boundary conditions at the interface, one obtains the law of reflection and Snell's Law from phase matching:

$$\theta_{\text{refl}} = \theta_{\text{inc}} \qquad \theta_{\text{trans}} = \text{sgn}(n_2) \sin^{-1}\left(\frac{n_1}{|n_2|} \sin\theta_{\text{inc}} \right) \qquad (1.19)$$

Note that if the index of refraction of a medium is negative, then the refracted angle, according to Snell's law, should also become "negative." This suggests that the refraction is anomalous, and the refracted angle is on the same side of the interface normal as the incident angle is. The wave and Poynting vectors associated with this oblique scattering problem are also obtained:

$$\mathbf{k}_{\text{inc}} = k_1(\cos\theta_{\text{inc}}\hat{z} + \sin\theta_{\text{inc}}\hat{x})$$
$$\mathbf{k}_{\text{refl}} = k_1(-\cos\theta_{\text{inc}}\hat{z} + \sin\theta_{\text{inc}}\hat{x}) \qquad (1.20)$$
$$\mathbf{k}_{\text{trans}} = k_2(\cos\theta_{\text{trans}}\hat{z} + \sin\theta_{\text{trans}}\hat{x})$$

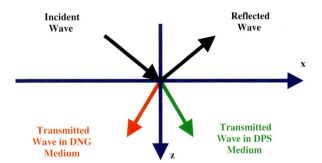

Figure 1.8 Geometry of the scattering of a wave obliquely incident upon a DPS–DNG interface.

$$\mathbf{S}_{\text{inc}} = \frac{1}{2} \frac{|E_0|^2}{\eta_1} (\cos\theta_{\text{inc}}\hat{z} + \sin\theta_{\text{inc}}\hat{x})$$

$$\mathbf{S}_{\text{refl}} = \frac{1}{2} \frac{|RE_0|^2}{\eta_1} (-\cos\theta_{\text{inc}}\hat{z} + \sin\theta_{\text{inc}}\hat{x}) \tag{1.21}$$

$$\mathbf{S}_{\text{trans}} = \frac{1}{2} \frac{|TE_0|^2}{\eta_2} (\cos\theta_{\text{trans}}\hat{z} + \sin\theta_{\text{trans}}\hat{x})$$

Assuming that the transmitted wave is propagating in a DPS medium, it is clear that the Poynting and wave vectors are in the same direction. However, if the transmitted wave is propagating in a DNG medium, the index is less than zero and one obtains immediately from Snell's law that

$$\mathbf{k}_{\text{trans}} = -|n_2| \frac{\omega}{c} (\cos|\theta_{\text{trans}}|\hat{z} - \sin|\theta_{\text{trans}}|\hat{x})$$

$$\mathbf{S}_{\text{trans}} = \frac{1}{2} \frac{|TE_0|^2}{\eta_2} (\cos|\theta_{\text{trans}}|\hat{z} - \sin|\theta_{\text{trans}}|\hat{x})$$

so that the wave and Poynting vectors point in opposite directions, the Poynting vector being directed in a causal direction away from the interface.

This negative-refraction behavior was verified with FDTD calculations [22]. The electric field intensity distributions were obtained with the 2D FDTD simulator when a $f_0 = 30$ GHz (needless to say, this choice is arbitrary; the numerical results presented here can be obtained at any frequency with a proper scaling of

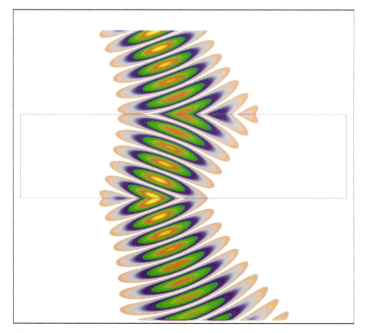

Figure 1.9   The NIR behavior has been confirmed with FDTD simulations. From [22]. Copyright © 2003 by The Optical Society of America.

the parameters) perpendicularly polarized CW Gaussian beam is incident on a DPS–DNG interface with $\theta_{\text{inc}} = 20°$. To reduce the effect of reflection and thus to observe the negative refraction more clearly, the parameters of these slabs were chosen such that the slabs are impedance matched to free space. Therefore, the electric and magnetic Drude models were selected to be identical, that is, $\omega_{pe} = \omega_{pm} = \omega_p$ and $\Gamma_e = \Gamma_m = \Gamma$. Only low loss values were considered by setting $\Gamma = 10^{+8}s^{-1} \ll \omega_p$. This means that the index of refraction had the form

$$n(\omega) = \sqrt{\frac{\varepsilon(\omega)}{\varepsilon_0}} \sqrt{\frac{\mu(\omega)}{\mu_0}} = 1 - \frac{\omega_p^2}{\omega(\omega - j\Gamma)} = 1 - \frac{\omega_p^2}{\omega^2 + \Gamma^2} - j\frac{\Gamma}{\omega}\frac{\omega_p^2}{\omega^2 + \Gamma^2}$$

$$\approx 1 - \frac{\omega_p^2}{\omega^2} - j\frac{\Gamma\omega_p^2}{\omega^3} \tag{1.22}$$

The DNG slab had $n_{\text{real}}(\omega_0) \approx -1$, by setting $\omega_p = 2\pi\sqrt{2}f_0 = 2.66573 \times 10^{11}$ rad/s and, hence, $\Gamma = 3.75 \times 10^{-4}\omega_p$. As can be seen in Figure 1.9, the negative angle of refraction is clearly seen. The refracted angle is equal and opposite to the angle of incidence. The discontinuities in the derivatives of the fields at the DPS–DNG interfaces (i.e., the so-called V-shaped patterns at both interfaces) are clearly seen. A simulation movie of this case is available in [22].

1.8 PHASE COMPENSATION WITH A DNG MEDIUM

As we reviewed in our recent paper [48], one of the interesting features of DNG media is their ability to provide phase compensation or phase conjugation due to their negative refraction. Here, we provide an illustrative example to highlight the insight behind this phenomenon. Consider a slab of conventional lossless DPS material with positive index of refraction n_1 and thickness d_1 and a slab of lossless DNG metamaterial with negative refractive index $-|n_2|$ and thickness d_2. Although not necessary, but for the sake of simplicity in the argument, we assume that each of these slabs is impedance matched to the outside region (e.g., free space). Let us take a monochromatic uniform plane wave normally incident on this pair of slabs. As this wave propagates through the slab, the phase difference between the exit and entrance faces of the first slab is obviously $n_1 k_0 d_1$, where $k_0 \equiv \omega\sqrt{\varepsilon_0\mu_0}$, while the total phase difference between the front and back faces of this two-layer structure is $|n_1|k_0d_1 - |n_2|k_0d_2$, implying that whatever phase difference is developed by traversing the first slab, it can be decreased and even compensated for by traversing the second slab. If the ratio of d_1 and d_2 is chosen to be $d_1/d_2 = |n_2|/|n_1|$ at the given frequency, then the total phase difference between the front and back faces of this two-layer structure will become zero. This means that the DNG slab acts as a phase compensator in this structure [19,27]. We should note that such phase compensation/conjugation does not depend on the *sum* of thicknesses, $d_1 + d_2$, rather it depends on the *ratio* of d_1 and d_2. So, in principle, $d_1 + d_2$ can be any value as long as d_1/d_2 satisfies the above condition. Therefore, even though this two-layer structure is present, the wave traversing this structure would not experience any phase difference between the input and output faces. This feature can lead to several interesting ideas in device and component designs, as will be discussed later.

Such phase compensation has been verified using the FDTD simulator, as shown in Figure 1.10. The FDTD predicted electric field intensity distribution for a perpendicularly polarized CW Gaussian beam incident on this DPS–DNG slab pair is shown. A Gaussian beam was launched toward the DPS–DNG slab pair, each slab having a thickness of $2\lambda_0$. The DPS slab had $n(\omega) = +3$, while the DNG slab had $n_{real}(\omega_0) \approx -3$. As is evident from Figure 1.10, the beam expands

Figure 1.10   FDTD predicted electric field intensity distribution for phase compensator–beam translator system DPS–DNG stacked pair. The Gaussian beam is normally incident on a stack of two slabs, the first being a DPS slab with $n_{real}(\omega) = +3$ and the second being a DNG slab with $n_{real}(\omega_0) \approx -3$. The initial beam expansion in the DPS slab is compensated by its refocusing in the DNG slab. The Gaussian beam is translated from the front face of the system to its back face with only -0.323 dB attenuation over the $4\lambda_0$ distance.

in the DPS slab and then the negative-refraction property refocuses it in the DNG slab, and the waist of the intensity of the input beam is recovered at the back face. The electric field intensity could, in principle, be maintained over the total thickness of $4\lambda_0$. There is only a -0.323-dB (7.17%) reduction in the peak value of the intensity of the beam when it reaches the back face. Moreover, the phase of the beam at the output face of the stack is the same as its value at the entrance face. Therefore, the beam emerges at the output of the slab pair in phase with the input beam and only slightly smaller in amplitude. Thus the DPS–DNG slab pair, in essence, translates the field from one location to another with low losses; that is, it acts as a beam translator.

Using multiple matched DPS–DNG stacks, one could produce a phase-compensated, time-delayed, waveguiding system. Each pair in the stack would act as shown in Figure 1.10. Thus the phase compensation–beam translation effects would occur throughout the entire system. Moreover, by changing the index of any of the DPS–DNG pairs, one changes the speed at which the beam traverses that slab pair. Consequently, one can change the time for the beam to propagate from the entrance face to the exit face of the entire DPS–DNG stack. In this manner one could realize a volumetric, low-loss time delay line for a Gaussian beam system.

This phase compensation can lead to a wide variety of potential applications that could have a large impact on a number of engineering systems. One such set of applications offers the possibility of having subwavelength, electrically small cavity resonators and waveguides with lateral dimension below diffraction limits. These ideas are briefly reviewed in Chapter 2.

1.9 DISPERSION COMPENSATION IN A TRANSMISSION LINE USING A DNG MEDIUM

The DNG medium, because of its dispersive nature, might also be used as an effective dispersion compensation device for time-domain applications. The dispersion produces a variance of the group speed of the signal components as they propagate in the DNG medium. Cheng and Ziolkowski have considered the use of volumetric DNG metamaterials for the modification of the propagation of signals along a microstrip transmission line [49]. If one could compensate for the dispersion along such transmission lines, signals propagating along them would not become distorted. This could lead to a simplification of the components in many systems. Microstrip dispersion can be eliminated by correcting for the frequency dependence of the effective permittivity associated with this type of transmission line. As shown in [50, 51], for a microstrip transmission line of width w and a conventional dielectric substrate height h one has the approximate result for the effective relative permittivity of the air–substrate–microstrip system:

$$\varepsilon_{\text{eff}}(f) = \varepsilon_r - \frac{\varepsilon_r - \varepsilon_{es}}{1 + G(f/f_d)^2} \tag{1.23}$$

where the constants

$$f_d = \frac{Z_c}{2\mu_0 h} \qquad G = 0.6 + 0.0009 Z_c$$

the characteristic impedance is

$$Z_c \cong \frac{1}{2\pi} \sqrt{\frac{\mu_0}{\varepsilon_{es}\varepsilon_0}} \log\left[F_1 \frac{h}{w} + \sqrt{1 + \left(2\frac{h}{w}\right)^2} \right]$$

with $F_1 = 6 + (2\pi - 6) \exp[-(30.666h/w)^{0.7528}]$, and the electrostatic relative permittivity is

$$\varepsilon_{es} \cong \frac{\varepsilon_r + 1}{2} + \left(\frac{\varepsilon_r - 1}{2}\right) \left[1 + 10\left(\frac{h}{w}\right)\right]^{-ab}$$

with

$$a = 1 + \frac{1}{49} \log\left[\frac{(w/h)^4 + (w/52h)^2}{(w/h)^4 + 0.432}\right] + \frac{1}{18.7} \log\left[1 + \left(\frac{1}{18.1}\frac{w}{h}\right)^3\right]$$

$$b = 0.564 \left(\frac{\varepsilon_r - 0.9}{\varepsilon_r + 3.0}\right)^{0.053}$$

The goal is to design a length of metamaterial-loaded transmission line that can be included in some manner with the same length of microstripline to make the paired system dispersionless; that is, we want to produce a dispersion-compensated segment of transmission line. This means we want to introduce a metamaterial with relative permittivity ε_{MTM} and permeability μ_{MTM} so that the overall relative permittivity and permeability of the system is

$$\frac{\varepsilon(f)}{\varepsilon_0} = \varepsilon_{eff}(f) + \varepsilon_{MTM}(f) \qquad \frac{\mu(f)}{\mu_0} = 1 + \mu_{MTM}(f) \qquad (1.24)$$

in such a manner that the wave impedance in the metamaterial remains the same as it is in the original substrate, that is,

$$Z = \sqrt{\frac{\mu(f)}{\varepsilon(f)}} = Z_0 \sqrt{\frac{1 + \mu_{MTM}(f)}{\varepsilon_{MTM}(f) + \varepsilon_{eff}(f)}} = Z_0 \sqrt{\frac{1}{\varepsilon_{eff}(f)}} \qquad (1.25)$$

and the index of refraction in the medium compensates for the dispersion effects associated with the microstrip geometry itself; that is, the effective index of the pair becomes that of free space,

$$n_{eff}(f) = \sqrt{\varepsilon_{eff}(f)} + \sqrt{\frac{\varepsilon(f)}{\varepsilon_0}} \sqrt{\frac{\mu(f)}{\mu_0}}$$

$$= \sqrt{\varepsilon_{eff}(f)} + \sqrt{\varepsilon_{eff}(f) + \varepsilon_{MTM}(f)} \sqrt{1 + \mu_{MTM}(f)} = 1 \qquad (1.26)$$

These conditions are satisfied if $\varepsilon_{eff}(f)[1 + \mu_{MTM}(f)] = \varepsilon_{MTM}(f) + \varepsilon_{eff}(f)$ so that

$$\mu_{MTM}(f) = \frac{1}{\sqrt{\varepsilon_{eff}(f)}} - 1 \qquad \varepsilon_{MTM}(f) = \varepsilon_{eff}(f)\mu_{MTM}(f) \qquad (1.27)$$

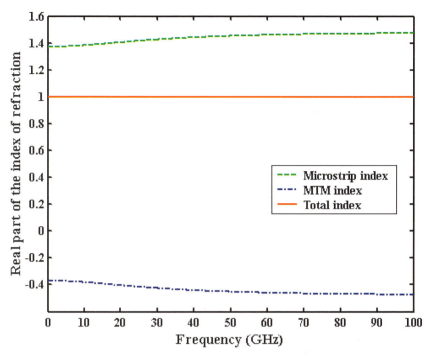

Figure 1.11   Real part of index of refraction of microstrip only, of metamaterial (MTM) only, and total MTM-dispersion-compensated transmission line.

We note that the effective permittivity and permeability of such a metamaterial should be negative, implying that a DNG material must be utilized for this purpose. [The range of validity of condition (1.27) should be consistent with that of the effective medium approximation (1.23).] A plot of the index of refraction of the uncompensated line, the metamaterial compensator, and the dispersion-compensated line is shown in Figure 1.11 for a microstrip transmission line at 10 GHz using Roger's Duroid 5880 substrate. The substrate had the relative permittivity $\varepsilon_r = 2.2$ and its height was $h = 31$ mils $= 0.7874$ mm. The width of the transmission line was $w = 2.428$ mm $= 95.6$ mils to achieve a 50 Ω impedance. As shown in Figure 1.11, in principle, complete dispersion compensation is theoretically possible.

1.10   SUBWAVELENGTH FOCUSING WITH A DNG MEDIUM

Another interesting potential application of a DNG medium that results from its negative-refraction properties was first theoretically suggested by Pendry [7]. It is the idea of a "perfect lens" or focusing beyond the diffraction limit. In his analysis of the image formation process in a flat slab of lossless DNG material, Pendry showed that the evanescent spatial Fourier components can be ideally reconstructed in addition to the faithful reconstruction of all the propagating spatial Fourier components. The evanescent wave reconstruction is due to the presence of the "growing exponential effect" in the DNG slab discussed in Section 1.5. This effect in theory leads to the formation of an image with a resolution higher than the conventional limit. His idea has motivated much interest in studying wave interactions with DNG media.

Various theoretical and experimental works by several groups have explored this possibility; they have shown the possibility and limitations of subwavelength focusing using a slab of DNG or negative-index metamaterials [52, 53]. The subwavelength focusing in the planar 2D structures made of negative-index transmission lines has also been investigated [31]. The presence of the growing exponential in the DNG slab has also been explained and justified using the equivalent distributed circuit elements in transmission line model [46]. It has also been shown that "growing evanescent envelopes" for the field distributions can be achieved in a suitably designed, periodically layered stacks of frequency-selective surfaces (FSSs) [54].

It was shown analytically in [12] that the perfect-focus solution exists only for the frequency-independent, lossless DNG slab case for which $\varepsilon_r = \mu_r = -1$. For all other cases, a line source will produce paraxial foci. If the line source is located z_0 away from the front face of the slab, the foci produced by a DNG slab have been shown analytically [12] to occur at the distances $z_{f1} = |n_2||z_0|$ and $z_{f1} = d(1 + 1/|n_2|) - |z_0|$ away from that face and the source. Thus for an $n_{\text{real}}(\omega_0) \approx -1$ planar slab of thickness d, the first focus is located at $z_{f1} = |z_0|$ and the second is at $z_{f2} = 2d - |z_0|$. The electric field intensity predicted by the FDTD simulator for a lossy Drude slab with $n_{\text{real}}(\omega_0) \cong -1$ and $\Gamma = 10^{-5}\omega_0$ is shown in Figure 1.12a. The line source is 50 cells in front of a 100-cell-deep slab. The location of the source and the expected locations of the two foci are indicated by the intersections of the vertical and horizontal black lines. The transverse profiles of the intensity along the horizontal lines are shown in Figure 1.12b. While the foci appear where expected, they are not perfect. As explained in [12], the perfect focus is not obtained because of the presence of dispersion and the large generation of surface waves which take energy away from the focusing mechanism. The transverse full width at half maximum (FWHM) of the input intensity and the intensities at $z_{f1} = 50$ cells and $z_{f2} = 150$ cells from the front interface are 29, 31, and 34 cells, respectively. Nonetheless, since a cell is $\lambda_0/100$, these distances are all subwavelength. The appearance of the paraxial foci is demonstrated in Figure 1.13a. The line source is 10 cells in front of a 120-cell-deep lossy Drude slab with $n_{\text{real}}(\omega_0) \cong -6$ and $\Gamma = 10^{-5}\omega_0$. The location of the source and the expected locations of the two foci are again indicated by the intersections of the vertical and horizontal black lines. The paraxial focusing (channeling) of the beam within the slab is readily apparent. The waist of the beam occurs where the focus within the slab is located. The transverse profiles of the intensity along the front interface and along the horizontal foci lines are shown in Figure 1.13b. The FWHM of these profiles are, respectively, 23, 28,

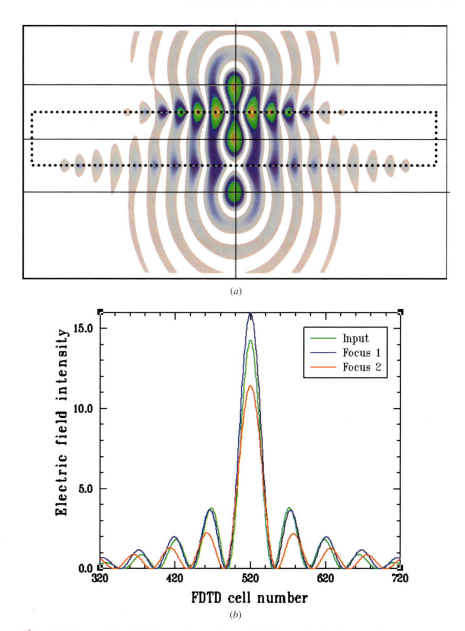

Figure 1.12 (*a*) The FDTD predicted electric field intensity distribution illustrates the focusing of the field generated by a line source in a $n_{real}(\omega_0) \approx -1$ DNG slab. (*b*) Transverse profiles of the electric field intensity distribution at various locations.

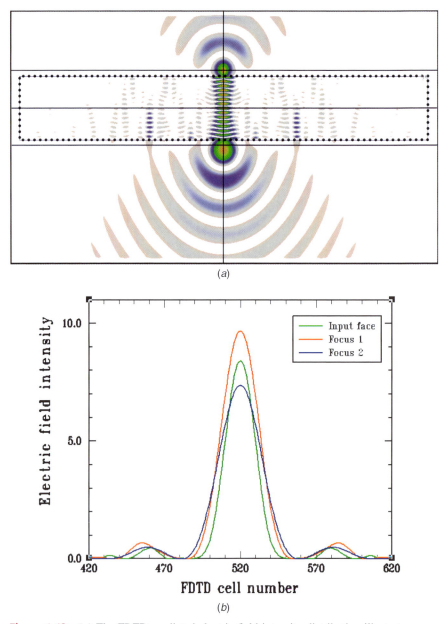

(a)

(b)

Figure 1.13   (a) The FDTD predicted electric field intensity distribution illustrates the focusing of the field generated by a line source in a $n_{real}(\omega_0) \approx -6$ DNG slab. (b) Transverse profiles of electric field intensity distribution at various locations. Channeling of the beam in the DNG slab is observed; the wings of the beam are seen to feed the center of the beam.

and 32 cells. Again, the FWHM of the intensity profile of the beam at the foci is subwavelength.

To emphasize the beam dynamics further, the electric field intensity distribution for a case for which the source is far from the slab is given in Figure 1.14a. The line source is 180 cells in front of a 360-cell-deep lossy Drude slab with $n_{real}(\omega_0) \cong -1$ and $\Gamma = 10^{-5}\omega_0$. The paraxial focusing of the beam within the slab and external to it is readily apparent. The waists of the beam within the slab and beyond it occur at the predicted locations of the point foci. The longitudinal profile of the intensity along the beam axis is given in Figure 1.14b. The location of the beam foci coincide with the locations of the predicted point foci. These results correlate nicely to those shown in [42] for the corresponding lossy Lorentz slab.

The use of a planar DNG slab as a lens is illustrated with an FDTD simulation of the focusing of a Gaussian beam. A diverging CW-modulated Gaussian beam is assumed to be normally incident on such a planar DNG slab $n_{real}(\omega_0) \approx -1$. The waist of the beam was $\lambda_0/2$ at the total field–scattered field plane from which it was launched into the simulation space. This source plane was $2\lambda_0$ away from the DNG interface so that there would be sufficient distance for the beam to diverge before it hit the interface. The DNG slab also had a depth of $2\lambda_0$. Thus the locations of the foci are degenerate at $z_{f1} = |z_0| = 2\lambda_0$ and $z_{f2} = 2d - |z_0| = 2\lambda_0$, which occur at the back face of the slab. The waist of the beam at this focus should be approximately the same as it is in the source plane. This behavior is illustrated in Figure 1.15. This result clearly shows that the planar DNG medium turns the diverging wave vectors toward the beam axis and, hence, acts as a lens to focus the beam. Since all angles of refraction are the negative of their angles of incidence for the $n_{real}(\omega_0) \approx -1$ slab, the initial beam distribution is essentially recovered at the back face of the slab; that is, as designed, the focal plane of the beam in the DNG medium is located at the back face of the DNG slab. From the electric field intensity obtained from the FDTD simulation, we note that the peak intensity is about 18 percent lower than its value at the original waist of the Gaussian beam. This variance stems from the presence of additional wave processes, such as surface wave generation, and from dispersion and loss in the actual Drude model used to define the DNG slab in the FDTD simulation.

The corresponding results for the Gaussian beam interacting with the matched DNG slab with $n_{real}(\omega_0) \approx -6$ reveals related but different results. In contrast to the $n_{real}(\omega_0) \approx -1$ case, when the beam interacts with the matched DNG slab with $n_{real}(\omega_0) \approx -6$, there is little focusing observed. The negative angles of refraction dictated by Snell's law are shallower for this higher magnitude of the refractive index, that is, $\theta_{trans} \approx -\sin^{-1}[\sin(\theta_{inc})/6]$. Rather than a strong focusing, the medium channels power from the wings of the beam toward its axis, hence maintaining its amplitude as it propagates into the DNG medium. This difference in behaviors between the two types of DNG slabs is illustrated in Figure 1.16. A Gaussian beam is launched from a source plane which is $2\lambda_0$ from the front of a pair of DNG slabs. Each slab is λ_0 deep. The first DNG slab has $n_{real}(\omega_0) \approx -1$ and the second has $n_{real}(\omega_0) \approx -6$, both having $\Gamma = 10^{-5}\omega_0$.

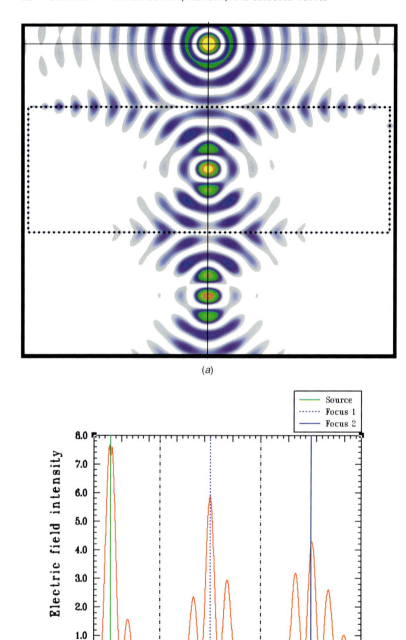

(a)

(b)

Figure 1.14 (a) The FDTD predicted electric field intensity distribution illustrates the focusing of the field generated by a line source in a $n_{real}(\omega_0) \approx -1$ DNG slab. (b) Longitudinal profile of electric field intensity distribution. The actual beam foci coincide with the predicted locations of the point foci.

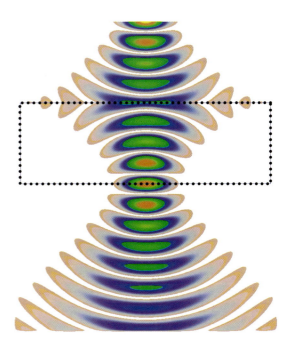

Figure 1.15 The FDTD predicted electric field intensity distribution illustrates the focusing of a diverging Gaussian beam with a $n_{real}(\omega_0) \approx -1$ DNG slab. The source and slab distances were selected to have the DNG slab focus the beam at its output face. From [22]. Copyright © 2003 by the Optical Society of America.

The FDTD predicted electric field intensity distribution given in Figure 1.16*a* shows the beam is initially focused by the $n_{real}(\omega_0) \approx -1$ slab, as expected. The $n_{real}(\omega_0) \approx -6$ slab then channels the beam through it with only minor focusing. The strong axial compression of the beam caused by the (factor of 6) decrease in the wavelength in the $n_{real}(\omega_0) \approx -6$ slab is apparent. The longitudinal profile of the electric field intensity along the beam axis is shown in Figure 1.16*b*. It shows the field is being focused throughout both slabs and the slab pair produces an output intensity that is larger than its input value.

We note that in all of the focusing cases considered the beam appears to diverge significantly once it leaves the DNG slab. The properties of the DNG medium hold the beam together as it propagates through the slab. Once it leaves the DNG slab, the beam must begin diverging; that is, if the DNG slab focuses the beam as it enters, the same physics will cause the beam to diverge as it exits. Moreover, there will be no focusing of the power from the wings to maintain the center portion of the beam. The rate of divergence of the exiting beam will be determined by its original value and the properties and size of the DNG medium. We also point out that a beam focused into a DNG slab will generate a diverging beam within the slab and a converging beam upon exit from the slab. This behavior has also been confirmed with the FDTD simulator.

It must be mentioned that a planar DNG slab is unable to focus a collimated beam (i.e., flat beam) or a plane wave, since the negative angle of refraction can occur only if there is oblique incidence. To focus a flat Gaussian beam (one with nearly an infinite radius of curvature), one must resort to a curved lens. In contrast to focusing (diverging) a plane wave with a convex (concave) DPS lens,

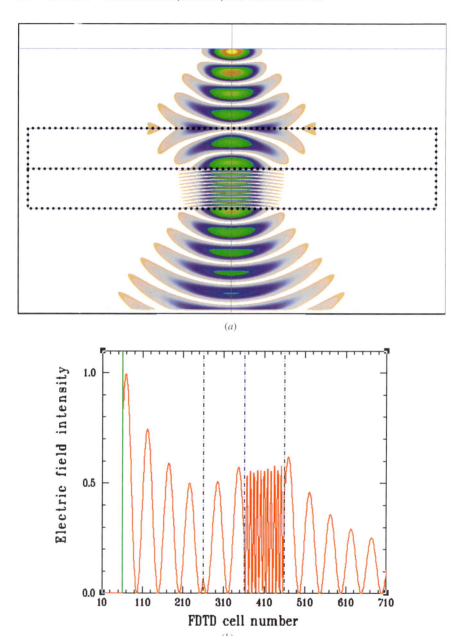

(a)

(b)

Figure 1.16   (a) The FDTD predicted electric field intensity distribution illustrates the focusing of a Gaussian beam in the first $n_{\text{real}}(\omega_0) \approx -1$ DNG slab and the channeling of the resulting beam in the second $n_{\text{real}}(\omega_0) \approx -6$ DNG slab. (b) Longitudinal profile of electric field intensity distribution. The slab pair focuses the beam and produces an output amplitude which is expectedly larger than its input value.

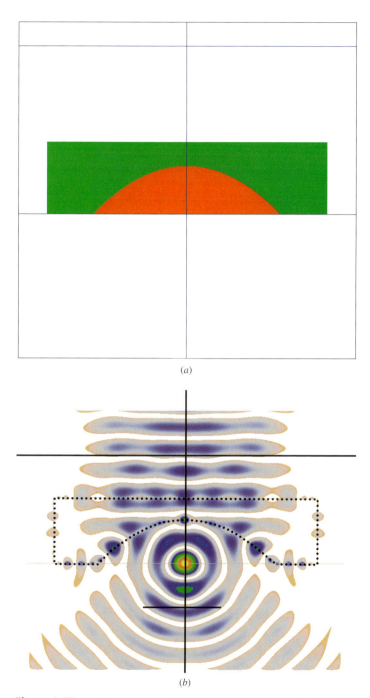

(a)

(b)

Figure 1.17   (*a*) Planoconcave DNG lens configuration. (*b*) The FDTD predicted electric field intensity distribution illustrates the focusing of a Gaussian beam with the planoconcave lens. The focal spot has both the longitudinal ($0.19\lambda_0$) and transverse ($0.17\lambda_0$) dimensions approximately equal.

one must use a concave (convex) DNG lens to achieve a focus (divergence). Such a planoconcave DNG lens with $n_{\text{real}}(\omega_0) \approx -1$ is shown in Figure 1.17a. It was formed by removing a parabolic section from the back side of a slab that was $1.5\lambda_0$ deep and $6.0\lambda_0$ wide. The focal length was chosen to be λ_0, and the location of the focus was chosen to be at the center of the back face of the slab. The full width of the removed parabolic section at the back face was $4\lambda_0$. A Gaussian beam with a waist of $2\lambda_0$ was launched $2\lambda_0$ distance away from the planar side of this lens and was normally incident on it. It is known that a DPS planoconvex lens of index n_{DPS} with a similar radius of curvature $R = 2\lambda_0$ (the red region in Fig. 1.17a) would have a focus located a distance $f_{\text{DPS}} = R/(n_{\text{DPS}} - 1) = 2\lambda_0/(n_{\text{DPS}} - 1)$ from its back face. Thus, to have the focal point within the very near field, as it is in the DNG case, the index of refraction would have to be very large. In fact, to have it located at the back face would require $n_{\text{DPS}} \rightarrow \infty$. This would also mean that very little of the incident beam would be transmitted through such a high-index lens because the magnitude of the reflection coefficient would approach 1. In contrast, the DNG lens achieves a greater bending of the incident waves with only moderate absolute values of the refractive index and is impedance matched to the incident medium. Moreover, since the incident beam waist occurs at the lens, the expected waist of the focused beam would be $w_{\text{focus}} \approx (\lambda_0 f_{\text{DPS}})/(\pi w_0) = \lambda_0/[\pi(n_{\text{DPS}} - 1)]$ [55]. For a normal glass lens $n_{\text{DPS}} \approx 1.5$; hence, the transverse waist at the focus would be $w_{\text{focus}} \approx \lambda_0/1.57$ and the corresponding intensity half-maximum waist would be $0.589 w_{\text{focus}}$. The longitudinal size of the focus is the depth of focus, which for the normal glass lens would be $2(\pi w_{\text{focus}}^2/\lambda_0)$. Again, to achieve a focus that is significantly subwavelength using a DPS lens, a very large index value would be required and would lead to similar disadvantages in comparison to the DNG lens. However, for the DNG planoconcave lens, one obtains more favorable results. Figure 1.17b shows a snapshot of the FDTD-predicted electric field intensity distribution when the intensity is peaked at the focal point. The radius of the focus along the beam axis (half-intensity radius) is measured to be about $\lambda_0/5$ and along the transverse direction it is about $\lambda_0/6$. This subwavelength focal region is significantly smaller than would be expected from the corresponding, traditional DPS lens. Moreover, even though the focal point is in the extreme near field of the lens, the focal region is nearly symmetrical and has a resolution that is much smaller than a wavelength.

1.11   METAMATERIALS WITH A ZERO INDEX OF REFRACTION

Metamaterials, in which the permittivity and/or permeability are near zero and thus the refractive index is much smaller than unity, can offer exciting potential applications. Their location on the $\varepsilon - \mu$ space diagram is represented by the red dot in Figure 1.18. Planar metamaterials that exhibit both positive and negative values of the index of refraction near zero have been realized experimentally by several research groups [8, 31, 32, 56–62] and are discussed in several chapters

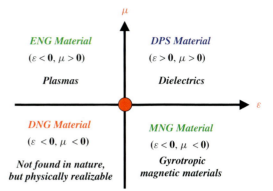

Figure 1.18 The zero-index media lie at the intersection of the various types of materials.

of this book. Within these studies, there have also been several demonstrations, both theoretically and experimentally, of planar metamaterials that exhibit a zero index of refraction within a specified frequency band. In particular, by matching the resonances in a series–parallel lumped-element circuit realization of a DNG metamaterial at a specified frequency, the propagation constant as a function of frequency continuously passes through zero (giving a zero index) with a nonzero slope (giving a nonzero group speed) in its transition from a DNG region of its operational behavior to a DPS region [56, 59, 63]. Several applications of these series–parallel metamaterials have been proposed and realized (e.g., phase shifters, couplers, and compact resonators).

Several investigations have also presented volumetric metamaterials that exhibit near-zero-index medium properties, for instance [64–68]. These zero-index electromagnetic bandgap (EBG) structure studies include working in a passband. By introducing a source into a zero-index EBG with an excitation frequency that lies within the EBG's passband, Enoch, Tayeb, and co-workers produced extremely narrow antenna patterns [66–68, Chapter 10]. Alù et al. have also shown theoretically that, by covering a subwavelength tiny aperture in a flat perfectly conducting screen with a slab of materials with $\mu \ll \mu_o$, one can significantly increase the power transmitted through such a hole, due to the coupling of the incident wave into the leaky wave supported by such a layer [69]. By covering both sides of the hole, not only can one increase the transmitted power through the hole but this power can be directed as a sharp beam in a given direction [69, 70].

These results stimulated a study by Ziolkowski [71] that details the propagation and scattering properties of a passive, dispersive metamaterial that is matched to free space and has an index of refraction equal to zero. One-, two-, and three-dimensional problems corresponding to source and scattering configurations have been treated analytically. The 1D and 2D results have been confirmed numerically with FDTD simulations. It has been shown that the electromagnetic fields in a matched zero-index medium [i.e., $\varepsilon_{\mathrm{real}}(\omega_0) \cong 0$, $\mu_{\mathrm{real}}(\omega_0) \cong 0$ so that $Z(\omega_0) = Z_0$ and $n_{\mathrm{real}}(\omega_0) \cong 0$] take on a static character in space, yet remain dynamic in time, in such a manner that the underlying physics remains associated with propagating fields.

To illustrate this behavior, consider Maxwell's equations:

$$\nabla \times \mathbf{E}_\omega = -j\omega\mu\mathbf{H}_\omega \qquad \nabla \cdot (\varepsilon\mathbf{E}_\omega) = \rho_\omega$$
$$\nabla \times \mathbf{H}_\omega = j\omega\varepsilon\mathbf{E}_\omega + \mathbf{J}_\omega \qquad \nabla \cdot (\mu\mathbf{H}_\omega) = 0 \tag{1.28}$$

When $\varepsilon_{\text{real}}(\omega_0) \cong 0$ and $\mu_{\text{real}}(\omega_0) \cong 0$, Maxwell's equations reduce to

$$\nabla \times \mathbf{E}_\omega = 0 \qquad \nabla \cdot (\varepsilon\mathbf{E}_\omega) = 0$$
$$\nabla \times \mathbf{H}_\omega = \vec{J}_\omega \qquad \nabla \cdot (\mu\mathbf{H}_\omega) = 0 \tag{1.29}$$

The equations on the right are automatically satisfied in the zero-index medium if the fields are finite. Thus one obtains staticlike equations for the fields within a zero-index medium. For an infinite cylindrical zero-index medium surrounded by free space, the solutions for a infinite line current

$$\mathbf{J}_\omega(\rho, \phi, z) = I_0 \frac{\delta(\rho)}{2\pi\rho} \hat{z} \tag{1.30}$$

are

$$\mathbf{E}_\omega(\rho, \phi, z) = -Z_0 \frac{I_0}{2\pi a} \frac{j H_0^{(2)}(k_0 a)}{H_1^{(2)}(k_0 a)} \hat{z}$$
$$\mathbf{H}_\omega(\rho, \phi, z) = \frac{I_0}{2\pi\rho} \hat{\phi} \tag{1.31}$$

for $r \leq a$ and

$$\mathbf{E}_\omega(\rho, \phi, z) = -Z_0 \frac{I_0}{2\pi a} \frac{j H_0^{(2)}(k_0 \rho)}{H_1^{(2)}(k_0 a)} \hat{z}$$
$$\mathbf{H}_\omega(\rho, \phi, z) = \frac{I_0}{2\pi a} \frac{H_1^{(2)}(k_0 \rho)}{H_1^{(2)}(k_0 a)} \hat{\phi} \tag{1.32}$$

for $r > a$, where $k_0 = \omega\sqrt{\varepsilon_0}\sqrt{\mu_0}$. Thus the solution inside the cylinder has a spatially constant electric field and magnetostatic magnetic field, whereas outside the cylinder, cylindrical waves propagate away from the source. Nonetheless, there is power flowing outward from the source in both regions; that is, the time-averaged Poynting's vector is

$$\langle \mathbf{S}_\omega \rangle = \frac{Z_0 I_0^2}{(2\pi a)^2} \frac{1}{|H_1^{(2)}(k_0 a)|^2} \frac{2}{\pi k_0 \rho} \hat{\rho} \tag{1.33}$$

If the spectrum of the time history driving the source is localized, one can easily take into account the time variations of the fields. In particular, one can approximately write

$$\mathbf{H}(\rho, \phi, z, t) = \frac{I_0}{2\pi\rho} f(t)\hat{\phi}$$
$$\mathbf{E}(\rho, \phi, z, t) \approx -Z_0 \frac{I_0}{2\pi a} f(t)\hat{z} \tag{1.34}$$

These 2D results have been confirmed with FDTD simulations. The FDTD predicted electric field distributions at various times are shown in Figure 1.19

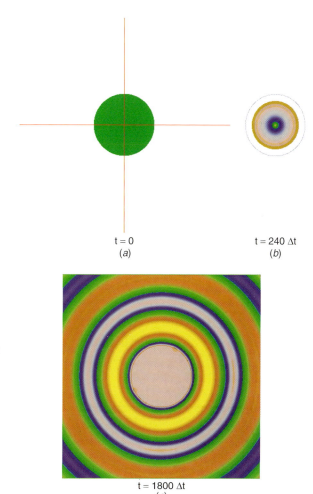

t = 0
(a)

t = 240 Δt
(b)

t = 1800 Δt
(c)

Figure 1.19 An infinite line source is located at the center of a matched zero-index infinite cylinder. The FDTD predicted electric field intensity distributions are shown for (a) $t = 0$, (b) $t = 240\Delta t$, and (c) $1800\Delta t$. A cylindrical wave propagates away from the cylinder while a uniform electric field intensity develops within the cylinder. From [71]. Copyright © 2004 by the American Physical Society.

at the initial time, early in the simulation, and at the end of the simulation. The infinite line source is driven at 30 GHz, is orthogonal to the plane, and is centered in the zero-index cylinder, which had a $0.6\lambda_0 = 60$-cell radius. The time sequence shows that the electromagnetic field energy propagates radially outward through the zero-index cylinder into the free-space region. The cross-sectional profiles of the electric and magnetic field distributions are given in Figures 1.20a and b, respectively. The electric field is clearly spatially constant throughout the entire cylinder but varies in time. On the other hand, the magnetic field component within the cylinder has taken on the predicted magnetostatic spatial characteristics and also varies in time. The electromagnetic field within the matched zero-index cylinder transitions to an oscillatory, propagating field once it exits the cylinder and enters free space. Comparisons of the time histories at various locations

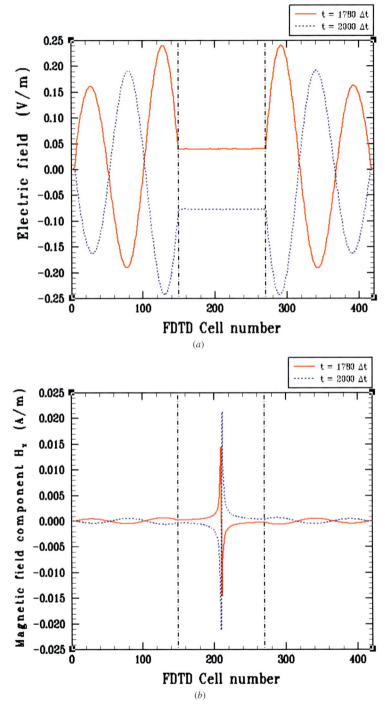

Figure 1.20 Cross-sectional profiles of (*a*) electric field and (*b*) magnetic field when a line source is driven at the center of a zero-index cylinder. From [71]. Copyright © 2004 by the American Physical Society.

within the cylinder show that there is zero phase difference between any two points once steady state has been achieved.

There have been related discussions [72–74] of metamaterials that exhibit "nihility," basically the zero-index properties described here. The generalization to more general bianisotropic media in which only the chiral parameters are nonzero has been considered in [74]. The latter suggests the intriguing possibility of force-free *electromagnetic* field configurations in metamaterials exhibiting such chiral nihility. Force-free magnetic fields are used, for example, to explain the behavior of plasmas associated with solar prominences and spheromaks (toroidal plasma states). All of these zero-index medium examples simply further illustrate how "exotic" the physical properties of metamaterials can be.

1.12 SUMMARY

In this chapter, we briefly reviewed the history and several selected topics associated with the fundamental properties of DNG metamaterials. We have shown how wave interaction with such materials can lead to interesting, unconventional features not observed in standard DPS media. A more comprehensive review can be found in our recent paper [48]. These physics characteristics can lead to exciting engineering concepts with future potential applications. Some of these concepts will be reviewed in Chapter 2.

REFERENCES

1. J. C. Bose, "On the rotation of plane of polarisation of electric waves by a twisted structure," *Proc. Roy. Soc.*, vol. 63, pp. 146–152, 1898.
2. I. V. Lindell, A. H. Sihvola, and J. Kurkijarvi, "Karl F. Lindman: The last Hertzian, and a Harbinger of electromagnetic chirality," *IEEE Antennas Propag. Mag.*, vol. 34, no. 3, pp. 24–30, 1992.
3. W. E. Kock, "Metallic delay lenses," *Bell Sys. Tech. J.*, vol. 27, pp. 58–82, 1948.
4. V. G. Veselago, "The electrodynamics of substances with simultaneously negative values of ε and μ," *Sov. Phys. Uspekhi*, vol. 10, no. 4, pp. 509–514, 1968. [*Usp. Fiz. Nauk*, vol. 92, pp. 517–526, 1967.]
5. D. R. Smith, W. J. Padilla, D. C. Vier, S. C. Nemat-Nasser, and S. Schultz, "Composite medium with simultaneously negative permeability and permittivity," *Phys. Rev. Lett.*, vol. 84, no. 18, pp. 4184–4187, May 2000.
6. R. A. Shelby, D. R. Smith, and S. Schultz, "Experimental verification of a negative index of refraction," *Science*, vol. 292, no. 5514, pp. 77–79, 6 Apr. 2001.
7. J. B. Pendry, "Negative refraction makes a perfect lens," *Phys. Rev. Lett.*, vol. 85, no. 18, pp. 3966–3969, Oct. 2000.
8. C. Caloz, C.-C. Chang, and T. Itoh, "Full-wave verification of the fundamental properties of left-handed materials in waveguide configurations," *J. Appl. Phys.*, vol. 90, no. 11, pp. 5483–5486, Dec. 2001.
9. A. K. Iyer and G. V. Eleftheriades, "Negative refractive index metamaterials supporting 2-D waves," in *2002 IEEE MTT International Microwave Symposium (IMS) Digest*, Seattle, WA, June 2–7, 2002, pp. 1067–1070.
10. C. Caloz, H. Okabe, T. Iwai, T. Itoh, "Transmission line approach of left-handed materials," paper presented at the 2002 IEEE AP-S International Symposium and USNC/URSI

National Radio Science Meeting, San Antonio, TX, June 16–21, 2002, abstract, *URSI Digest*, p. 39.

11. I. V. Lindell, S. A. Tretyakov, K. I. Nikoskinen, and S. Ilvonen, "BW media—Media with negative parameters, capable of supporting backward waves," *Microwave Opt. Tech. Lett.*, vol. 31, no. 2, pp. 129–133, Oct. 2001.

12. R. W. Ziolkowski and E. Heyman, "Wave propagation in media having negative permittivity and permeability," *Phys. Rev. E*, vol. 64, no. 5, 056625, Oct. 2001.

13. R. W. Ziolkowski and F. Auzanneau, "Passive artificial molecule realizations of dielectric materials," *J. Appl. Phys.*, vol. 82, no. 7, pp. 3195–3198, Oct. 1997.

14. F. Auzanneau and R. W. Ziolkowski, "Microwave signal rectification using artificial composite materials composed of diode loaded, electrically small dipole antennas," *IEEE Trans. Microwave Theory Tech.*, vol. 46, no. 11, pp. 1628–1637, Nov. 1998.

15. D. C. Wittwer and R. W. Ziolkowski, "Two time-derivative Lorentz material (2TDLM) formulation of a Maxwellian absorbing layer matched to a lossy media," *IEEE Trans. Antennas Propag.*, vol. 48, no. 2, pp. 192–199, Feb. 2000.

16. R. W. Ziolkowski, "Superluminal transmission of information through an electromagnetic metamaterial," *Phys. Rev. E*, vol. 63, 046604, Apr. 2001.

17. R. W. Ziolkowski and C.-Y. Cheng, "Existence and design of trans-vacuum-speed metamaterials," *Phys. Rev. E*, vol. 68, 026612, Aug. 2003.

18. D. R. Smith and N. Kroll, "Negative refractive index in left-handed materials," *Phys. Rev. Lett.*, vol. 85, pp. 2933–2936, Oct. 2000.

19. N. Engheta, "An idea for thin subwavelength cavity resonators using metamaterials with negative permittivity and permeability," *IEEE Antennas Wireless Propag. Lett.*, vol. 1, pp. 10–13, 2002.

20. J. A. Kong, B.-I. Wu, and Y. Zhang, "A unique lateral displacement of a Gaussian beam transmitted through a slab with negative permittivity and permeability," *Microwave Opt. Tech. Lett.*, vol. 33, pp. 136–139, Mar. 2002.

21. P. Kolinko and D. R. Smith, "Numerical study of electromagnetic waves interacting with negative index materials," *Opt. Express*, vol. 11, pp. 640–648, Apr. 2003.

22. R. W. Ziolkowski, "Pulsed and CW Gaussian beam interactions with double negative metamaterial slabs," *Opt. Express*, vol. 11, pp. 662–681, Apr. 2003.

23. R. W. Ziolkowski, "Pulsed and CW Gaussian beam interactions with double negative metamaterial slabs: Errata," *Opt. Express*, vol. 11, no. 13, pp. 1596–1597, June 2003.

24. S. Foteinopoulou, E. N. Economou, and C. M. Soukoulis, "Refraction in media with a negative refractive index," *Phys. Rev. Lett.*, vol. 90, 107402, Mar. 2003.

25. A. Alù, and N. Engheta, "Guided modes in a waveguide filled with a pair of single-negative (SNG), double-negative (DNG), and/or double-positive (DPS) layers," *IEEE Trans. Microwave Theory Tech.*, vol. MTT-52, no. 1, pp. 199–210, Jan. 2004.

26. A. Alù and N. Engheta, "An overview of salient properties of guided-wave structures with double-negative and single-negative metamaterials," in *Negative Refraction Metamaterials: Fundamental Properties and Applications* (G. V. Eleftheriades and K. G. Balmain, Eds.), IEEE Press/Wiley, Hoboken, NJ, 2005, Chap. 15, pp. 339–380.

27. N. Engheta, "Ideas for potential applications of metamaterials with negative permittivity and permeability," *Advances in Electromagnetics of Complex Media and Metamaterials*, NATO Science Series (S. Zouhdi, A. H. Sihvola, and M. Arsalane, Eds.), Kluwer Academic, Dordrecht, The Netherlands, 2002, pp. 19–37.

28. I. S. Nefedov and S. A. Tretyakov, "Waveguide containing a backward-wave slab," e-print in arXiv:cond-mat/0211185 v1, at http://arxiv.org/pdf/cond-mat/0211185, Nov. 10, 2002.

29. B.-I. Wu, T. M. Grzegorczyk, Y. Zhang, and J. A. Kong, "Guided modes with imaginary transverse wave number in a slab waveguide with negative permittivity and permeability," *J. Appl. Phys.*, vol. 93, no. 11, pp. 9386–9388, June 2003.

30. A. Topa, "Contradirectional interaction in a NRD waveguide coupler with a metamaterial slab," paper presented at XXVII General Assembly of International Union of Radio Science (URSI GA'02), Maastricht, The Netherlands, Aug. 17–24, 2002, *CD Digest*, paper no. 1878.

31. A. Grbic and G. V. Eleftheriades, "Overcoming the diffraction limit with a planar left-handed transmission line lens," *Phys. Rev. Lett.*, vol. 92, no. 11, 117403, Mar. 2004.
32. A. A. Grbic and G. V. Eleftheriades, "Experimental verification of backward-wave radiation from a negative refractive index metamaterial," *J. Appl. Phys.*, vol. 92, pp. 5930–5935, Nov. 2002.
33. J. Lu, T. M. Grzegorczyk, Y. Zhang, J. Pacheco, Jr., B.-I. Wu, J. A. Kong, and M. Chen, "Čerenkov radiation in materials with negative permittivity and permeability," *Opt. Express*, vol. 11, pp. 723–734, Apr. 2003.
34. Z. M. Zhang and C. J. Fu, "Unusual photon tunneling in the presence of a layer with negative refractive index," *Appl. Phys. Lett.*, vol. 80, pp. 1097–1099, Feb. 2002.
35. L. Wu, S. He, and L. Chen, "On unusual narrow transmission bands for a multi-layered periodic structure containing left-handed materials," *Opt. Express*, vol. 11, pp. 1283–1290, June 2003.
36. R. W. Ziolkowski and A. Kipple, "Application of double negative metamaterials to increase the power radiated by electrically small antennas," *IEEE Trans. Antennas Propag.*, vol. 51, no. 10, pp. 2626–2640, Oct. 2003.
37. R. W. Ziolkowski, "Gaussian beam interactions with double negative (DNG) metamaterials," in *Negative Refraction Metamaterials: Fundamental Properties and Applications* (G. V. Eleftheriades and K. G. Balmain, Eds.), IEEE Press/Wiley, Hoboken, NJ, 2005, Chapter 4, pp. 171–211.
38. A. Taflove, *Computational Electrodynamics: The Finite-Difference Time-Domain Method*, Artech House, Norwood, MA, 1995.
39. D. C. Wittwer and R. W. Ziolkowski, "Maxwellian material based absorbing boundary conditions for lossy media in 3D," *IEEE Trans. Antennas Propag.*, vol. 48, pp. 200–213, Feb. 2000.
40. R. A. Shelby, D. R. Smith, S. C. Nemat-Nasser, and S. Schultz, "Microwave transmission through a two-dimensional, isotropic, left-handed metamaterial," *Appl. Phys. Lett.*, vol. 78, pp. 489–491, Jan. 2000.
41. M. M. I. Saadoun, and N. Engheta, "Theoretical study of electromagnetic properties of non-local omega media" in *Progress in Electromagnetic Research (PIER) Monograph Series*, Vol. 9 (A. Priou, Guest Ed.), EMW Publishing, Cambridge, MA, 1994, Chapter 15, pp. 351–397.
42. P. F. Loschialpo, D. L. Smith, D. W. Forester, and F. J. Rachford, "Electromagnetic waves focused by a negative-index planar lens," *Phys. Rev. E*, vol. 67, 025602(R), Feb. 2003.
43. M. K. Kärkkäinen, "Numerical study of wave propagation in uniaxially anisotropic Lorentzian backward-wave slabs," *Phys. Rev. E*, vol. 68, 026602, Aug. 2003.
44. R. W. Ziolkowski and A. Kipple, "Causality and double-negative metamaterials," *Phys. Rev. E*, vol. 68, 026615, Aug. 2003.
45. A. Alù, and N. Engheta, "Pairing an epsilon-negative slab with a mu-negative slab: Anomalous tunneling and transparency," *IEEE Trans. Antennas Propag., Special Issue on Metamaterials*, vol. AP-51, no. 10, pp. 2558–2570, Oct. 2003.
46. A. Alù and N. Engheta, "A physical insight into the 'growing' evanescent fields of double-negative metamaterial lens using its circuit equivalence," *IEEE Trans. Antennas Propag.*, vol. 54, pp. 268–272, Jan. 2006.
47. A. Alù and N. Engheta, "Radiation from a traveling-wave current sheet at the interface between a conventional material and a material with negative permittivity and permeability," *Microwave Opt. Tech. Lett.*, vol. 35, no. 6, pp. 460–463, Dec. 2002.
48. N. Engheta and R. W. Ziolkowski, "A positive future for double-negative metamaterials," *IEEE Trans. Microwave Theory Tech.*, Special Issue on Metamaterial Structures, Phenomena, and Applications, vol. 53, no. 4 (part II), pp. 1535–1556, Apr. 2005.
49. C.-Y. Cheng and R. W. Ziolkowski, "Tailoring double negative metamaterial responses to achieve anomalous propagation effects along microstrip transmission line," *IEEE Trans. Microwave Theory Tech.*, vol. 51, pp. 2306–2314, Dec. 2003.
50. F. Gardiol, *Microstrip Circuits*, Wiley, New York, 1994, pp. 48–50.

51. W. J. Getsinger, "Microstrip dispersion model," *IEEE Trans. Microwave Theory Tech.*, vol. MTT-21, pp. 34–39, Jan. 1973.
52. D. R. Smith, D. Schurig, M. Rosenbluth, S. Schultz, S. Anantha Ramakrishna, and J. B. Pendry, "Limitation on subdiffraction imaging with a negative refractive index slab," *Appl. Phys. Lett.*, vol. 82, no. 10, pp. 1506–1508, Mar. 2003.
53. C. Luo, S. G. Johnson, and J. D. Joannopoulos, "Subwavelength imaging in photonic crystals," *Phys. Rev. B*, vol. 68, 045115, July 2003.
54. A. Alù and N. Engheta, "Tunneling and 'growing evanescent envelopes' in a pair of cascaded sets of frequency selective surfaces in their band gaps," in *Proceedings of the 2004 URSI International Symposium on Electromagnetic Theory*, vol. 1, Pisa, Italy, May 24–27, 2004, pp. 90–92.
55. B. E. A. Saleh and M. C. Teich, *Fundamentals of Photonics*, Wiley, New York, 1991, pp. 94–95.
56. C. Caloz and T. Itoh, "Microwave applications of novel metamaterials," in *Proceedings of the International Conference on Electromagnetics in Advanced Applications, ICEAA'03*, Torino, Italy, Sept. 2003, pp. 427–430.
57. C. Caloz, A. Sanada, and T. Itoh, "A novel composite right/left-handed coupled-line directional coupler with arbitrary coupling level and broad bandwidth," *IEEE Trans. Microwave Theory Tech.*, vol. 52, pp. 980–992, Mar. 2004.
58. C. Caloz and T. Itoh, "A novel mixed conventional microstrip and composite right/left-handed backward wave directional coupler with broadband and tight coupling characteristics," *IEEE Microwave Wireless Components Lett.*, vol. 14, no. 1, p. 31–33, Jan. 2004.
59. G. G. V. Eleftheriades, A. K. Iyer, and P. C. Kremer, "Planar negative refractive index media using periodically *L-C* loaded transmission lines," *IEEE Trans. Microwave Theory Tech.*, vol. 50, pp. 2702–2712, Dec. 2002.
60. R. Islam, F. Eleck, and G. V. Eleftheriades, "Coupled-line metamaterial coupler having co-directional phase but contra-directional power flow," *Electron. Lett.*, vol. 40, no. 5, pp. 315–317, Mar. 2004.
61. A. A. Oliner, "A periodic-structure negative-refractive-index medium without resonant elements," paper presented at 2002 IEEE AP-S Int. Symp./USNC/URSI National Radio Science Meeting, San Antonio, TX, June 16–21, 2002, *URSI Digest*, p. 41.
62. A. A. Oliner, "A planar negative-refractive-index medium without resonant elements," in *MTT Int. Microwave Symp. (IMS'03) Digest*, Philadelphia, PA, June 8–13, 2003, pp. 191–194.
63. R. W. Ziolkowski and C.-Y. Cheng, "Lumped element models of double negative metamaterial-based transmission lines," *Radio Sci.*, vol. 39, RS2017, doi: 10.1029/2003RS002995, Apr. 2004.
64. B. Gralak, S. Enoch, and G. Tayeb, "Anomalous refractive properties of photonic crystals," *J. Opt. Soc. Am. A*, vol. 17, no. 6, pp. 1012–1020, June 2000.
65. M. Notomi, "Theory of light propagation in strongly modulated photonic crystals: Refractionlike behavior in the vicinity of the photonic bandgap," *Phys. Rev. B*, vol. 62, no. 16, 10696, Oct. 2000.
66. S. Enoch, G. Tayeb, P. Sabouroux, N. Guerin, and P. Vincent, "A metamaterial for directive emission," *Phys. Rev. Lett.*, vol. 89, 213902, Nov. 2002.
67. S. Enoch, G. Tayeb, and B. Gralak, "The richness of dispersion relation of electromagnetic bandgap materials," *IEEE Trans. Antennas Propag.*, vol. 51, no. 10, pp. 2659–2666, Oct. 2003.
68. G. Tayeb, S. Enoch, P. Vincent, and P. Sabouroux,, "A compact directive antenna using ultrarefractive properties of metamaterials," in *Proceedings of the International Conference on Electromagnetics in Advanced Applications, ICEAA'03*, Torino, Italy, Sept. 2003, pp. 423–426.
69. A. Alù, F. Bilotti, N. Engheta, and L. Vegni, "How metamaterials may significantly affect the wave transmission through a sub-wavelength hole in a flat perfectly conducting

screen," paper present at Workshop on Metamaterials for Microwave and (Sub) millime-tre Wave Applications: Photonic Bandgap and Double Negative Designs, Components and Experiments, London, Nov. 24, 2003.

70. A. Alù, F. Bilotti, N. Engheta, and L. Vegni, "Metamaterial bilayers for enhancement of wave transmission through a small hole in a flat perfectly conducting screen," in *2004 IEEE Antennas and Propagation Society (AP-S) International Symposium Digest*, Monterey, CA, June 20–26, 2004, vol. 3, pp. 3163–3166.

71. R. W. Ziolkowski, "Propagation in and scattering from a matched metamaterial having a zero index of refraction," *Phys. Rev. E*, vol. 70, 046608, Oct. 2004.

72. A. Lakhtakia, "On perfect lenses and nihility," *Int. J. Infrared Millim. Waves*, vol. 23, no. 3, pp. 339–343, Mar. 2002.

73. A. Lakhtakia, "An electromagnetic trinity from 'negative permittivity' and 'negative permeability,' " *Int. J. Infrared Millim. Waves*, vol. 23, no. 6, pp. 813–818, June 2002.

74. S. Tretyakov, I. Nefedov, A. Sihvola, S. Maslovski, and C. Simovski, "Waves and energy in chiral hihility," *J. Electromagnetic Waves Appl.*, vol. 17, no. 5, pp. 695–706, May 2003.

CHAPTER 2

FUNDAMENTALS OF WAVEGUIDE AND ANTENNA APPLICATIONS INVOLVING DNG AND SNG METAMATERIALS

Nader Engheta, Andrea Alù, Richard W. Ziolkowski, and
Aycan Erentok

2.1 INTRODUCTION

In this chapter, we discuss some of the salient and unconventional features of cavity resonators, waveguides, scatterers, and antennas loaded or covered with double-negative and/or double-positive (DNG, DPS) metamaterials. Before starting this review, we should point out an important and general observation, which applies to each of the applications presented here: *The unconventional electromagnetic characteristics of metamaterials are exhibited when these materials are paired with other materials with at least one oppositely signed constitutive parameter.* In other words, when we pair a DNG material with a DPS, epsilon-negative (ENG), or mu-negative (MNG) layer, we may obtain interesting wave propagation properties that may be absent if we paired one DNG layer with another one. If the entire universe had been filled with DNG materials, we would not have had the possibility of exploiting these unusual features. (That is, in such a DNG world, the main difference we would have experienced is the Poynting vector being antiparallel with the phase velocity, but since this property would have been present at all points, it would not have had any practical utility.)

We also note that indeed the *interface* between two media with at least one pair of oppositely signed parameters can play a major role in offering anomalous behaviors for the combined structure. At the boundary between such two media, using the Maxwell equations one can write the continuity of the tangential electric and magnetic field components as

$$\frac{1}{-j\omega\mu_1}\frac{\partial E_{1,\text{tan}}}{\partial n}\bigg|_{\text{Interface}} = \frac{1}{-j\omega\mu_2}\frac{\partial E_{2,\text{tan}}}{\partial n}\bigg|_{\text{Interface}}$$

$$\frac{1}{j\omega\varepsilon_1}\frac{\partial H_{1,\text{tan}}}{\partial n}\bigg|_{\text{Interface}} = \frac{1}{j\omega\varepsilon_2}\frac{\partial H_{2,\text{tan}}}{\partial n}\bigg|_{\text{Interface}}$$

where $\partial/\partial n$ represents the normal derivative and ε_i and μ_i, $i = 1, 2$, are the permittivity and permeability in the two media, respectively. It is clear that the normal derivatives of these tangential components are not necessarily continuous, and furthermore, if μ_1 and μ_2 and/or ε_1 and ε_2 have opposite signs, then the derivatives of the tangential fields on both sides of the interface will have opposite signs. As a result, we observe a "V-shaped" discontinuity (see Fig. 1.9) for the tangential components of fields at the interface between such media [1], which may imply a concentrated resonant phenomenon at that interface (similar to the current and voltage distributions at the junction between an inductor and a capacitor at the resonance of an $L-C$ circuit). As will be described later in this chapter, this feature can lead to interesting characteristics for wave interaction in devices and components containing metamaterials.

It is also important to note that this "interface resonance" is essentially independent of the total thickness of the paired layers, since it arises along the discontinuity between two such conjugate materials [2]. The mechanism behind this resonance can be described in several ways, one of which is the equivalent circuit approach [2]. This resonant characteristic, which may occur in subwavelength structures formed by pairing such media, has provided us with ideas for cavities, waveguides, scatterers, antennas, and lenses that may operate with dimensions below the conventional diffraction limits. Here, we review some of these ideas and concepts.

2.2 SUBWAVELENGTH CAVITIES AND WAVEGUIDES

As anticipated in the previous paragraph, the interface resonance at the junction between two conjugate materials with at least one pair of oppositely signed constitutive parameters would provide a concentrated resonant phenomenon. In some of our works, we have utilized this effect to design thin, subwavelength cavity resonators and parallel-plate waveguides in which a layer of DNG material is paired with a layer of DPS material or in which ENG and MNG materials are paired together or with DPS materials [1, 3–5]. By exploiting the antiparallel nature of the phase velocity and Poynting vectors in a DNG slab, indeed we found the possibility of resonant modes in electrically thin parallel-plate structures containing such bilayered structures.

The geometry in this simple case is depicted in Figure 2.1 and consists of two parallel plates, perfect electric conductors (PECs), filled by two stacked planar slabs of homogeneous and isotropic materials with constitutive parameters

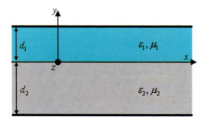

Figure 2.1 Geometry of the 1D parallel-plate waveguide filled with a pair of layers made of any two of ENG, MNG, DNG, and DPS materials. From [3], copyright © 2004 by the Institute of Electrical and Electronics Engineers (IEEE).

ε_1, ε_2, μ_1, μ_2 and thicknesses d_1, d_2. In [3] we have shown how the judicious choice of these parameters, involving pairing of DPS and DNG or ENG and MNG materials (or in other words the choice of oppositely signed values for the permittivities and/or permeabilities) provides interesting possibilities for having waveguides with no cutoff modes supporting fast and slow waves, independent of the total thickness of the waveguide, or also electrically large waveguides supporting only one single propagating mode.

In the thin-waveguide limit we have shown how, unlike the standard DPS–DPS case, the dispersion relation for the supported modes depends not on the total thickness of the waveguide $d = d_1 + d_2$ but instead on the *ratio* of the two slab thicknesses d_1/d_2, leading to the theoretical possibility of having waveguides supporting a resonant mode even when the total thickness tends to electrically small values [3]. This is evident in Figure 2.2, where the two cases of a DPS–DPS and a DPS–DNG waveguide are compared. Their dispersion plots for transverse electric (TE$_x$) polarization are sketched in the figure, showing the minimum total thickness of the waveguide, $d_1 + d_2$, required to have a supported mode propagating with the factor $e^{-j\beta x}$ for any pair (d_1, β). The dispersion relation for TE$_x$ polarization can be written in an implicit form as

$$\frac{\mu_1}{k_{t1}^{\text{TE}}} \tan(k_{t1}^{\text{TE}} d_1) = -\frac{\mu_2}{k_{t2}^{\text{TE}}} \tan(k_{t2}^{\text{TE}} d_2) \tag{2.1}$$

with $k_{t1} = \sqrt{\omega^2 \varepsilon_1 \mu_1 - \beta^2}$ and ω being the monochromatic radian frequency of excitation.

The figure clearly highlights the main difference between the two cases: In a standard DPS–DPS waveguide the minimum required thickness to have a TE propagating mode inside a parallel-plate waveguide has a finite value (i.e., the cutoff thickness), whereas in a DPS–DNG waveguide, independent of the value of β, a mode can be supported if the ratio d_1/d_2 is sufficiently close to the value of $-\mu_2/\mu_1$. It is interesting to emphasize that this phenomenon may be interpreted in terms of the compact resonance present at the interface between the two conjugate materials composing the slabs (in the DPS–DNG case): The resonance that happens "spatially" in a standard DPS–DPS waveguide, for which the transverse dimension has to be comparable with the wavelength of operation, may be substituted in the DPS–DNG waveguide by the *interface* resonance present at the junction between the two slabs, which effectively acts as a compact resonant "circuit" loading the waveguide. This allows, in principle, a marked reduction of the waveguide lateral dimension, which has also been shown experimentally (see, e.g., [4]). It is also worth noting how Figure 2.2 predicts the special case of a subwavelength cavity resonator, whose dispersion plot is shown on the $\beta = 0$ cut. As anticipated in [1], the possibility of a resonant cavity with subwavelength dimension loaded with conjugate materials may open up interesting venues and applications. Along the same line, 2D and 3D subwavelength cavities or Fabry–Perot components may be envisioned, which offer the realistic possibilities of going beyond the diffraction limitations in several applications [5,6].

Another interesting case arises from the anomalous behavior of ENG–MNG waveguides. Their interface resonance again can provide the possibility of subwavelength resonant modes, but it can also be combined with the rapidly decaying

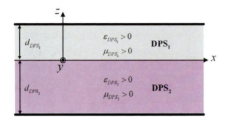

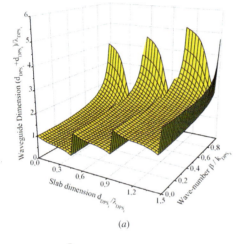

(a)

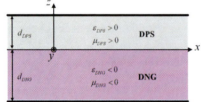

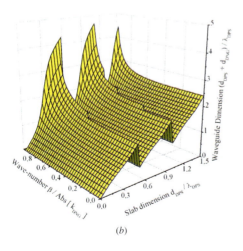

(b)

Figure 2.2 Dispersion plots for (a) DPS–DPS and (b) DPS–DNG waveguides. From [7]. Copyright © 2003 by the Institute of Electrical and Electronics Engineers (IEEE).

property of the field away from the interface due to the evanescent behavior of the electromagnetic fields in ENG and MNG media. This property, as proposed in [3], allows designing electrically thick waveguides with monomodal behavior, which may be useful in those applications where monomodality is a key issue.

In [7] we have analyzed in more detail the mode excitation in such waveguides filled with conjugate materials, showing its peculiar flow of energy that consists of two antiparallel flows in the two material slabs. Every mode in such a waveguide is in fact characterized by a *forward* powerflux flowing in the region with the positive constitutive parameter (permittivity or permeability depending on the polarization of interest) and a *backward* power flux in the other region, meaning that the Poynting vector and the phase flow are parallel in the "positive" medium and antiparallel in the "negative" one. This implies that the *net* power flow, the one actually carried by the mode, may be forward moving (when the first flow dominates), backward moving (when the second one is dominant), or even zero (when the two fluxes are the same but with opposite directions), suggesting the possibility of unusual resonant modes with a nonzero phase velocity (i.e., they would be similar to a standing wave in a standard resonator but with incident and reflected waves spatially separated in the two different materials).

Figure 2.3, as an example, shows the results of a mode-matching analysis of a discontinuity between a DPS–DNG parallel-plate waveguide (on the left) and an empty waveguide (on the right). The two waveguides support only one propagating mode each, whose real parts of the longitudinal Poynting vector distributions are described by the solid lines on each side of the discontinuity. The black arrows represent the real part of the Poynting vector at each point of the waveguide. As can be clearly seen from the figure, far enough away from the discontinuity the arrows show the power flow is longitudinal as given by

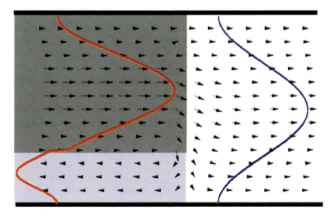

Figure 2.3 Mode-matching analysis of a discontinuity. The black arrows represent the real part of the Poynting vector. On the left we have a DPS–DNG waveguide with one propagating mode with power distribution described by the red line. On the right there is an empty parallel-plate waveguide that supports again only one propagating mode with field distribution given by the blue line. From [7]. Copyright © 2005 by the Institute of Electrical and Electronics Engineers (IEEE).

the propagating modes only. In the DPS–DNG waveguide portion, as mentioned above, they are oppositely directed and well separated from each other; that is, there is no transverse exchange of power for each mode, as required by the orthogonality relations. In the vicinity of the discontinuity, however, the distribution of the real part of the Poynting vectors (the black arrows) is rearranged by the presence of the evanescent nonpropagating modes, and they clearly reconstruct the transmitted mode (on the right of the discontinuity) and interestingly feed the backward flow (on the left of the discontinuity).

In [8] we have also studied this waveguide problem with a circuit model approach, showing how this inherent resonance at the interface between such two *conjugate* materials may be explained with an *LC* tank circuit analogy. To discuss the details of this analysis, we first need to review the circuit equivalence for wave propagation in general isotropic media.(The reader is referred to [3] for more details.)

Considering the propagation of a transverse magnetic (TM$_z$) plane wave in a homogeneous isotropic medium, one can write from Maxwell's equations

$$\frac{\partial E_x}{\partial z} = -j\omega\tilde{\mu}_{eq}H_y \qquad \frac{\partial H_y}{\partial z} = -j\omega\tilde{\varepsilon}_{eq}E_x \tag{2.2}$$

where $\tilde{\mu}_{eq}$ and $\tilde{\varepsilon}_{eq}$ are shorthand notations for $\tilde{\mu}_{eq} \equiv \mu\left[1 - k_x^2/(\omega^2\mu\varepsilon)\right]$ and $\tilde{\varepsilon}_{eq} \equiv \varepsilon$ and ε, μ are the material permittivity and permeability [9]. (By duality, one can easily write the corresponding terms for the TE case as well, which are $\tilde{\mu}_{eq} \equiv \mu$ and $\tilde{\varepsilon}_{eq} \equiv \varepsilon\left[1 - k_x^2/(\omega^2\mu\varepsilon)\right]$.) These expressions may obviously be viewed as analogs to the transmission line (TL) equations $\partial V/\partial z = -j\omega L_{eq}I$, $\partial I/\partial z = -j\omega C_{eq}V$, with the equivalent series inductance per unit length L_{eq} and equivalent shunt capacitance per unit length C_{eq} being proportional to $\tilde{\mu}_{eq}$ and $\tilde{\varepsilon}_{eq}$, that is,

$$L_{eq} \propto \tilde{\mu}_{eq} \qquad C_{eq} \propto \tilde{\varepsilon}_{eq} \tag{2.3}$$

Clearly the TL analogy may in general offer an interesting physical interpretation and alternative insight, effectively linking the voltage and current distributions along a circuit network to their local counterparts represented by the electric and magnetic fields. This is well known in the DPS case [9], but it can be easily extended to the metamaterial parameters, as it has been shown in [2, 8]. We note that even in a conventional DPS material, where μ and ε are positive, the value of L_{eq} in the TM case and C_{eq} in the TE case may become negative when $k_x^2 > \omega^2\mu\varepsilon$, that is, for an evanescent wave. As is well known, a negative equivalent inductance or capacitance at a given frequency may be interpreted effectively as a positive capacitance or inductance at that frequency, respectively [2]. Therefore, for the TM case the evanescent plane-wave propagation in a DPS medium may be modeled using a TL with a negative series inductance per unit length and a positive shunt capacitance per unit length, which effectively implies a positive series capacitance per unit length and a positive shunt capacitance per unit length. In such a C–C line, which is a ladder network made of capacitors, the currents and voltages cannot "propagate" along the line, but instead they have an evanescent behavior, consistent with their electromagnetic counterpart. When a DNG material or an ENG or MNG medium is used, their suitable equivalent TL

TABLE 2.1 Effective TL Models in Lossless DPS, DNG, ENG, MNG Slabs for TE and TM Propagating and Evanescent Waves

	DPS ($\mu > 0,\ \varepsilon > 0$)	DNG ($\mu < 0,\ \varepsilon < 0$)	ENG ($\mu > 0,\ \varepsilon < 0$)	MNG ($\mu < 0,\ \varepsilon > 0$)
TE propagating, $k_x^2 < \omega^2\mu\varepsilon$	$L_{eq} > 0 \quad \kappa \in \Re$ $C_{eq} > 0 \quad Z_t \in \Re$ L-C TL	$L_{eq} < 0 \quad \kappa \in \Re$ $C_{eq} < 0 \quad Z_t \in \Re$ C-L TL	Not applicable, since for $k_x \in \Re$ we always have $k_x^2 > \omega^2\mu\varepsilon$ in lossless ENG	Not applicable, since for $k_x \in \Re$ we always have $k_x^2 > \omega^2\mu\varepsilon$ in lossless MNG
TE evanescent, $k_x^2 > \omega^2\mu\varepsilon$	$L_{eq} > 0 \quad \kappa \in \Im$ $C_{eq} < 0 \quad Z_t \in \Im$ L-L TL	$L_{eq} < 0 \quad \kappa \in \Im$ $C_{eq} > 0 \quad Z_t \in \Im$ C-C TL	$L_{eq} < 0 \quad \kappa \in \Im$ $C_{eq} > 0 \quad Z_t \in \Im$ C-C TL	$L_{eq} > 0 \quad \kappa \in \Im$ $C_{eq} < 0 \quad Z_t \in \Im$ L-L TL
TM propagating, $k_x^2 < \omega^2\mu\varepsilon$	$L_{eq} > 0 \quad \kappa \in \Re$ $C_{eq} > 0 \quad Z_t \in \Re$ L-C TL	$L_{eq} < 0 \quad \kappa \in \Re$ $C_{eq} < 0 \quad Z_t \in \Re$ C-L TL	Not applicable, since for $k_x \in \Re$ we always have $k_x^2 > \omega^2\mu\varepsilon$ in lossless ENG	Not applicable, since for $k_x \in \Re$ we always have $k_x^2 > \omega^2\mu\varepsilon$ in lossless MNG
TM evanescent, $k_x^2 > \omega^2\mu\varepsilon$	$L_{eq} < 0 \quad \kappa \in \Im$ $C_{eq} > 0 \quad Z_t \in \Im$ C-C TL	$L_{eq} > 0 \quad \kappa \in \Im$ $C_{eq} < 0 \quad Z_t \in \Im$ L-L TL	$L_{eq} > 0 \quad \kappa \in \Im$ $C_{eq} < 0 \quad Z_t \in \Im$ L-L TL	$L_{eq} < 0 \quad \kappa \in \Im$ $C_{eq} > 0 \quad Z_t \in \Im$ C-C TL

Source: From [8]. Copyright © 2003 by the Society of Photo-Optical Instrumentation Engineers (SPIE).

models may exhibit anomalous properties consistent with the features of wave propagation in such media. In general, one may consider Table 2.1 showing the equivalent TL model for plane waves in lossless homogeneous isotropic media, with all of the possibilities for signs of the real part of their permittivity and permeability, both for the cases of propagating and evanescent waves. When losses are present, μ and/or ε have complex values, which translates into positive series resistances and/or shunt conductances in the TL model.

Returning to the waveguide problem, a standard transverse resonance technique [9] may be employed that considers the possible resonance between the parallel plates as a mode which bounces from one plate to the other. Obviously the way in which each medium should be considered as a TL may be obtained from Table 2.1, by considering the transverse wavenumbers in the two media for the guided mode with waveguide wavenumber κ. A sketch of this analysis is given in Figure 2.4. The DPS–DPS waveguide is here represented by two standard L–C segments closed by two short circuits representing the metallic plates. It is well known that such a circuit structure can resonate only when its length is comparable with the wavelength of operation. For a DPS–DNG waveguide the situation is different: At the interface between the two layers, conjugate elements from the L–C and C–L lines may go into resonance, justifying the anomalous behavior of the metamaterial waveguides and the peculiar "V-shaped" field distributions. The same technique may be exploited to study waveguides filled with other different pairs: ENG–MNG waveguides may support subwavelength modes when their equivalent subwavelength L–L and C–C TL segments go into resonance together. Similarly, DPS–ENG or DPS–MNG waveguides may show similar properties for the polarizations for which DPS materials behave respectively as MNG or ENG materials for the evanescent waves propagating into them (see Table 2.1). The choice of the guided wavenumber β therefore should ensure an appropriate evanescent tail in the transverse distribution of the mode in the DPS material.

Similar anomalous behavior may be verified in open waveguides, based on surface wave propagation along DNG slabs [10]. Here the pairing with DPS materials is ensured by the presence of the surrounding free space or the possible presence of a dielectric cover/cladding. The interface between the two media again provides a resonant guide for anomalous modes propagating along the surface. Figure 2.5 shows the dispersion diagram for the surface wave propagation along a DNG slab (black lines), as compared with the case of a standard DPS slab (red lines). It is important to note the differences in the dispersion plots, clearly evident in the lower branch of the DNG modes (i.e., the lowest order mode). In this plot, $\Delta k = \sqrt{\omega^2 \mu \varepsilon - \omega^2 \mu_0 \varepsilon_0}$ and $2d$ is the thickness of the planar slab. In a standard DPS slab, the surface wave propagation is possible when $\omega\sqrt{\varepsilon_0 \mu_0} = k_0 < \beta < k_{\text{DPS}} = \omega\sqrt{\varepsilon_{\text{DPS}} \mu_{\text{DPS}}}$, as confirmed by all of the branches for the DPS case (red lines). In a DNG slab, on the other hand, the lowest branch has no cutoff thickness, since its wavenumber β unconventionally increases as the thickness of the slab is decreased. This implies that a subwavelength planar slab would be characterized by a slow surface plasmon mode (with a large value of β for its $e^{-j\beta x}$ propagation factor, possibly even

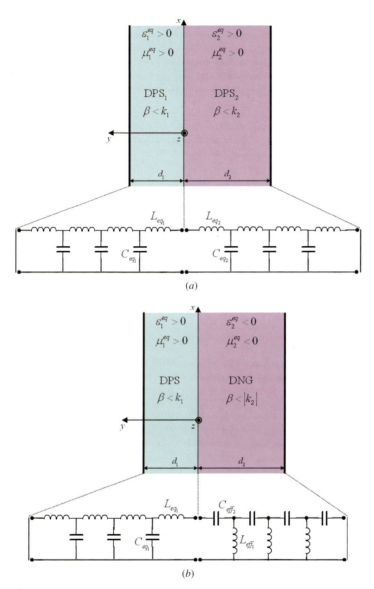

Figure 2.4 Parallel-plate waveguides: (*a*) filled with standard materials; (*b*) filled with DPS–DNG pair and their equivalent distributed circuit elements and TL models. In both cases, when the waveguide supports a given mode or the cavity is resonating, the corresponding equivalent circuit should be resonant at that frequency. The behavior of the voltages and currents along the circuit gives an insight to the corresponding behavior of the fields in the waveguide, and moreover, the intrinsic resonances between conjugate elements in (*b*) suggest an explanation for the anomalous behavior of metamaterial waveguides. From [8]. Copyright © 2003 by the Society of Photo-Optical Instrumentation Engineers (SPIE).

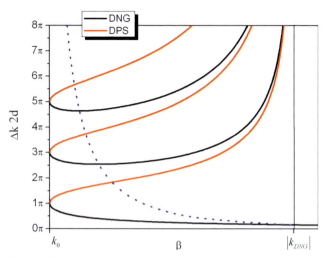

Figure 2.5 Dispersion plots of odd surface modes supported by slabs of DPS (red lines) and DNG (black lines) materials. The blue dotted line joins the minima of the black lines, which are the loci for which the surface wave does not carry any net power. From [7], Copyright © 2005 by the Institute of Electrical and Electronics Engineers (IEEE).

larger than $|k_{\text{DNG}}| = \omega\sqrt{|\varepsilon_{\text{DNG}}\mu_{\text{DNG}}|}$). Its field distribution would be highly concentrated around the two surfaces of the slab (i.e., at the DPS–DNG interfaces), allowing the effective cross section of the guided mode to become narrower than the conventional diffraction limits. Also in this case the peculiar power flow in this open waveguide would see opposite power flows inside and outside the slab. The net flow may be forward, that is, parallel with the phase flow, backward (antiparallel), or zero, for which we are at the minimum points in the dispersion curves shown in Figure 2.5. The blue dotted line in the figure joins all of these minima, and it can be shown that it is tangent at infinity with the lowest order mode of the DNG slab.

To describe the difference between a DNG and a DPS slab waveguide, let us consider the following situation. In a standard open dielectric waveguide one may be able to find a mode with zero cutoff thickness; that is, this mode can still be guided even when the DPS slab thickness becomes electrically very small. However, in this case the lateral field distribution of such a mode is spread out in the space around the slab when the slab section is electrically too thin, and the effective cross section of the guided mode may become much larger than the slab's lateral dimension (see Figs. 2.6a and b). On the other hand, when a DNG slab is considered, a highly concentrated guided mode (lowest order odd mode) propagating along the material slab may still be found even for very thin DNG slabs. In fact, the thinner the DNG slab is, the more confined the lowest order odd mode becomes (see Figs. 2.6c and d). This can lead to the possibility of building ultrathin open waveguides overcoming the standard diffraction limitation in energy transport [10] and may offer potential applications for miniaturization of interconnects.

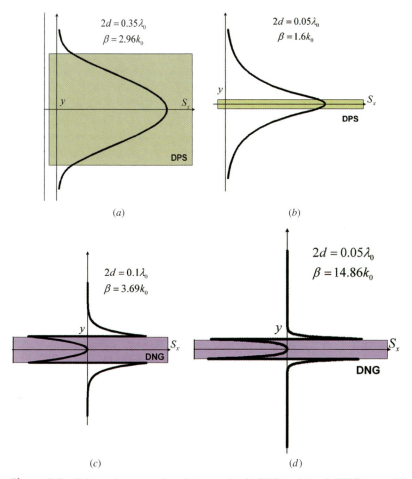

Figure 2.6 Schematic comparison between (a, b) DPS and (c, d) DNG open slab waveguides. For the standard DPS case, as the slab waveguide becomes thinner, the lateral distribution of the guided mode will be spread out (i.e., the mode will be weakly guided). [Compare (a) with (b).] However, in the DNG open slab, the lowest order odd mode will be more confined as the DNG slab gets thinner [compare (c) and (d)].

The unusual power flow properties of a DNG open slab waveguide led us to the conceptualization of "antidirectional" couplers when two of these open waveguides (one DNG and one DPS) are placed at the proximity of each other and thus are coupled together [11]. When we juxtapose two open waveguides, one with forward and the other with backward behavior, it becomes possible to provide an anomalous coupling with the backward feeding of port 1 from port 2, as depicted in Figure 2.7. In this figure a sketch of the power flows in each region of such an idealized coupler is also shown for a given modal distribution (with fixed β of propagation), highlighting the anomalous Poynting vector distribution in such complex configurations.

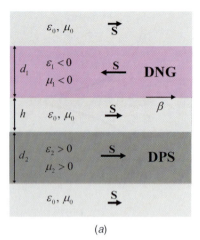

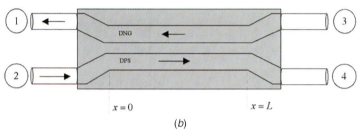

Figure 2.7 (*a*) Geometry and possible application of antidirectional coupler with DNG and DPS open slab waveguides placed in parallel and in proximity of each other. From [11]. Copyright © 2002 by the Institute of Electrical and Electronics Engineers. (*b*) This can lead to the "backward" coupling of energy from port 2 to port 1.

2.3 SUBWAVELENGTH CYLINDRICAL AND SPHERICAL CORE–SHELL SYSTEMS

In the previous sections we highlighted how a judicious pairing of materials with oppositely signed parameters may provide a means to realize compact resonant structures and subwavelength guided-wave systems through the presence of the interface resonances. When concentric spherical (or coaxial cylindrical) shells of such pairs of materials are considered, similar interface resonances may be induced with appropriate designs [12–14]. In this way one may form subwavelength (3D or 2D) resonant structures that support resonant modes effectively independent of the total width of the object. When the coaxial cylindrical shells are covered by a metallic cylinder, for instance, a thin subwavelength metallic waveguide may be designed with a propagating mode independent of its total size; when instead the two shells are surrounded by free space, we may have an open subwavelength waveguide or a resonant scatterer [10, 12, 13]. In the case of the scatterer, we have analyzed the possibility of a large scattering width (or cross section) from electrically small objects, showing that when the

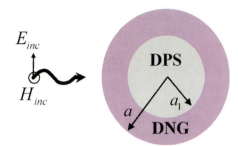

Figure 2.8 Cross section of spherical structure composed of two concentric layers of different isotropic materials (DPS–DNG in figure).

corresponding material polariton (i.e., natural mode) becomes resonant (caused by the interface resonance when subwavelength objects are considered), a high scattering is expected from small DPS–DNG or ENG–MNG coaxial cylindrical or spherical shells [12].

Similar to the planar geometry discussed in the previous section, here also in the limit of electrically small objects the condition for such resonances depends not as much on the total size of the small scatterers—as is the case when standard DPS materials are considered—but rather on the ratio of the core–shell radii. In particular, consider the spherical geometry depicted in Figure 2.8, where a two-layered spherical scatterer is surrounded by free space and excited by a plane wave. The scatterer is composed of a core material with constitutive parameters ε_1, μ_1 and radius a_1 and a cover with parameters ε_2, μ_2 and radius a. Again, a resonant condition for the polariton of order n (corresponding to an $e^{jn\varphi}$ angular variation) may be achieved independent of the total size of the scatterer, provided that the ratio of the core–shell radii a_1/a satisfies the following conditions for the two polarizations:

$$\text{TE}: \quad \gamma \equiv \frac{a_1}{a} \simeq {}^{[2n+1]}\sqrt{\frac{[(n+1)\mu_0 + n\mu_2][(n+1)\mu_2 + n\mu_1]}{n(n+1)(\mu_2 - \mu_0)(\mu_2 - \mu_1)}}$$

$$\text{TM}: \quad \gamma \equiv \frac{a_1}{a} \simeq {}^{[2n+1]}\sqrt{\frac{[(n+1)\varepsilon_0 + n\varepsilon_2][(n+1)\varepsilon_2 + n\varepsilon_1]}{n(n+1)(\varepsilon_2 - \varepsilon_0)(\varepsilon_2 - \varepsilon_1)}}$$

(2.4)

These conditions have been obtained in the quasi-static limit and are therefore valid only in the limit of subwavelength spheres, but a general theory valid for larger objects has been presented in [12]. It is interesting to note, however, again in the subwavelength limit, that here the resonant condition is represented by a geometric "filling" ratio of the structure rather than an overall spatial limit, as it occurs with DPS materials. As anticipated, the interface resonance here again plays a key role, similar to what happens in a lumped circuit where the judicious choice of inductors and capacitors are sufficient to induce a resonance at the required frequency.

It should be noted that the physical constraint $0 \leq a_1/a \leq 1$ should be satisfied in (2.4), which restricts the possible range of material parameters for existence of these resonances. These ranges of parameters for the case of the TM scattering mode are shown in Figure 2.9. We can see from this figure that, for

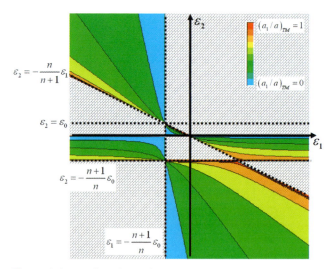

Figure 2.9 Regions for which the resonant condition (2.4) for the TM-polarized scattered wave is satisfied, with the corresponding values for a_1/a between zero and unity. The "forbidden" regions indicated with the "brick" symbols present values of ε_1 and ε_2 for which the condition (2.4) for the TM case cannot be fulfilled. From [12]. Copyright © 2005 by the American Institute of Physics.

such a TM resonance to exist, we should at least have one of the parameters ε_1 or ε_2 negative; that is, we should have SNG or DNG materials for one of the layers.

Another interesting feature of this configuration is the possibility of exciting higher order resonant polaritons in subwavelength scatterers. When considering a conventional DPS small scatterer, generally the scattering field is weak and dominated by the dipolar ($n = 1$) term. By increasing its size, this first-order polariton may be brought into resonance. However, higher order terms also start to contribute significantly in the scattering field. Equation (2.4) then ensures the possibility of choosing the material parameters appropriately, and their filling ratio, so that a higher order resonant scattering condition may be obtained by a subwavelength conjugate pair of shells. In other words, one would be able to build a subwavelength open-cavity resonator that may "hit" higher order resonances instead of the dipolar one.

This phenomenon, as reported in [14], may be particularly interesting if we consider the higher directivity in the scattering (or radiation) pattern of the higher order multipoles. In other words, instead of the usual broad beam of a dipolar scattering, the resonant scattering induced by these subwavelength scatterers may be modeled by properly choosing the filling ratio of the DPS–DNG pair [following Eq. (2.4)] in order to have scattered beams with higher directivity toward specific directions. For the standard DPS scatterers, it is known that the resonances of higher order scattering terms may be achieved only with electrically large objects (which correspond to the classical relation between the beam directivity and the size of the aperture from which they are generated), and their radiation/scattering patterns would then contain not only the desired resonant order but also contributions from other scattering multipoles. However, for a subwavelength small scatterer made of DPS–DNG bilayer or a DPS–ENG bilayer, one may achieve a scattering pattern that is dominated by a higher order multipole, even though the size of the object is small. Clearly, this ideal phenomenon may be limited by the presence of ohmic losses and the sensitivity of this phenomenon to the design parameters, as discussed in [12, 14].

Figure 2.10a gives the absolute values of the first-order scattering coefficient in terms of the ratio of the radii for various values of the outer radius for a DPS–ENG concentric spherical bilayer. It can be seen how for a smaller outer radius of this scatterer the scattering coefficient remains the same when the ratio of the two shell radii is chosen properly. As already mentioned, here indeed the *ratio* of the two radii, rather than the outer radius, plays a more important role for having a resonating mode supported by this system. This implies the presence of resonant modes or polaritons in subwavelength scatterers. In Figure 2.10b, the resonant peaks of the scattering coefficient c_1^{TM} are compared, in a logarithmic scale, for a DPS–ENG (black) and a DPS–DPS (red) scatterer. This clearly shows how this subwavelength scatterer, which is partially composed of materials with a negative parameter, may act as a "compact resonator," occupying a very small volume but exhibiting a large resonant scattering cross section.

Analogous results may also be found for tiny cylindrical objects made of a core–shell pair with oppositely signed parameters, which may be employed not only as a scatterer but also as a leaky-wave radiator or as an open waveguide, hinting again to the possibility of decreasing the working dimensions of such devices. A similar analysis may be applied to more complex geometries, and in this quasi-static limit similar conditions on the ratio of geometric parameters may be straightforwardly derived for such resonances.

It is worth noting that another set of ratio of radii may be obtained for which an "opposite" effect occurs; that is, the total scattering cross section of this object may be drastically reduced and thus the object becomes essentially "transparent". This condition, as a counterpart to Eq. (2.4), reads as

$$\text{TE:} \quad \gamma \equiv \frac{a_1}{a} \simeq \sqrt[2n+1]{\frac{(\mu_2 - \mu_0)[(n+1)\mu_2 + n\mu_1]}{(\mu_2 - \mu_1)[(n+1)\mu_2 + n\mu_0]}}$$

$$\text{TM:} \quad \gamma \equiv \frac{a_1}{a} \simeq \sqrt[2n+1]{\frac{(\varepsilon_2 - \varepsilon_0)[(n+1)\varepsilon_2 + n\varepsilon_1]}{(\varepsilon_2 - \varepsilon_1)[(n+1)\varepsilon_2 + n\varepsilon_0]}} \tag{2.5}$$

for the two polarizations. This implies that for a given dielectric or metallic cylinder or sphere it is possible to choose a suitable metamaterial "cover" layer with proper material parameters in order to make the object effectively transparent, leading to interesting applications and concepts, as discussed in [15].

As an example, we show in Figure 2.11 the comparison between the radial component of the scattered electric field in the near zone of a standard metallic perfectly conducting sphere and that of the same sphere covered with a suitable

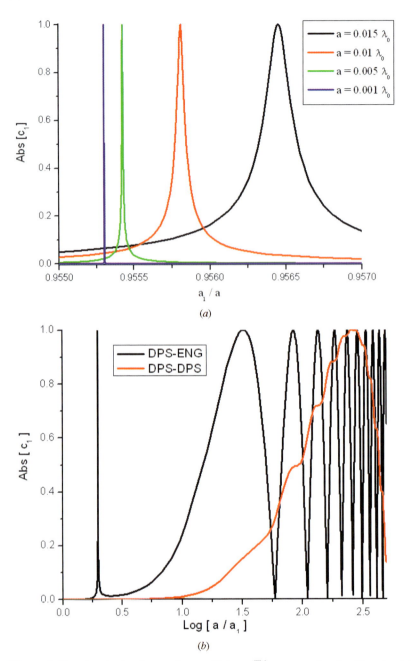

Figure 2.10   Magnitude of scattering coefficient c_1^{TM} versus ratio of radii a_1/a: (a) for the ENG–DPS spherical scatterer with $\varepsilon_1 = -3\varepsilon_0$, $\varepsilon_2 = 10\varepsilon_0$, $\mu_1 = \mu_2 = \mu_0$ with the outer radius a as a parameter; (b) for $a_1 = 0.01\lambda_0$, $\varepsilon_1 = 10\varepsilon_0$, $\varepsilon_2 = \pm1.2\varepsilon_0$, $\mu_1 = \mu_2 = \mu_0$, as a comparison with a DPS–DPS case (logarithmic scale). From [12]. Copyright © 2005 by the American Institute of Physics.

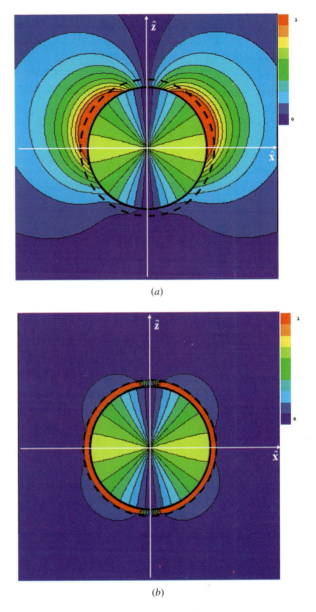

Figure 2.11 Contour plots of distribution of magnitude of radial component of scattered electric field in x–z plane induced by plane wave traveling along z direction with electric field along x axis: (a) for perfectly conducting sphere with $a_1 = \lambda_0/5$; (b) for the same sphere, covered with $\varepsilon_2 = 0.1\varepsilon_0$, $\mu_2 = \mu_0$, $a = 1.087a_1$.

metamaterial cover satisfying condition (2.5). From this figure one can clearly see the drastic reduction of the scattered field after the cover is placed over the original sphere. In the contour plots in Figure 2.11 the red colors correspond to lower values of the field and the blue colors to higher values. It is interesting to see that the field distribution in the metal is similar in the two cases, but the scattered field in the exterior region is drastically reduced when a thin cover is introduced that is suitably designed.

2.4 ENG–MNG AND DPS–DNG MATCHED METAMATERIAL PAIRS FOR RESONANT ENHANCEMENTS OF SOURCE-GENERATED FIELDS

We have mentioned that if one forms a pair of complementary SNG materials, the resulting structure can be regarded as the join of two reactive impedances with opposite signs. Such a pairing can be designed to produce a resonance phenomenon. Consider, as an example, the simple geometry shown in Figure 2.12. An infinite planar electric current sheet $\mathbf{J}_s = I_0 \delta(z)\hat{x}$ is sandwiched between two semi-infinite material slabs with the indicated thicknesses. The bottom slab is terminated with a PEC; the region above the slabs is free space. It is straightforward to obtain the electromagnetic fields in the region above the slabs. One finds

$$E_{\omega x,1}(x, y, z) = A e^{-jk_1 z} \qquad H_{\omega y,1}(x, y, z) = \frac{A}{\eta_0} e^{-jk_1 z} \tag{2.6}$$

$$A = -\eta_2 \eta_3 e^{+jk_1 d_1} \frac{I_0}{Z_{\text{Total}}} \left(\frac{T_{12} e^{-jk_2 d_1}}{1 - R_{12} e^{-j2k_2 d_1}} \right) \left(\frac{1 - e^{-j2k_3 d_2}}{1 + e^{-j2k_3 d_2}} \right) \tag{2.7}$$

where the wavenumbers $k_i = \omega \sqrt{\varepsilon_i} \sqrt{\mu_i}$ and the wave impedances $\eta_i = \sqrt{\mu_i}/\sqrt{\varepsilon_i}$ for $i = 1, 2, 3$ and the terms

$$R_{12} = \frac{\eta_1 - \eta_2}{\eta_1 + \eta_2}$$

$$Z_{\text{Total}} = \eta_2 \frac{1 + R_{12} e^{-j2k_2 d_1}}{1 - R_{12} e^{-j2k_2 d_1}} + \eta_3 \frac{1 - e^{-j2k_3 d_2}}{1 + e^{-j2k_3 d_2}} \tag{2.8}$$

The total impedance term Z_{Total} represents the sum of the input impedances seen by the source looking toward the PEC and toward the free-space interface. A resonance occurs when this total impedance is zero. The corresponding emitted fields will become resonantly large.

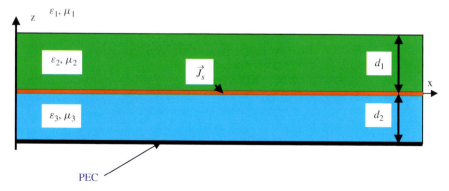

Figure 2.12 Current source sandwiched between two slabs, one terminated with a PEC and the other having an interface with free space, $\varepsilon_1 = \varepsilon_0$, $\mu_1 = \mu_0$.

Configurations for which this resonance can occur include those cases where the two input impedances are equal in magnitude and opposite in sign. This can be achieved by setting $\eta_3 = -\eta_2 = -\eta$ and $2k_2d_1 = 2k_3d_2 = \psi$ so that

$$Z_{\text{Total}} = \eta_2 \frac{1 + R_{12}e^{-j2k_2d_1}}{1 - R_{12}e^{-j2k_2d_1}} + \eta_3 \frac{1 - e^{-j2k_3d_2}}{1 + e^{-j2k_3d_2}} = \eta \frac{1 + R_{12}e^{-j\psi}}{1 - R_{12}e^{-j\psi}} - \eta \frac{1 - e^{-j\psi}}{1 + e^{-j\psi}}$$

$$R_{12} = \frac{\eta_1 - \eta_2}{\eta_1 + \eta_2} = \frac{\eta_0 - \eta}{\eta_0 + \eta} \tag{2.9}$$

The total impedance can then be made to be zero if $|\eta| \to \infty$ so that $R_{12} = -1$. This means that the remaining interface should act like a PEC to form an effective resonant cavity.

One way to achieve these conditions is to set $d_1 = d_2 = d$ and then have $\mu_2 = |\mu| \gg +1$ and $\varepsilon_2 \to -\delta$, $\delta \ll 1$, so that $\eta_2 = \sqrt{\mu_2/\varepsilon_2} = \sqrt{|\mu|}/\sqrt{-\delta} \to +j\infty$ in the top slab and $\mu_3 = -|\mu| \ll -1$ and $\varepsilon_3 \to +\delta$, $\delta \ll 1$, so that $\eta_3 = \sqrt{\mu_3/\varepsilon_3} = \sqrt{-|\mu|}/\sqrt{\delta} \to -j\infty$ in the bottom slab, that is, to have the slabs be a matched ENG–MNG pair. These choices give the impedance conditions $\eta_3 = -\eta_2 = -\eta$, and hence $R_{12} \to -1$ and the wavenumber conditions $2k_2d_1 = 2k_3d_2 = \psi = -2j\omega\sqrt{|\delta|}\sqrt{|\mu|}d$, independent of the size of d. Note that because these phase terms are imaginary, one would actually want d to be much smaller than a wavelength so that the amount of decay of the fields in regions 2 and 3 will be minimal, even with $\delta \ll 1$. A naturally small resonant cavity situation is thus obtained. These material conditions could be achieved, for example, with the capacitively loaded loop (CLL) metamaterial block introduced in [16, 17] for artificial magnetic conductor (AMC) studies. The top slab would be constructed with CLL elements having their resonance frequency above the operating frequency (slightly smaller elements than those designed to operate at the operating frequency) while the bottom slab would have different CLL elements with their resonance frequency below the operating frequency (slightly larger elements than those designed to operate at the operating frequency).

The resonance condition can also be achieved with a DPS–DNG matched pair. In particular, if one sets $\eta_3 = \eta_2 = \eta$ and $2k_2d_1 = -2k_3d_2 = \psi$ so that

$$R_{12} = \frac{\eta_1 - \eta_2}{\eta_1 + \eta_2} = \frac{\eta_0 - \eta}{\eta_0 + \eta}$$

$$Z_{\text{Total}} = \eta_2 \frac{1 + R_{12}e^{-j2k_2d_1}}{1 - R_{12}e^{-j2k_2d_1}} + \eta_3 \frac{1 - e^{-j2k_3d_2}}{1 + e^{-j2k_3d_2}} = \eta \frac{1 + R_{12}e^{-j\psi}}{1 - R_{12}e^{-j\psi}} + \eta \frac{1 - e^{+j\psi}}{1 + e^{+j\psi}}$$

$$= \eta \frac{1 + R_{12}e^{-j\psi}}{1 - R_{12}e^{-j\psi}} - \eta \frac{1 - e^{-j\psi}}{1 + e^{-j\psi}} \tag{2.10}$$

then the total impedance can again go to zero if $R_{12} \to -1$. Thus, resonance occurs if the lower slab is a DNG metamaterial of thickness $d_2 = d$ that has $\mu_3 \to -\infty$ and $\varepsilon_3 \to -\delta$, $\delta \ll 1$, while the upper slab is a DPS metamaterial of thickness $d_1 = d$ that has $\mu_2 \to +\infty$ and $\varepsilon_2 \to +\delta$, $\delta \ll 1$, so that $\eta_3 = \eta_2 = \eta = \sqrt{|\mu|/\delta} \to +\infty$ and $2k_2d_1 = -2\omega\sqrt{|\delta|}\sqrt{|\mu|}d = -2k_3d_2 = -\psi$.

Note that the actual distances d_1, d_2 in either the ENG–MNG or the DNG–DPS configuration could again be scaled along with the actual material parameters. In particular, the impedance and phase conditions are satisfied if $d_1/d_2 = |\mu_3|/|\mu_2| = |\varepsilon_3|/|\varepsilon_2|$. Thus, the slab thicknesses could be scaled to match their permeability and permittivity ratios.

2.5 EFFICIENT, ELECTRICALLY SMALL DIPOLE ANTENNAS: DNG NESTED SHELLS

One can then ask the following question: Can a DNG (or SNG) layer be used as an alternative paradigm to modify the input impedance of an antenna or even match it to free space, thus providing the possibility of improving the antenna performance? This problem has been studied analytically and numerically by Ziolkowski and Kipple [18]. They have considered the possibility of matching an electrically small electric dipole antenna to free space by surrounding it with a DNG shell and have successfully demonstrated that this dipole–DNG shell system produces a much larger radiated power when compared to that produced by the same antenna in free space [18].

Consider an ideal electrically small electric dipole antenna of length ℓ that is driven by the current I_0 at the frequency f_0 corresponding to the wavelength $\lambda = 1/(f_0\sqrt{|\varepsilon\|\mu|})$ and that is embedded in a DPS medium so that (as explained in Chapter 1) the wavenumber $k = \omega\sqrt{\varepsilon}\sqrt{\mu} > 0$ and the wave impedance $\eta = \sqrt{\mu}/\sqrt{\varepsilon} > 0$. It produces the complex power at the radius r given by [19]

$$P = \oiint_S \left(\frac{1}{2}\mathbf{E}_\omega \times \mathbf{H}_\omega^*\right) \cdot \hat{r} \; dS = \eta\left(\frac{\pi}{3}\right)\left|\frac{I_0\ell}{\lambda}\right|^2 \left[1 - j\frac{1}{(kr)^3}\right] = P_{\text{rad}} + jP_{\text{reac}}$$

(2.11)

[using the engineering convention for time-harmonic signals: $\exp(+j\omega t)$]. The (well-known) capacitive reactive power component in Eq. (2.11) is very large near an electrically small antenna (e.g., for the distance $r = a$, where a is the radius of the smallest sphere that can surround the antenna) and severely limits its efficiency as a radiator, that is, the reactance ratio $P_{\text{reac}}/P_{\text{rad}} \gg 1$ for $ka \ll 1$ indicating that the radiated power is much smaller than the reactive power for an electrically small radiator. The dipole's reactive power is dominated by the electric field energy, that is, $P_{\text{reac}} = \omega(W_m - W_e) \approx -\omega W_e$, where W_e and W_m are the corresponding time-averaged electric and magnetic field energies. It was then noticed that this capacitive reactance becomes an inductive reactance when the same antenna is embedded in a DNG medium for which (as explained in Chapter 1) the wavenumber $k = \omega\sqrt{\varepsilon}\sqrt{\mu} = -\omega\sqrt{|\varepsilon|}\sqrt{|\mu|} < 0$ while the wave impedance $\eta = \sqrt{\mu}/\sqrt{\varepsilon} = \sqrt{|\mu|}/\sqrt{|\varepsilon|} > 0$ [18]. Basically, the negative permittivity loading the capacitor makes it act as an inductor. This behavior suggested a possible means of resonantly matching these capacitive and inductive behaviors. Consequently, the possibility of naturally matching an electrically small electric dipole antenna to free space by surrounding it with a hollow DNG shell has been considered.

We have demonstrated with analytical and numerical investigations [18] that the capacitance exhibited by an electrically small electric dipole antenna in free space can be matched by the inductance of a surrounding lossless DNG shell, creating an effective *LC* resonator configuration which significantly increases the real power radiated by the dipole with a corresponding decrease in the total reactance ratio. In particular, the three-region (two-nested-sphere) geometry used in the majority of our investigations is shown in Figure 2.13. The dipole antenna is located along the vertical z axis at the center of the nested spheres. Relations for the electric and magnetic vector potentials, and for the resulting electric and magnetic fields, were obtained in a straightforward manner for each region. The unknown coefficients in those relations were found by applying the appropriate electromagnetic boundary conditions, that is, by making the nonzero tangential fields E_θ and H_ϕ continuous across each shell interface. The resulting equations, presented in [18], were straightforward to obtain and solve numerically. They have been generalized to include passive losses in the medium parameters; that is, the medium parameters were set to $\varepsilon = \varepsilon_r \varepsilon_0 - j\varepsilon''$ and $\mu = \mu_r \mu_0 - j\mu''$, where $\varepsilon_r, \mu_r < 0$ and $\varepsilon'', \mu'' > 0$. An electrically small dipole antenna $\ell = 100\ \mu\text{m} = \lambda_0/300$ driven at $f_0 = 10$ GHz (free-space wavelength

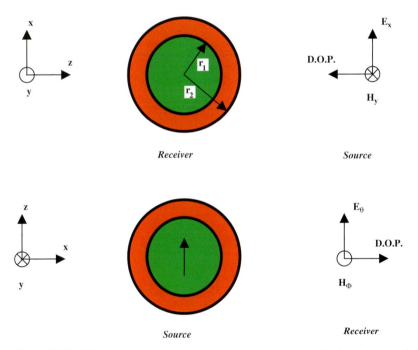

Figure 2.13 Three-region (two-nested-sphere) geometries used to investigate reciprocity of scattering (upper) and source (lower) problems. For both cases, region 1 is given by (ε_1, μ_1) for $r \le r_1$, region 2 has (ε_2, μ_2) for $r_1 < r \le r_2$, and region 3 has (ε_3, μ_3) for $r_2 < r$. The direction of propagation (D.O.P.) of the wave leaving the source and incident on the receiver is shown. From [20]. Copyright © 2005 by the American Physical Society.

$\lambda_0 = 3.0$ cm) was embedded in a small sphere of DPS material (free space) with a radius $r_1 = 100$ μm, which in turn was surrounded by a DNG shell of outer radius r_2 with the permittivity and permeability values $(\varepsilon_{2r}, \mu_{2r}) = (-3, -3)$ and equal electric and magnetic loss tangents LT $= \varepsilon''/|\varepsilon_r|\varepsilon_0 = \mu''/|\mu_r|\mu_0$ set to the fixed values LT $= 0, 0.00001, 0.0001, 0.001$. The external region $r > r_2$ was assumed to be free space.

The radiated power gain, that is, the power radiated by the dipole–DNG shell system normalized by the power radiated by an ideal dipole antenna in free space, whose length is equal to the diameter of the outer sphere that produces the maximum radiated power, was obtained. Both the dipole antenna in the DNG shell and the free-space reference dipole antenna were ideal and were driven with 1.0 A current at their terminals. These radiated power gain results are plotted as a function of the DNG shell's outer radius in Figure 2.14. As reported in [18], the maximum in the lossless case occurs at $r_{2,\text{max}} = 185.8$ μm. The fact that this configuration is a natural mode of this system and produces a resonant enhancement despite the fact that it is much smaller than a free-space wavelength $(r_{2,\text{max}} \approx \lambda_0/161)$ is surprising but in complete agreement with the subwavelength resonator concepts introduced earlier in this chapter. The occurrence of this natural mode was confirmed mathematically by demonstrating

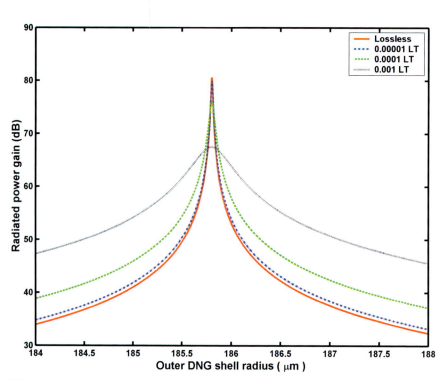

Figure 2.14   Radiated power gain for electrically small electric dipole antenna located at the center of various lossless and lossy DNG shells with inner radius $r_1 = 100$ μm. From [20]. Copyright © 2005 by the American Physical Society.

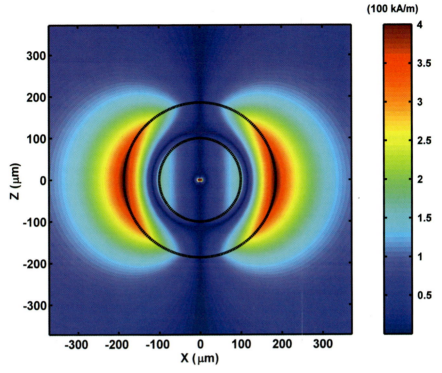

Figure 2.15 Contour plot of magnitude of the real part of magnetic field distribution produced by the z-oriented resonant electrically small electric dipole–lossless DNG shell system. From [20]. Copyright © 2005 by the American Physical Society.

that the determinant of the solution system approaches zero for these material and geometry parameters. Because the enhanced radiated power gain occurs when the dipole–DNG shell interactions are resonant, the losses simply reduce the peak of the response and cause a broadening of the resonance region. They do not make the effect disappear. The contour plot of the magnitude of the real part of the magnetic field distribution for the dipole–DNG shell system with $r_1 = 100$ μm and $r_2 = 185.8$ μm given in Figure 2.15 illustrates the natural mode that is excited by the electrically small dipole antenna.

Recent work by Alù and Engheta has shown a direct correlation between the behavior of nested DPS–DNG and nested ENG–MNG systems [2] and has additionally demonstrated resonant scattering from nested metamaterial shells [12], including various combinations of DPS, DNG, ENG, and MNG shells. Their findings led us to investigate the possibility of reciprocity between the radiation and scattering resonances for a variety of nested metamaterial shells [20]. Should reciprocity be observed, we then hypothesized that the SNG nested sphere geometries, composed of various combinations of ENG and MNG metamaterial shells and found to support these lowest order TM (dipole) scattering resonances, would also maximize the power radiated by an electrically small electric dipole antenna.

We thus anticipated that the electrically small electric dipole antenna could be matched with a nested set of MNG–ENG shells—or even a single ENG shell—in place of the DNG shell used previously. The possibility of matching an electrically small antenna with an ENG material alone is especially appealing, since ENG materials are found in nature or have been manufactured artificially.

The reciprocal scattering problem is also shown in Figure 2.13. The source and receiver have been interchanged. A plane wave polarized along the x direction is incident from infinity along the $-z$ axis onto the nested two-sphere configuration. However, because the nested spheres are electrically small, only the transverse magnetic with respect to the r-direction TM_r ($n = 1, m = 1$) mode (mode number n for the polar angle θ, m for the azimuthal angle ϕ) of this plane wave causes any significant response and was used in the calculations below. The scattering can thus be viewed in terms of an induced dipole antenna oriented along the x axis. This mode also provided the best match to the above z-oriented electrically small electric dipole antenna analysis, which corresponds to radiation of the TM_r ($n = 1, m = 0$) mode. Because the scattering of this lowest order mode from the nested spheres can be described as the excitation of an effective dipole antenna located at their center, the nested electrically small spheres act as a dipole receiver. The dipole antenna strength is directly related to the energy

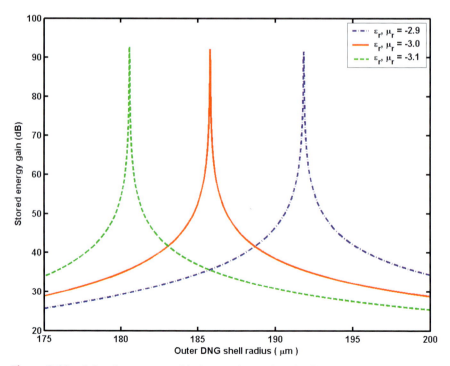

Figure 2.16 Gain of energy stored in inner sphere when the TM_r ($n = 1, m = 1$) mode scatters from various lossless DPS–DNG shell systems with inner radius $r_1 = 100\ \mu m$. From [20]. Copyright © 2005 by the American Physical Society.

captured by the spheres; resonant scattering occurs when the electromagnetic energy captured by the spheres is resonant. We thus represent the signal at the receiver for the reciprocal problem as the energy captured by the interior DPS sphere.

The electric and magnetic field relations for the three-region scattering geometry and the corresponding derived scattering parameters were obtained and calculated. The simulator was validated using well-known cases, including dielectric Mie scattering from each shell interface with a DPS interior; its output was also compared to data provided by Alù and Engheta [12,13]. The DPS–DNG systems were then analyzed numerically for scattering resonances.

The energy received and stored in the inner DPS sphere as a result of an incident 10-GHz TM_r ($n = 1$, $m = 1$) wave scattering from the lossless nested DPS–DNG $r_1 = 100$-µm configuration used in the source problem is shown in Figure 2.16, plotted as a function of the DNG shell's outer radius. The energy stored in the inner DPS sphere is normalized relative to the energy that would be stored in that sphere if all three regions were composed of free space. The peak in the normalized stored energy for the $(\varepsilon_2, \mu_2) = (-3.0\varepsilon_0, -3.0\mu_0)$ case occurs at $r_{2,\text{max}} = 185.8$ µm $\approx \lambda_0/161$, the same radius at which the source problem gain

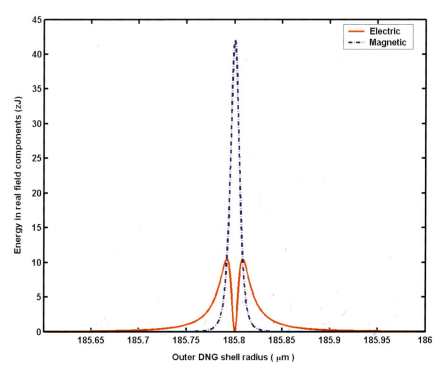

Figure 2.17 Energy stored in real components of the electric and magnetic fields in inner DPS sphere when TM_r ($n = 1$, $m = 1$) mode scatters from the lossless DPS–DNG shell system with inner radius $r_1 = 100$ µm.

was maximized. The determinant of the TM_r ($n = 1, m = 1$) scattering coefficient matrix produces a sharp minimum at this radius, indicating the presence of a natural mode. The ratio $r_{2,max}/r_1$ agrees with the value predicted by the resonant scattering expression given in [12, 13], obtained from a quasi-static ($kr_i \ll 1$, $i = 1, 2$) approximation of the scattering matrices. When considering the energy stored in the real and imaginary components of the electric and magnetic fields within the inner sphere, it was found that the energy stored in the real part of the magnetic field goes to a local maximum while the energy stored in the real part of the electric field goes to local minimum for the resonant geometry. Put another way, the magnetic fields in the DPS–DNG scattering system become completely real at the resonant configuration, while the electric fields become completely imaginary. The same behavior is exhibited in the dipole–DNG shell system. This result is summarized in Figure 2.17.

This indicates that the resonant DPS–DNG system is behaving as a TM resonator with matched capacitance and inductance. The magnitude of the real part of the magnetic field distribution generated for this case is shown in Figure 2.18. Clearly, the same natural mode of the nested sphere system found in the source

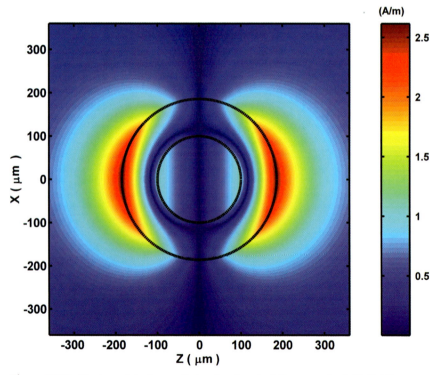

Figure 2.18   Contour plot of magnitude of real part of the magnetic field distribution produced when TM_r ($n = 1, m = 1$) mode is scattered from the resonant lossless DPS–DNG shell system. From [20]. Copyright © 2005 by the American Physical Society.

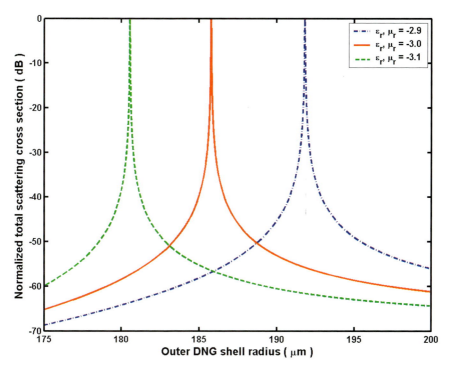

Figure 2.19 Total scattering cross section normalized by unity cross-section value when $TM_r (n = 1, m = 1)$ mode scatters from various lossless DPS–DNG shell systems with inner radius $r_1 = 100$ µm. From [20]. Copyright © 2005 by the American Physical Society.

problem was excited in the scattering problem. Reciprocity of the source and scattering problems in the presence of the DNG shell is thus established.

To further emphasize this reciprocity, the total scattering cross section data for the TM_r ($n = 1$, $m = 1$) mode normalized to its unity value $\sigma_{\text{unity}} = 3 \times (2\pi/k_0^2)$, as seen in Figure 2.19, also exhibits resonant enhancements for the same geometry. In fact, the dipole scattering coefficient is seen to go to one at the critical radius, $r_{2,\text{max}} \approx \lambda_0/161$, again despite the extremely small size of the scatterer. As noted above, the incident wave induces a dipole moment over the inner DPS sphere, which then reradiates as an electrically small antenna. A resonant response results from the presence of the inductive DNG shell that matches this capacitive element to free space.

The energy (detector) and the total scattering cross section for the $r_1 = 100$-µm DPS–DNG shell systems with $(\varepsilon_2, \mu_2) = (-2.9\varepsilon_0, -2.9 \mu_0)$ and $(\varepsilon_2, \mu_2) = (-3.1\varepsilon_0, -3.1 \mu_0)$ are also given, respectively, in Figures 2.16 and 2.19 for comparison. The nested sphere configuration is resonant for a smaller outer shell radius when the index of refraction of the shell is more negative. Reciprocity between these scattering results and the corresponding dipole source results has been verified.

2.6 EFFICIENT, ELECTRICALLY SMALL DIPOLE ANTENNAS: ENG NESTED SHELLS—ANALYSIS

Along with the DPS–DNG analyses, source and scattering resonances were analyzed for a wide variety of DPS–SNG and MNG–ENG systems. They were also studied [21] for related four-region (three-nested-sphere) configurations, for example, for a nested DPS–MNG–ENG–DPS configuration. For example, several source and scattering resonances for configurations with similar sizes and material parameters as the "supergain" case presented above were observed, including (1) inner MNG sphere with $(\varepsilon_r, \mu_r) = (1, -1)$ and radius $r_1 = 100.0$ μm and surrounding ENG shell with $(\varepsilon_r, \mu_r) = (-3, 3)$ and $r_2 = 185.8$ μm, (2) inner DPS, free-space sphere with radius $r_1 = 99.9$ μm and surrounding ENG shell with $(\varepsilon_r, \mu_r) = (-3, 1)$ and $r_2 = 185.8$ μm, and (3) an example four-region case with an inner DPS, free-space sphere with radius $r_1 = 100.0$ μm surrounding MNG shell with $(\varepsilon_r, \mu_r) = (1, -5)$ and $r_2 = 133.9$ μm and outer ENG shell with $(\varepsilon_r, \mu_r) = (-5, 1)$ and $r_3 = 185.8$ μm.

The DPS–ENG case is of particular interest due to the relative ease of manufacturing a DPS–ENG system while still exploiting the resonance properties emphasized by the DPS–DNG scattering cases. In particular, the fact that the shell must be inductive to achieve a resonant match with an electrically small dipole antenna infers that an ENG shell should be adequate to achieve similar results. In particular, the inner electrically small DPS sphere with the electrically small dipole antenna driving it acts as a capacitive element. Similarly, the electrically small ENG shell also acts as a capacitive dipole element but with a negative permittivity. Hence, the ENG shell effectively acts as an inductive element. A

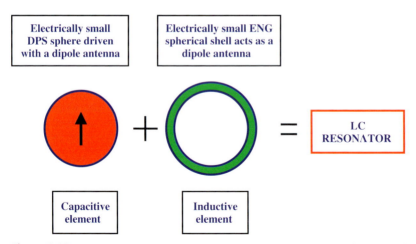

Figure 2.20 An electrically small ENG shell combined with an electrically small dipole antenna can also lead to a resonant system.

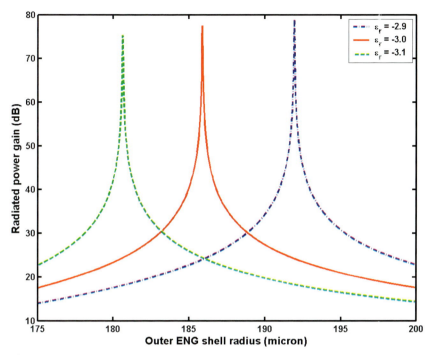

Figure 2.21   Radiated power gain for electrically small electric dipole antenna located at the center of various lossless ENG shells with inner radius $r_1 = 100$ μm. From [20]. Copyright © 2005 by the American Physical Society.

properly designed LC pair will be resonant at the desired operating frequency. This behavior is summarized in Figure 2.20.

   With a dipole antenna of length $\ell = 100$ μm in free space surrounded by a lossless ENG shell with inner radius $r_1 = 100$ μm and material properties $(\varepsilon_r, \mu_r) = (-3.0, 1)$, the radiated power gain as a function of the outer radius is shown in Figure 2.21. The resonant peak occurs at $r_{2,\text{max}} = 185.9$ μm. The resonant behavior for lossless ENG shells with material properties $(\varepsilon_r, \mu_r) = (-2.9, 1)$ and $(\varepsilon_r, \mu_r) = (-3.1, 1)$ are also shown in Figure 2.21 for comparison. The radiated power gain for all of these systems is comparable in magnitude to that of the electrically small dipole–lossless DNG shell case shown in Figure 2.14. Potentially more manufacturable cases have also been considered; the results indicate that an ENG shell may indeed provide a practical alternative to a DNG shell for purposes of resonantly matching a TM_r-producing antenna to free space.

   The distributions of the magnitude of the real parts of the electric and magnetic fields at the resonance are shown in Figures 2.22a and b, respectively. Clearly, the same TM_r resonance that was found in the dipole–DNG system is excited in the dipole–ENG case.

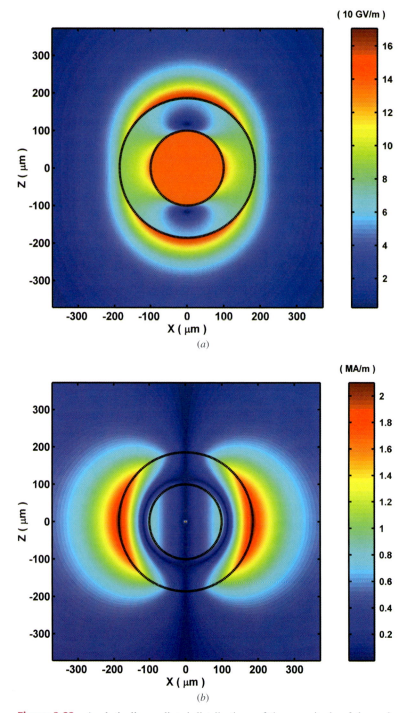

Figure 2.22 Analytically predicted distributions of the magnitude of the real parts of (*a*) electric and (*b*) magnetic fields for the *z*-oriented resonant electrically small dipole−ENG shell system.

2.7 EFFICIENT, ELECTRICALLY SMALL DIPOLE ANTENNAS: HFSS SIMULATIONS OF DIPOLE–ENG SHELL SYSTEMS

To investigate several practical issues associated with the dipole–ENG shell system, we have developed an appropriate numerical model using Ansoft's High Frequency Structure Simulator (HFSS). This package is a full-wave, vector Maxwell equation solver that utilizes a finite-element approach. The 3D objects and the simulation region are discretized with tetrahedra; thus, even a curved boundary is approximated locally with straight segments. The dipole antenna is modeled with two cylindrical wires with a gap source. Consequently, the physical size and shape of the antenna is included in the model and there is a discrete approximation of the spherical shell. The HFSS model thus includes several perturbations that in fact could impact the resonant mode of the system that we are trying to excite.

The HFSS-predicted radiated power gain is shown in Figure 2.23. There has been a shift in the resonance frequency due to the extra capacitance introduced by the finite radius of the dipole antenna. Nonetheless, the shape and peak of the gain curve match the analytical results very well. The HFSS-predicted electric field values for the resonant configuration along the broadside direction from the middle of the gap outward are compared to the corresponding analytical results in Figure 2.24. The differences between the results near the origin are due to the

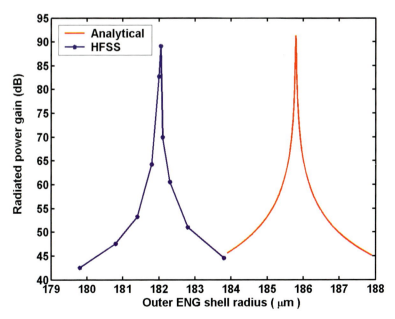

Figure 2.23 Comparison of HFSS and analytically predicted radiated power gains for the resonant electrically small dipole–ENG shell system.

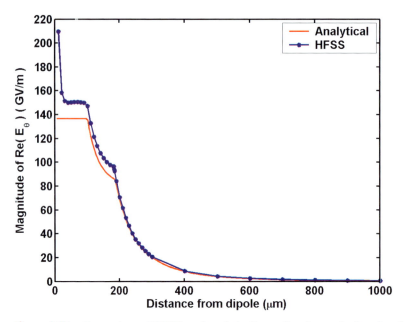

Figure 2.24 Comparison of HFSS and analytically predicted magnitudes of real part of electric field along the broadside direction from the electric dipole antenna in the resonant electrically small dipole–ENG shell system.

presence of the gap source in the HFSS model. The HFSS-predicted electric and magnetic field distributions are shown in Figures 2.25a and b, respectively. Very good agreement with the analytical results is obtained. Moreover, the actual field distributions near the dipole antenna are now apparent.

In summary, the resonant enhancement of the power radiated by an electrically small electric dipole antenna when it is surrounded by a properly designed DNG shell is found to correlate with the scattering enhancements produced by the same DPS–DNG shell configuration. These resonant enhancements occurred even though the sizes of these configurations were very small compared to the radiation wavelength. Fundamental TM scattering resonances that have been observed for various nested MNG–ENG and DPS–ENG configurations have additionally been found to correlate with resonances in the radiation produced by an electrically small electric dipole antenna centered within those systems. The DPS–ENG results are of particular interest, due to the known possibility of manufacturing ENG materials. The DPS–ENG system produces radiated and scattered power gains of the same order of magnitude as the DNG-based system. Numerical solutions of the dipole–ENG shell system obtained with the commercial software package HFSS agreed very well with the analytical results.

Current work is focusing on the development of an approach for determining the "best" shell configuration given specific material and size constraints. We are also investigating the effects of nonideal dipole antennas and their associated

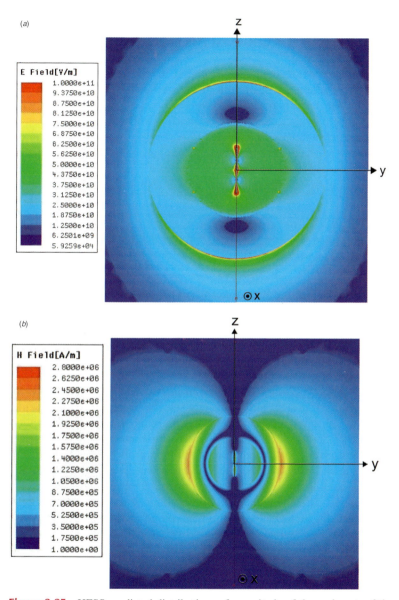

Figure 2.25   HFSS-predicted distributions of magnitude of the real parts of the (a) electric and (b) magnetic fields for the z-oriented resonant electrically small electric dipole–ENG shell system.

matching networks on these dipole–metamaterial shell systems. Experimental designs to realize the predicted match between a real TM_r-producing antenna and an ENG coating, as well as between other types of antennas and the appropriate metamaterials, and to demonstrate that metamaterials can significantly increase an antenna's radiated power are also currently under investigation.

2.8   METAMATERIAL REALIZATION OF AN ARTIFICIAL MAGNETIC CONDUCTOR FOR ANTENNA APPLICATIONS

The SNG medium parameters predicted to achieve the resonant source configurations considered in Section 2.4 dealt with metamaterial slabs that had extremely low permittivities and large permeabilities to make the magnitude of their wave impedances approach infinity. Such high-impedance metamaterial slabs act as AMCs and have many antenna applications. In particular, consider the scattering of a normally incident plane wave from a semi-infinite slab of thickness d having a wave impedance η and wave number k, as shown in Figure 1.6. The reflection S_{11} and transmission S_{21} coefficients from such a slab are readily derived and are found to be

$$
\begin{aligned}
S_{11} &= \frac{\eta - \eta_0}{\eta + \eta_0} \frac{1 - e^{-j2kd}}{1 - [(\eta - \eta_0)/(\eta + \eta_0)]^2 e^{-j2kd}} \\
S_{21} &= \frac{4\eta\eta_0}{(\eta + \eta_0)^2} \frac{e^{-jkd}}{1 - [(\eta - \eta_0)/(\eta + \eta_0)]^2 e^{-j2kd}}
\end{aligned}
\tag{2.12}
$$

It is straightforward to see that if $\eta \to \infty$, an in-phase reflection then occurs,

$$
\lim_{|\eta| \to \infty} S_{11} = +1 \qquad \lim_{|\eta| \to \infty} S_{21} = 0
\tag{2.13}
$$

independent of the thickness of the slab. This zero-phase or in-phase reflection condition characterizes an AMC.

Several planar and volumetric metamaterial structures have been investigated that act as AMCs [22–28]. As shown in later chapters, these include the Sievenpiper mushroom surfaces and frequency selective surfaces (FSSs) such as the University of California Los Angeles (UCLA) uniplanar, compact photonic bandgap (UC-PBG) surfaces. It has been shown by Erentok et al. [17] that a volumetric metamaterial constructed from a periodic arrangement of CLL elements acts as an AMC when the incident wave first interacts with the capacitor side of the CLLs and as an AEC from the opposite direction. This behavior is illustrated in Figure 2.26. The associated HFSS simulations of the scattering of a plane wave normally incident on such a CLL-based metamaterial block have been used to demonstrate that such a block has effective material properties that exhibit a two-time-derivative Lorentz material (2TDLM) behavior for the permeability and a Drude behavior for the permittivity. This effective material behavior is illustrated in Figure 2.27. The resonance of the real part of the 2TDLM model and the zero crossing of the real part of the Drude model occur at the same frequency at which the in-phase reflection occurs. This concurrence of the critical frequencies of both models produces a metamaterial slab with a high-impedance state at that frequency, that is, $\lim_{\omega \to \omega_0} \sqrt{\mu(\omega)/\varepsilon(\omega)} \to \infty$. Numerical simulation and experimental results for the CLL-based metamaterial slab have shown good agreement [17].

The use of the finite two-CLL-deep metamaterial AMC block for antennas has also been considered [17]. Numerical simulations of the interaction of a dipole antenna with such a metamaterial block have shown the expected AMC enhancements of the radiated fields. The dipole–AMC block configuration is

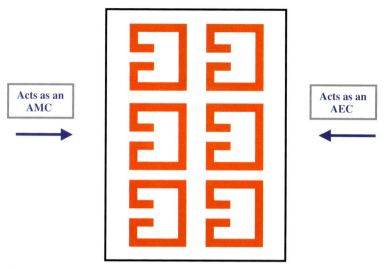

Figure 2.26   The two-CLL-deep based metamaterial block acts as an artificial magnetic conductor (AMC) when the plane wave is incident upon the capacitive gaps and acts as an artificial electric conductor (AEC) when the plane wave is incident from the opposite direction.

Figure 2.27   Effective relative permittivity $\varepsilon_r = \varepsilon_r' + j\varepsilon_r''$ and permeability $\mu_r = \mu_r' + j\mu_r''$ extracted from HFSS simulation results of scattering of a plane wave that is normally incident on a two-CLL-deep-based AMC block. From [17]. Copyright © 2005 by the Institute of Electrical and Electronics Engineers.

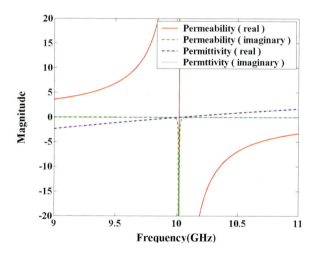

shown in Figure 2.28*a*. The behaviors of this system as a function of the antenna length ℓ and the distance of the antenna from the block h have been studied. As shown in Figure 2.28*b*, it has been found that resonant responses are obtained when the distance between the dipole and the metamaterial block is optimized. Significantly enhanced electric field values in the reflected field region and front-to-back ratios have been demonstrated. The E-plane and H-plane patterns of the dipole–AMC block system and of the free-space dipole are compared in

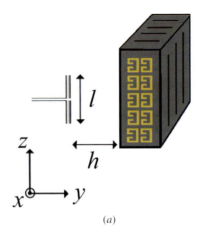

(a)

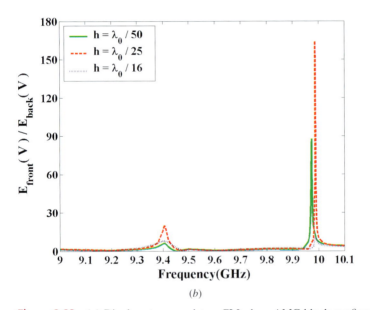

(b)

Figure 2.28   (a) Dipole antenna and two-CLL-deep AMC block configuration.
(b) HFSS-predicted resonant interaction between the dipole antenna and the two-CLL-deep-based AMC block results in very large front-to-back ratios.

Figures 2.29a and b, respectively, for the optimized case of a $\ell = 0.325\lambda_0$ antenna driven at 10 GHz ($\lambda_0 = 30$ mm) near the two-CLL-deep-based AMC block with dimensions 5.5 mm $\times$ 6.6 mm $\times$ 25.4 mm. The broadside power from the dipole antenna is more than doubled in the presence of this AMC block. The realized front-to-back ratio, as shown in Figure 2.28b for this case, is 44.3 dB

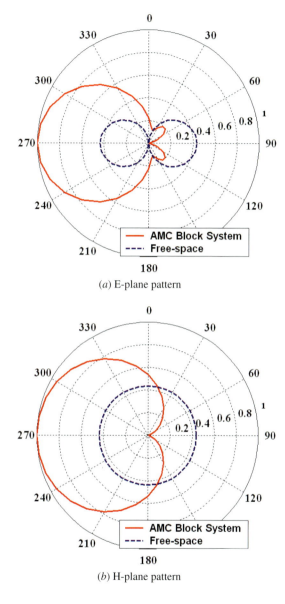

Figure 2.29 The far-field (*a*) E-plane and (*b*) H-plane patterns of the dipole antenna and two-CLL-deep-based AMC block system (solid line), shown in Figure 2.28*a*, are compared to those produced by a free-space dipole antenna (dotted line).

(*a*) E-plane pattern

(*b*) H-plane pattern

(164.25). The corresponding HFSS-predicted near-field electric field distributions in the E and H planes for the resonant frequency 9.987 GHz are shown in Figures 2.30*a* and *b*, respectively. These field distributions show that there is very good isolation between the front and back sides of the two-CLL-deep-based AMC block even in the near field of the system.

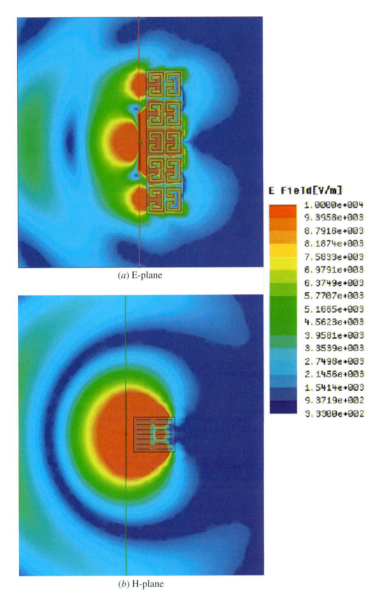

(a) E-plane

(b) H-plane

Figure 2.30   Near-field HFSS-predicted electric field distributions of a $0.325\lambda_0$ dipole antenna when it is in resonance at 9.987 GHz with the two-element-deep CLL-based AMC block: (a) E plane; (b) H plane.

2.9   ZERO-INDEX METAMATERIALS FOR ANTENNA APPLICATIONS

An example of the use of a zero-index slab in antenna applications is to obtain a highly directive beam (e.g., [29–31] and Chapters 1 and 11). Consider a line source that is located in the center of a slab of matched zero-index metamaterial, that is, $\varepsilon_{\text{real}}(\omega_0) \approx 0$ and $\mu_{\text{real}}(\omega_0) \approx 0$. The cylindrical wave generated by the line source will create a spatially constant, time-varying electric field distribution in the slab [32]. From Snell's law one knows that the waves transmitted out of the slab will have a transmitted angle of zero for any angle of incidence since the index of the incident medium is zero,

$$\theta_{\text{exterior}} = \sin^{-1}\left(\frac{n_{\text{interior}}}{n_{\text{exterior}}}\sin\theta_{\text{interior}}\right) = 0 \qquad n_{\text{interior}} = 0$$

This means the field radiated from any zero-index slab will propagate away from it in a direction orthogonal to the face of the slab from which it is emitted. Consequently, the cylindrical wave generated by the line source will be converted into a wave with a planar wave front as the wave emerges from the matched zero-index slab. Moreover, because the entire face of the slab will be driven in phase, it will then act as a uniformly driven aperture. Such an aperture will generate the most directive beam; that is, the far-field pattern of the aperture has the total divergence angle $\theta_{\text{divergence}} = \lambda_0/D$, where D is the largest dimension in the aperture. This also means that the output beam will become narrower as the slab, and hence the aperture, is made wider.

To convert the maximum amount of power from the source into a directive beam, one long face of the slab should be terminated with a perfect magnetic conductor (PMC). The PMC will cause an in-phase reflection of the wave being generated in the slab. A PEC should not be used because it will short out the constant electric field that is generated in the slab. Other configurations with PMC and PEC walls are possible if the walls are located at particular distances away from the slab so that the resulting reflected waves interact constructively with those being generated by the source toward the observer. Because the electric field will become constant throughout the slab, any thickness slab will generate the same directive beam for the same width. Thus, the slab can be designed to be very thin. Consequently, the PMC configuration will generate the largest radiated power into a directive beam with the most compact design.

Such an electrically thin PMC-backed, line-source-driven, matched zero-index slab has been modeled with a finite-difference time-domain (FDTD) simulator [33]. In particular, an electric line source is centered in a $10.2\lambda_0 \times 0.1\lambda_0$ metamaterial slab. The metamaterial is modeled as a low-loss Drude material having $\varepsilon_{\text{real}}(\omega_0) \approx 0$ and $\mu_{\text{real}}(\omega_0) \approx 0$ at $f_0 = \omega_0/2\pi = 30$ GHz. The long side of the slab away from the observer was terminated with a finite PMC ground plane (i.e., to avoid shorting out the electric field in the slab, a PMC rather than a PEC ground plane was used). The FDTD-predicted electric field intensity distributions at several times are shown in Figure 2.31. As predicted, the line source generates a cylindrical wave that forms a spatially constant electric field, which in turn causes the slab to emit a beam with planar wave fronts. The zero-index slab causes the entire output face of the slab to have the same phase, thus resulting in a highly directive output beam.

Finally, because a matched zero-index slab transforms cylindrical waves generated by an internal line source into planar wave fronts, it was anticipated

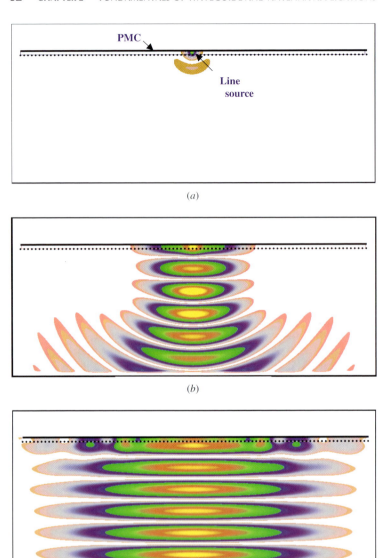

Figure 2.31 Electric field intensity radiated by a line source centered in a $\lambda_0/10$ thick zero-index slab that is terminated in a PMC sheet: (a) $t = 167\ \Delta t$; (b) $t = 1000\ \Delta t$; (c) $t = 4833\ \Delta t$.

that such a slab would do the same to the fields radiated by an external line source. This wave front transformer behavior was demonstrated in [32]. It was characterized with the FDTD simulator and the FDTD results confirmed the expected behavior.

2.10 SUMMARY

In this chapter, we reviewed some of the interesting features of DNG and SNG metamaterials, particularly when they are juxtaposed with complementary metamaterials and/or conventional DPS media. The phenomenon of the interface resonance was one of the reasons behind some of the unusual resonant properties of such a combination of metamaterials. A variety of unusual and interesting cavity, waveguide, and antenna properties that rely on such interface resonances were discussed. These features may lead to a number of exciting potential applications of metamaterials with impacts on the performance of various engineering systems. A further overview of these metamaterial properties and their engineering applications can be found in [34].

REFERENCES

1. N. Engheta, "An idea for thin subwavelength cavity resonators using metamaterials with negative permittivity and permeability," *IEEE Antennas Wireless Propag. Lett.*, vol. 1, no. 1, pp. 10–13, 2002.
2. A. Alù, and N. Engheta, "Pairing an epsilon-negative slab with a mu-negative slab: anomalous tunneling and transparency," *IEEE Trans. Antennas Propag.*, Special Issue on Metamaterials, vol. 51, no. 10, pp. 2558–2570, Oct. 2003.
3. A. Alù, and N. Engheta, "Guided modes in a waveguide filled with a pair of single-negative (SNG), double-negative (DNG), and/or double-positive (DPS) layers," *IEEE Trans. Microwave Theory Tech.*, vol. 52, no. 1, pp. 199–210, Jan. 2004.
4. S. Hrabar, J. Bartolic, and Z. Sipus, "Experimental investigation of sub-wavelength resonator based on backward metamaterials," in *2004 IEEE Antennas and Propagation Society International Symposium Digest*, Monterey, CA, June 20–26, 2004, vol. 3, pp. 2568–2571.
5. A. Alù and N. Engheta, "Sub-wavelength resonant structures containing double-negative (DNG) or single-negative (SNG) media: Planar, cylindrical and spherical cavities," presented at the *Progress in Electromagnetics Research Symposium (PIERS'03)*, Honolulu, Waikiki, HI, Oct. 13–16, 2003, p. 12, abstract.
6. N. Engheta and A. Alù, "May cavities and waveguides be ultra-thin and still support resonant modes when they contain double-negative (DNG) or single-negative (SNG) media?" presented at the *Progress in Electromagnetics Research Symposium (PIERS'03)*, Honolulu, HI, Oct. 13–16, 2003, p. 381, abstract.
7. A. Alù and N. Engheta, "Mode excitation by a line source in a parallel-plate waveguide filled with a pair of parallel double-negative and double-positive slabs," in *2003 IEEE Antennas and Propagation Society International Symposium Digest*, Columbus, OH, June 22–27, 2003, vol. 3, pp. 359–362. Also in A. Alù and N. Engheta, "An overview of salient properties of planar guided-wave structures with double-negative (DNG) and single-negative (SNG) metamaterials," in *Negative Refraction Metamaterials: Fundamental Properties and Applications* (G. V. Eleftheriades and K. G. Balmain, Eds.), IEEE Press/Wiley, Hoboken, NJ, 2005, pp. 339–380.
8. A. Alù and N. Engheta, "Distributed-circuit-element description of guided-wave structures and cavities involving double-negative or single-negative media," in *Proceedings of the SPIE Annual Meeting 2003, Complex Mediums IV: Beyond Linear Isotropic Dielectrics*, vol. 5218, San Diego, CA, August 3–8, 2003, pp. 145–155.
9. R. E. Collin, *Field Theory of Guided Waves*, IEEE Press, New York, 1991.

10. A. Alù and N. Engheta "Anomalies in the surface wave propagation along double-negative and single-negative cylindrical shells," presented at the *Progress in Electromagnetics Research Symposium (PIERS'04)*, Pisa, Italy, March 28–31, 2004, *CD Digest of Abstracts*.

11. A. Alù and N. Engheta, "Anomalous mode coupling in guided-wave structures containing metamaterials with negative permittivity and permeability," in *Proceedings of IEEE-Nano'2002 Conference on Nanotechnology*, Washington, DC, August 26–28, 2002, pp. 233–234.

12. A. Alù and N. Engheta, "Polarizabilities and effective parameters for collections of spherical nano-particles formed by pairs of concentric double-negative (DNG), single-negative (SNG) and/or double-positive (DPS) metamaterial layers," *J. Appl. Phys.*, vol. 97, 094310, May 1, 2005.

13. A. Alù and N. Engheta, "Resonances in sub-wavelength cylindrical structures made of pairs of double-negative and double-positive or ε-negative and μ-negative coaxial shells," in *Proceedings of the International Conference on Electromagnetics in Advanced Applications (ICEAA'03)*, Turin, Italy, Sept. 8–12, 2003, pp. 435–438.

14. A. Alù and N. Engheta, "Strong quadrupole scattering from ultra small metamaterial spherical nano-shells," in *Proceedings of USNC/CNC/URSI National Radio Science Meeting*, Monterey, CA, June 20–26, 2004, p. 210.

15. A. Alù and N. Engheta, "Achieving transparency with plasmonic coatings," *Phys. Rev. E*, vol. 72, 016623, July 26, 2005.

16. R. W. Ziolkowski, "Design, fabrication, and testing of double negative metamaterials," *IEEE Trans. Antennas Propagat.*, vol. 51, no. 7, pp. 1516–1529, July 2003.

17. A. Erentok, P. Luljak, and R. W. Ziolkowski, "Antenna performance near a volumetric metamaterial realization of an artificial magnetic conductor," *IEEE Trans. Antennas Propagat.*, vol. 53, pp. 160–172, Jan. 2005.

18. R. W. Ziolkowski and A. Kipple, "Application of double negative metamaterials to increase the power radiated by electrically small antennas," *IEEE Trans. Antennas Propagat.*, vol. 51, no. 10, pp. 2626–2640, Oct. 2003.

19. C. A. Balanis, *Antenna Theory Analysis and Design*, 2nd ed., Wiley, New York, 1997.

20. R. W. Ziolkowski and A. D. Kipple, "Reciprocity between the effects of resonant scattering and enhanced radiated power by electrically small antennas in the presence of nested metamaterial shells," *Phys. Rev. E*, vol. 72, 036602, Sept. 2005.

21. A. D. Kipple, "Fundamental investigations of double-negative (DNG) metamaterials including applications for antenna systems," Ph.D. dissertation, University of Arizona, Tucson, AZ, Oct. 2004.

22. D. Sievenpiper, L. Zhang, R. F. Jimenez Broas, N. G. Alexopolous, and E. Yablonovitch, "High-impedance electromagnetic surfaces with a forbidden frequency band," *IEEE Trans. Microwave Theory Tech.*, vol. 47, pp. 2059–2074, Nov. 1999.

23. D. Sievenpiper, H.-P. Hsu, J. Schaffner, G. Tangonan, R. Garcia, and S. Ontiveros, "Low-profile, four-sector diversity antenna on high-impedance ground plane," *Elect. Lett.*, vol. 36, pp. 1343–1345, Aug. 2000.

24. R. Coccioli, F.-R. Yang, K.-P. Ma, and T. Itoh, "A novel TEM waveguide using uniplanar compact photonic-bandgap (UC-PBG) structure," *IEEE Trans. Microwave Theory Tech.*, vol. 47, pp. 2092–2098, Nov. 1999.

25. F.-R. Yang, K.-P. Ma, Y. Qian, and T. Itoh, "Aperture-coupled patch antenna on UC-PBG substrate," *IEEE Trans. Microwave Theory Tech.*, vol. 47, pp. 2123–2130, Nov. 1999.

26. C. Caloz and T. Itoh, "A super-compact super-broadband tapered uniplanar PBG structure for microwave and milli-meter wave applications," in *2002 IEEE MTT-S International Microwave Symposium (IMS'02) Digest*, vol. 2, Seattle, WA, June 2002, pp. 1369–1372.

27. J. McVay, N. Engheta, and A. Hoorfar, "Peano high-impedance surfaces," in *Proceedings of the 2004 URSI International Symposium on Electromagnetic Theory, Pisa, Italy,*

vol. 1, May 23–27, 2004, pp. 284–286. The expanded full manuscript is *Radio Science*, vol. 40, RS6503, 2005.

28. J. McVay, N. Engheta, and A. Hoorfar, "High-impedance metamaterial surfaces using Hilbert-curve inclusions," *IEEE Microwave Wireless Components Lett.*, vol. 14, no. 3, pp. 130–132, Mar. 2004.

29. S. Enoch, G. Tayeb, P. Sabouroux, N. Guérin, and P. Vincent, "A metamaterial for directive emission," *Phys. Rev. Lett.*, vol. 89, p. 213902, 2002.

30. S. Enoch, G. Tayeb, and D. Maystre, "Dispersion diagrams of Bloch modes applied to the design of directive sources," *Progress in Electromagnetics Research*, PIER 41, EMW Publishing, Cambridge, MA, 2003, pp. 61–81.

31. G. Tayeb, S. Enoch, P. Vincent, and P. Sabouroux, "A compact directive antenna using ultrarefractive properties of metamaterials," in *Proceedings of the International Conference on Electromagnetics in Advanced Applications, ICEAA'03, Torino, Italy*, Sept. 2003, pp. 423–426.

32. R. W. Ziolkowski, "Propagation in and scattering from a matched metamaterial having a zero index of refraction," *Phys. Rev. E*, vol. 70, 046608, Oct. 2004.

33. R. W. Ziolkowski, "Antennas and propagation in the presence of metamaterials and other complex media: Computational electromagnetic advances and challenges," *IEICE Trans. Electron.*, vol. E88-B, pp. 2230–2238, June 2005.

34. N. Engheta and R. W. Ziolkowski, "A positive future for double-negative metamaterials," *IEEE Trans. Microwave Theory Tech.*, Special Issue on Metamaterial Structures, Phenomena, and Applications, vol. 53, no. 4 (part II), pp. 1535–1556, Apr. 2005.

CHAPTER 3

WAVEGUIDE EXPERIMENTS TO CHARACTERIZE PROPERTIES OF SNG AND DNG METAMATERIALS

Silvio Hrabar

3.1 INTRODUCTION

Effective permittivity and effective permeability are basic engineering parameters of metamaterials. The metamaterials possessing negative values for the real parts of one of these parameters [single-negative (SNG) material] and metamaterials possessing negative values for the real parts of both parameters [double-negative (DNG) material] were introduced in 2000 [1]. Since then there has been a need for experimental characterization of these metamaterials. The first choice might be a direct use (or modification) of a standard free-space method widely used in characterization of continuous materials. This method involves measurement of the transmission and reflection coefficients of a slab sample illuminated by a plane wave emanated from a highly directive antenna. One should use a rather large slab (with typical transversal dimensions of 10 wavelengths) in order to avoid diffraction at the edges. On the other hand, a bulk metamaterial usually comprises an array of small scatterers (the inclusions) embedded into a host material at a small mutual distance (a fraction of a wavelength). Manufacturing of such a large metamaterial sample is a rather time-consuming and expensive task at the present state of the art. Thus, the free-space technique does not appear to be a convenient method of testing the bulk metamaterials.

  The first experimental investigations of SNG and DNG metamaterials were performed in a so-called scattering chamber [1, 2]. Briefly, a scattering chamber is a thin (thinner than half of a wavelength) metallic box containing a small metamaterial sample surrounded by electromagnetic absorbers. If one launches a transverse electric (TE) polarized wave into such a chamber, its small thickness will prevent excitation of any standing waves (e.g., the excitation of waveguide modes) in the vertical direction. In addition, absorbers prevent the excitation of waveguide modes in the horizontal direction. Since the parallel metallic plates (a bottom and a top of the box) impose periodic boundary conditions, the whole

Metamaterials: Physics and Engineering Explorations, Edited by N. Engheta and R. W. Ziolkowski
Copyright © 2006 the Institute of Electrical and Electronics Engineers, Inc.

setup simulates a scenario in which a plane transverse electromagnetic (TEM) wave impinges on an infinite metamaterial slab at normal incidence.

Contrary to the methods with plane-wave excitation, one may turn to the waveguide methods. The waveguide environment is well defined and completely closed, the diffraction is not present, and the testing space is rather small, relaxing the requirements on the sample size. Of course, this environment is obviously different than free space, and one should be very careful in any interpretation of the results, particularly in the case of an anisotropic metamaterial. It is interesting that the literature is sparse in experimental investigations of metamaterials in waveguide environments. Therefore, the purpose of the research presented in this chapter is twofold: the characterization of bulk SNG and DNG metamaterials in a waveguide environment and an investigation of the properties of a waveguide filled with a metamaterial as a new guiding structure.

3.2 BASIC TYPES OF BULK METAMATERIALS WITH INCLUSIONS

Almost all bulk metamaterials used at the present state of the art are based on only two structures: a dense array of thin wires and an array of split-ring resonators (SRRs).

3.2.1 Thin-Wire Epsilon-Negative (ENG) Metamaterial

It has been shown [3–5] that an array of parallel wires (Fig. 3.1) exhibits a high-pass behavior for an incoming plane wave whose electric field is parallel to the wires.

Below a special frequency (a cutoff frequency of the array) there is no propagation and an electromagnetic wave will experience total reflection. This behavior is similar to the propagation of the electromagnetic waves in plasma. If a lattice constant (a) is much smaller than a wavelength ($a \ll \lambda$), the wire array can be thought of as a continuous plasma like material described by an equivalent macroscopic relative permittivity ($e^{j\omega t}$ time dependence is assumed

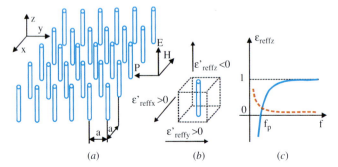

(a)   *(b)*   *(c)*

Figure 3.1 (*a*) Array of thin conducting wires. (*b*) Unit cell. (*c*) Effective permittivity of array: solid lines, real part; dashed lines, imaginary part.

with $\omega = 2\pi f$ being the angular frequency [3–5]):

$$\varepsilon_{\text{reff},z} = \varepsilon'_{\text{reff},z} - j\varepsilon''_{\text{reff},z} = 1 - \frac{f_p^2}{f^2 - j\gamma f} \tag{3.1}$$

Here, $\varepsilon_{\text{reff},z}$ denotes the effective relative permittivity in the z direction. The symbols f and f_p represent the frequency of the signal and the cutoff frequency of the array ("plasma frequency"), respectively, while the factor γ represents the losses. The plasma frequency generally depends on the geometry of the system (a lattice constant and wire radius), and different equations for the prediction of its value can be found in the literature [3–5]. The relative permittivity in the transversal directions (x direction and y direction) is always positive and, in a case of thin wires, is approximately equal to that of the vacuum ($\varepsilon_{\text{reff},x} \approx 1$, $\varepsilon_{\text{reff},y} \approx 1$). Strictly speaking, the permittivity in the direction parallel to the wires (z direction) also depends on the component of the wave vector in the z direction [5]. Thus Eq. (3.1) applies only if there is no component of the wave vector in the z direction, that is, if the propagation takes place only in the transversal (x–y) plane. This requirement was met in the scattering chamber [1,2], and it can also be met in the waveguide environment with the fundamental TE mode. In this case the wire medium can be considered as an isotropic 2D ENG metamaterial described by a scalar relative permittivity $\varepsilon_{\text{reff},z}$.

3.2.2 SRR Array Mu-Negative (MNG) Metamaterial

Since the first theoretical introduction in [6], an array of SRR inclusions (Fig. 3.2) has been widely used for the synthesis of MNG metamaterials [1,2,7–12]. A single SRR can be thought of as a small, capacitively loaded loop antenna [8,9]. If this antenna operates slightly above the resonant frequency, the local scattered magnetic field will be almost out of phase with the incident field. Thus, the resultant local magnetic field will be lower than that of the incident field. It leads to the negative magnetic polarization and negative effective permeability of the resulting metamaterial. It was shown [1,6,7,11] that the effective permeability of this metamaterial has the form given by

$$\mu_{\text{eff}} = \mu'_{\text{eff}} - j\mu''_{\text{eff}} = 1 - \frac{f_{mp}^2 - f_0^2}{f^2 - f_0^2 - j\gamma f} \tag{3.2}$$

where f is the frequency of the signal, f_{mp} denotes the frequency at which (in the lossless case) $\mu_{\text{eff}} = 0$ ("magnetic plasma frequency"), the symbol f_0 stands for the frequency at which μ_{eff} diverges (the resonant frequency of the SRR), and γ represents the losses. The dependence of μ_{eff} on frequency is qualitatively sketched in Figure 3.2c. In general, f_{mp} and f_0 depend both on the lattice constant and the inherent geometric parameters of the SRR itself (inner and outer radii of the rings, the width of the gap between the rings, and the slit width [6]).

Equation (3.2) describes the simplified SRR model that does not take into account the minor electrical polarization [6], which obviously influences the effective permittivity. Furthermore, the SRR shows some small bi-anisotropic effects [7]. All these effects are neglected throughout this experimental investigation, and the SRR is treated as a purely magnetic particle.

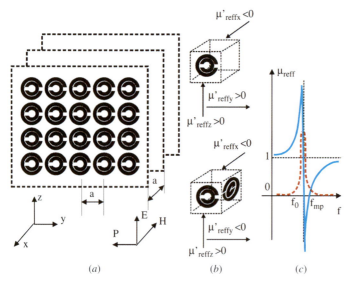

$\mu'_{reffx} < 0$

$\mu'_{reffy} > 0$

$\mu'_{reffz} > 0$

$\mu'_{reffx} < 0$

$\mu'_{reffy} < 0$

$\mu'_{reffz} > 0$

μ_{reff}

f_0 f_{mp} f

(a)   (b)   (c)

Figure 3.2 (*a*) Array of SRRs. (*b*) Unit cell: upper, 1D case; lower, 2D case. (*c*) Effective permeability of array: solid lines, real part; dashed lines, imaginary part.

It is obvious from the upper part of Figure 3.2*b* that an SRR is inherently an anisotropic particle. If the magnetic field vector of the incident plane wave is perpendicular to the SRR, it will give rise to the induced currents that eventually will yield the negative permeability. On the contrary, if the magnetic field vector is parallel to the SRR, it cannot give rise to the induced currents and the presence of the SRR does not affect the effective permeability. Due to this, the pioneering experimental study [1] actually dealt with an anisotropic metamaterial. That metamaterial supported backward-wave propagation only for the magnetic field direction perpendicular to the SRRs (it can be loosely said that the DNG metamaterial used in [1] was actually one dimensional). If one wants to achieve a nearly isotropic 2D MNG metamaterial, one should use at least two SRRs per unit cell (Fig. 3.2*b*, lower). Similar to the case of a thin-wire ENG metamaterial, the experiments performed to date in the scattering chamber [1,2] have dealt with either 1D or 2D metamaterials. Thus, one concludes that a general SRR-array-based MNG metamaterial can be described by a 2×2 uniaxial permeability tensor:

$$\overline{\mu} = \mu_0 \begin{bmatrix} \mu_{tr} & 0 \\ 0 & \mu_{lr} \end{bmatrix} = \mu_0(\overline{\mu}'_r - j\overline{\mu}''_r) = \mu_0\overline{\mu}_r \qquad (3.3)$$

where μ_0 is the absolute permeability and μ_{tr} and μ_{lr} are the relative permeabilities in the transversal (x) and longitudinal (y) directions, respectively. If one deals with an anisotropic SRR-based metamaterial that contains one SRR per unit cell (Fig. 3.2*b*, upper), the longitudinal (y-directed) permeability will be approximately equal to that of the vacuum ($\mu_{lr} \approx 1$). In contrast, if one deals with a nearly isotropic SRR-based MNG metamaterial (Fig. 3.2*b*, lower), the transversal

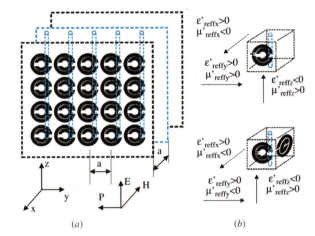

Figure 3.3 (*a*) DNG metamaterial based on thin wires and SRRs. (*b*) Unit cell: upper, 1D case; lower, 2D case.

permeability (*x* directed) and the longitudinal permeability (*y* directed) will be equal ($\mu_{tr} = \mu_{lr}$).

3.2.3 DNG Metamaterial Based on Thin Wires and SRRs

The first DNG metamaterial reported in the literature [1] was actually a combination of the thin-wire-based ENG structure and the SRR-based MNG structure (Fig. 3.3). It was assumed that the new composite material exhibited a macroscopic permittivity equal to that of the thin-wire ENG medium and a macroscopic permeability equal to the permeability of the SRR-based MNG medium. (This is a simplified model which neglects any interactions between the SRRs and the wires [14]). In this chapter, it is assumed that a general 2D bulk metamaterial available at the present state of the art can be fully described by a scalar macroscopic permittivity inherited from the thin-wire-based ENG medium (3.2) whereas the macroscopic permeability has the form of a uniaxial 2 × 2 tensor (3.3) associated with the SRR-array-based MNG medium. This general 2D bulk DNG metamaterial can be either magnetically anisotropic (Fig. 3.3*b*, upper) or magnetically isotropic (Fig. 3.3*b*, lower).

3.3 THEORETICAL ANALYSIS OF RECTANGULAR WAVEGUIDE FILLED WITH GENERAL METAMATERIAL

Let us analyze a rectangular waveguide filled with the general metamaterial described in the previous section (the metamaterial having an isotropic scalar permittivity and a uniaxial anisotropic permeability; Fig. 3.4). The wave equation for such a waveguide [11] reads as

$$\nabla \times \overline{\mu}_r^{-1}(\nabla \times \mathbf{E}) = k_0^2 \varepsilon_r \mathbf{E} \qquad k_0^2 = \omega^2 \mu_0 \varepsilon_0 \qquad (3.4)$$

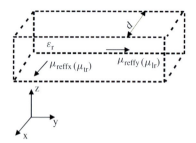

Figure 3.4 Rectangular waveguide filled with general metamaterial.

where **E** is the electric field, k_0 the free-space propagation factor, and ε_0 and ε_r the free-space permittivity and relative permittivity in the medium, respectively. Assuming that the propagating waves are TE modes, one derives [11] the dispersion equation

$$\frac{k_x^2}{\mu_{lr}} + \frac{k_y^2}{\mu_{tr}} = \varepsilon_r k_0^2 \qquad (3.5)$$

From (3.5) one easily finds the expression for the longitudinal propagation factor:

$$k_y = \pm\sqrt{\varepsilon_r \mu_{tr}\left(k_0^2 - \frac{k_x^2}{\varepsilon_r \mu_{lr}}\right)} = \beta_y - j\alpha_y \qquad k_x = \frac{m\pi}{d} \qquad m = 1, 2, 3, \ldots$$
$$(3.6)$$

where d is the waveguide width (Fig. 3.4), m is an integer, and k_x and k_y are the propagation factors in the transversal and longitudinal directions, respectively. The symbols β_y and α_y are the phase factor and attenuation factor, respectively. One should note that there are always two different solutions for the square root in (3.6) and, therefore, two different solutions for the longitudinal propagation factor k_y. The proper, physically meaningful solution is chosen by the requirement that $\alpha_y > 0$ in the lossy case [11], since every physical flow of energy must decay away from the source. If the filling material is lossless ($\varepsilon_r'' = 0$, $\mu_r'' = 0$), Eq. (3.6) can be rearranged into the following compact, more convenient form:

$$k_y = \pm k_0 \sqrt{\varepsilon_r' \mu_{tr}' \left[1 - \left(\frac{f_c}{f}\right)^2\right]} = \beta_y \qquad f_c = \frac{f_{c0}}{\sqrt{\varepsilon_r' \mu_{lr}'}} \qquad f_{c0} = \frac{mc}{2d} \quad (3.7)$$

where f is the frequency of the signal, whereas f_{c0} and f_c are, respectively, the cutoff frequencies of an empty waveguide and the same waveguide filled with material. Now one can analyze the influence of different types of fillings on the propagation of waves in the waveguide. In the familiar case of an isotropic DPS material ($\varepsilon_r' > 0$, $\mu_{tr}' > 0$, $\mu_{lr}' > 0$), the detailed analysis [11] shows that a physically meaningful solution ($\alpha_y > 0$) for $f > f_c$ is given by the positive sign in front of the square roots in (3.6) and (3.7) for the lossy and lossless cases, respectively. These solutions have a positive phase factor ($\beta_y > 0$); hence, the propagation is described by forward waves. Below the cutoff frequency ($f < f_c$), the wave vector component k_y becomes imaginary in the limiting lossless case (3.7); hence, there is no propagation. One notices that a waveguide filled with an

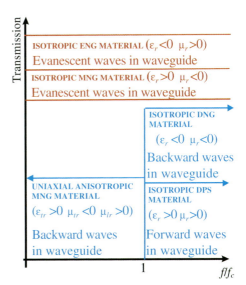

Figure 3.5 Influence of type of meta-material filling rectangular waveguide on wave propagation in waveguide.

isotropic double-positive (DPS) material shows the very well known high-pass behavior (Fig. 3.5).

A similar analysis shows that there is no wave propagation for all frequencies (both for $f < f_c$ and $f > f_c$) if a waveguide is filled either with an isotropic ENG material ($\varepsilon'_r < 0, \mu'_r > 0$) or with an isotropic MNG material ($\varepsilon'_r > 0, \mu'_r < 0$) (Fig. 3.5).

In the case of an isotropic DNG filling ($\varepsilon'_r < 0, \mu'_r < 0$) one must choose the solution of the wave vector component with the negative sign in front of the square root in both the lossy case (3.6) and the lossless case (3.7) for frequencies above the cutoff frequency ($f > f_c$). Now, one notices that $\beta_y < 0$ in (3.6), indicating that the propagation is described by backward waves. Below the cutoff frequency ($f < f_c$) one must choose the positive sign in front of the square root in the lossy case (3.6), which yields an imaginary k_y in the limiting lossless case (3.7). Thus, there is no wave propagation below the cutoff frequency and the waveguide again shows the familiar high-pass behavior (but with backward-wave propagation above the cutoff) (Fig. 3.5).

A very interesting but counterintuitive case occurs if the waveguide is filled with a uniaxial MNG material with negative transversal permeability ($\varepsilon'_r > 0, \mu'_{tr} < 0, \mu_{lr} > 0$). In the lossy case, the physically meaningful solution above the cutoff frequency is given by the positive sign in front of the square root in (3.6), which now yields a negative sign in front of the square root in the lossless case (3.7). The wave vector component k_y in (3.6) becomes imaginary, but now there is no wave propagation above the cutoff frequency. On the contrary, below the cutoff frequency one must choose the solution with a negative sign in front of the square root in (3.6) in the lossy case and again the solution with a negative sign in front of the square root in the lossless case (3.7). Consequently, in this case one becomes aware of the peculiar fact that wave propagation is now possible below the cutoff frequency [10, 11]. This propagation is in the form of backward waves [due to $\beta_y < 0$ in (3.6) and (3.7)]. From Figure 3.5, one concludes that now all of the TE modes can propagate below the cutoff frequency. Hence, a waveguide filled with a uniaxial MNG material with negative transversal permeability exhibits a low-pass behavior, and it can be considered as the dual of an ordinary waveguide.

The analysis presented so far has dealt only with dispersionless (nonphysical) material. In reality, every passive material has dispersion, due to energy conservation [15]. Therefore, any passive metamaterial can exhibit negative permeability (or negative permittivity) only within a limited frequency band. Due to this, all of the explained effects and propagation properties sketched in Figure 3.5 can occur only within a finite frequency band dictated by the dispersion properties of the particular metamaterial used to fill the waveguide.

The propagation of waves in a waveguide filled with a general metamaterial can be analyzed in a simple intuitive way using transmission line theory [11]. For the sake of simplicity let us, for the moment, assume that the metamaterial filling is lossless. The rectangular waveguide can be thought of as an ordinary, two-wire TEM transmission line loaded with an infinite number of short-circuited stubs (Fig. 3.6a).

The main two-wire transmission line can be modeled with a distributed series inductance and a distributed shunt capacitance (Fig. 3.6b). The distributed capacitance of the main line represents the permittivity of the metamaterial filling (ε). Let us first consider the energy flow along the waveguide (represented by the longitudinal component of the Poynting vector $\mathbf{P}_\ell$ in Fig. 3.6b). This longitudinal flow is responsible for the existence of the nonzero transversal magnetic field vector $\mathbf{H}_t$ and therefore the existence of the distributed inductance associated with the transversal permeability (μ_t). The energy flow down the stub is responsible for the existence of the nonzero longitudinal magnetic field vector $\mathbf{H}_l$ and the distributed series inductance of the stub associated with the longitudinal permeability (μ_l). Bearing in mind that the input admittance of a short-circuited stub can be modeled as a parallel tank circuit, one derives the equivalent circuit of the differential section of the waveguide filled with a general metamaterial (Fig. 3.6c).

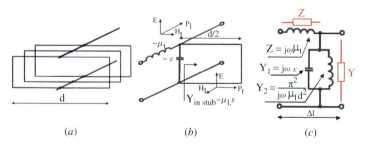

(a) (b) (c)

Figure 3.6 (a) Transmission line representation of rectangular waveguide. (b) Explanation of distributed inductance and capacitance. (c) Equivalent transmission line circuit. From [11] Copyright © 2005 by the Institute of Electrical and Electronics Engineers, Inc.

In the equivalent circuit in Figure 3.6c, the symbol Z represents the impedance per unit length of the waveguide, Y_1 and Y_2 the shunt admittances, and d the transverse dimension of the waveguide. The wave impedance (Z_w) of the waveguide filled with a general lossless metamaterial is given by [11]

$$Z_w = \pm\sqrt{\frac{Z}{Y}} = \pm\sqrt{\frac{\mu_0\mu_{tr}}{\varepsilon_0\varepsilon_r[1-(f_c/f_0)^2]}} \qquad Y = Y_1 + Y_2 \qquad f_c = \frac{f_{c0}}{\sqrt{\varepsilon_r\mu_{lr}}}$$

(3.8)

Now, one can easily understand the propagation properties of the different cases shown in Figure 3.5. If the waveguide is filled with an ordinary DPS material, the series reactance Z will always be positive (since $\mu_{tr} > 0$). Below the cutoff frequency, the stubs are shorter than one-quarter of the wavelength, and the shunt admittance Y_2 shows an inductive character (since $\mu_{tr} > 0$). The tank circuit formed by the shunt capacitance and the shunt inductance operates below the resonant frequency (f_c) and the overall shunt admittance [Y in (3.8)] has an inductive character. In that case, the ratio Z/Y in (3.8) is negative, and the wave impedance Z_w is an imaginary number (a pure reactance), indicating total reflection and therefore the absence of wave propagation. The waveguide behaves as the transmission line, for which both the series impedance and the shunt admittance have an inductive character (L–L transmission line, see the first row in Table 3.1). This type of transmission line obviously supports only evanescent waves and wave propagation is not possible. Above the resonant frequency (f_c), the admittance of the tank circuit exhibits a capacitive behavior and the ratio Z/Y becomes a positive real number. The transmission line now has the usual form (series inductance and parallel capacitance), an L–C line (see the first row in Table 3.1); and the propagation of waves is possible in the form of ordinary forward waves.

In the case of an isotropic ENG filling, the capacitance in the tank circuit in Figure 3.6b will have a negative sign (due to $\varepsilon_r < 0$). This negative capacitance

TABLE 3.1 Comparison of Equivalent Transmission Line Circuits Representing Waveguide Filled with Different Types of Metamaterials

Type of Waveguide Filling	μ_r	ε_r	Type of Equivalent Transmission Line (Series Element–Shunt Element)	
			$f < f_c$	$f > f_c$
Isotropic DPS	$\mu_{rl} = \mu_{rt} = \mu_r$, $\mu_r > 0$	$\varepsilon_r > 0$	L–L	L–C
Isotropic ENG	$\mu_{rl} = \mu_{rt} = \mu_r$, $\mu_r > 0$	$\varepsilon_r < 0$	L–L	L–L
Isotropic MNG	$\mu_{rl} = \mu_{rt} = \mu_r$, $\mu_r < 0$	$\varepsilon_r > 0$	C–C	C–C
Isotropic DNG	$\mu_{rl} = \mu_{rt} = \mu_r$, $\mu_r < 0$	$\varepsilon_r < 0$	C–C	C–L
Uniaxial MNG	$\mu_{rl} > 0$, $\mu_{rt} < 0$	$\varepsilon_r > 0$	C–L	C–C

actually behaves as an ordinary (positive) inductance at a single frequency, yielding an $L-L$ type of transmission line (the second row in Table 3.1). Similarly, in the case of an isotropic MNG filling both the series and the shunt inductances are negative (due to $\mu_r < 0$), and they can be modeled as equivalent capacitances yielding a $C-C$ type of transmission line that does not support propagating waves (the third row in Table 3.1).

In the case of an isotropic DNG filling, a series negative inductance ($\mu_r < 0$) behaves as a capacitance whereas a negative capacitance ($\varepsilon_r < 0$) behaves as an inductance. One ends up with a $C-L$ transmission line (for $f > f_c$) that supports backward-wave propagation.

In the peculiar case of a uniaxial MNG filling, one realizes that the series and shunt inductances are independent. The series inductance is negative (due to $\mu_{tr} < 0$), that is, it behaves as capacitance, whereas the shunt inductance has a positive sign (due to $\mu_{lr} > 0$). This leads to the $C-L$ type of transmission line for frequencies below the cutoff frequency, indicating backward-wave propagation [11].

Above the cutoff frequency ($f > f_c$), one has a $C-C$ type of transmission line (due to the capacitive behavior of the tank circuit), which explains its low-pass behavior.

To test the theory presented here, several different rectangular waveguides were fabricated using copper, and they were filled with different metamaterials. The scattering parameters of the experimental waveguides were measured using an HP 8720B network analyzer. These experiments and their results are discussed in the following sections.

3.4 INVESTIGATION OF RECTANGULAR WAVEGUIDE FILLED WITH 2D ISOTROPIC ENG METAMATERIAL

A standard X-band waveguide (cross section of 22.5 mm $\times$ 10 mm, $f_{c0} = 6.6$ GHz) was filled with a 2D isotropic ENG wire-based metamaterial designed to have $f_p = 15$ GHz (the equations given in [4] were used in this design). The metamaterial was comprised of a 20×4 array of thin (diameter of 0.4 mm) copper wires (Fig. 3.7a). The wires were stretched between the upper and lower walls of the waveguide and were soldered to the waveguide body. The wires formed a rectangular grid with lattice constant $a = 4.5$ mm. The rows next to the waveguide walls were located at a distance 2.25 mm from the wall (Fig. 3.7b). It was shown in Section 3.2.1 that one can consider the wire medium as a 2D isotropic dielectric medium if there is only the fundamental TE_{01} mode in the waveguide. It was also shown (Section 3.3) that a waveguide filled with a dispersionless (nonphysical) 2D isotropic ENG material does not support propagation of the electromagnetic waves. In this experiment, the plasma frequency of the wire-based ENG metamaterial was higher than the cutoff frequency of an empty waveguide ($f_p > f_{c0}$). Therefore, one concludes that the presence of the wire medium should shift the waveguide cutoff to a higher frequency due to the negative value of the effective permittivity below the plasma frequency. It can be seen

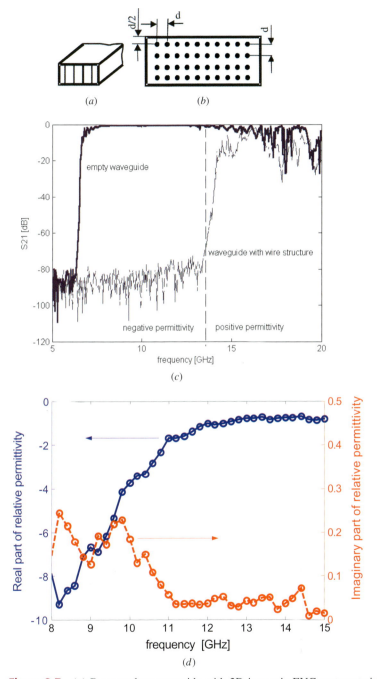

Figure 3.7 (*a*) Rectangular waveguide with 2D isotropic ENG metamaterial. (*b*) Top view of waveguide. (*c*) Measured transmission coefficient (S_{21}). (*d*) Extracted effective relative permittivity: solid line, real part; dashed line, imaginary part.

that the measured transmission coefficients (S_{21}) of the experimental waveguides (Fig. 3.7c) are in good agreement with the theoretical predictions. Irregularities in the S_{21} curve at frequencies above 15 GHz are primarily caused by the excitation of higher order waveguide modes, which have not been taken into account in the presented theoretical analysis.

It was also attempted to determine the effective permittivity of the ENG wire-based metamaterial. Since the operating frequency band of the ENG material is located below the plasma frequency, one assumes that the influence of the negative permittivity on the propagation in a waveguide will be much more pronounced than the influence of the losses. Let us, for the moment, assume that losses are so small that they can be neglected. In that case, the effective permittivity can be determined in a very simple manner using a transmission line analogy. A waveguide filled with an isotropic ENG material can be thought of as an $L-L$ transmission line (the second row in Table 3.1). From basic transmission line theory it is very well known that the input impedance of any infinitely long transmission line is equal to its characteristic impedance. However, the $L-L$ transmission line is essentially a reactive structure that supports only evanescent waves and the fields decay very rapidly with distance along the line (the waveguide). The manufactured waveguide was 80 mm long (approximately two guiding wavelengths at 10 GHz), and it was expected that all of the fields would die off before the electromagnetic waves reached the far end of the waveguide. Therefore, the input impedance of the waveguide was equal to the wave impedance (3.8).

Now, let us analyze the influence of the losses. One can model losses with additional resistance connected in series with a shunt inductance L; thus the wave impedance would become a complex number. The fields along this new $L-RL$ line would decay even more rapidly due to losses, and the input impedance would be again equal to the wave impedance.

The waveguide filled with the isotropic ENG metamaterial was terminated with a standard X-band waveguide matched load, and the input impedance was measured with a network analyzer for frequencies below and above the cutoff frequency. At first, it was noted that the input impedance at frequencies below the cutoff frequency was almost pure reactance, confirming that the hypothesis of small losses was correct. Assuming that the input impedance was equal to the wave impedance of the waveguide, the effective permittivity of the metamaterial filling was calculated from (3.8), which was modified by introducing the complex effective permittivity $\varepsilon_r = \varepsilon_r' - j\varepsilon_r''$. The ambiguity of the sign in front of the square root in (3.8) was resolved by a simple intuitive fact that the imaginary part of wave impedance of an $L-RL$ type of transmission line must have an inductive character (a positive sign of $\mathrm{Im}\{Z_w\}$). It can be seen that the real part of the extracted relative permittivity (Fig. 3.5d) is indeed negative and qualitatively obeys a parabolic dependence on frequency, as was theoretically predicted [3–5]. It also can be noticed that the imaginary part of the permittivity is rather small (less than 0.25); thus the experimental wire-based ENG metamaterial had low losses.

3.5 INVESTIGATION OF RECTANGULAR WAVEGUIDE FILLED WITH 2D ISOTROPIC MNG METAMATERIAL

The double ring [7], which is a special case of the SRR [6], was used for the fabrication of a metamaterial with negative permeability. It consists of two conductive rings placed back to back on a thin dielectric substrate with their slots oriented in opposite directions (Fig. 3.8a). The rings were designed to have their resonant frequency [f_0 in (3.2)] at 7.8 GHz, and they were fabricated using a standard etching process applied to the Cu-clad substrate, which has a double copper cladding (thickness of 0.7 mm, $\varepsilon_r = 2.6$). The rings had an outer diameter of 4 mm with trace widths equal to 1.0 mm and slit widths equal to 0.5 mm. To achieve a 2D isotropic metamaterial, four double rings were used in an arrangement of the unit cell shown in Figure3.8b. A standard X-band waveguide ($f_{c0} = 6.6$ GHz) was filled with 40 double rings (10×2 unit cells, each with the dimensions 7×7 mm) as depicted in Figure 3.8c. Since the resonant frequency of the double ring (and therefore the band associated with the negative permeability) is located above the cutoff frequency of the empty waveguide ($f_p > f_{c0}$), it was expected that a stop band would occur. The measured transmission coefficient of the experimental waveguide (Fig. 3.8d) indeed revealed the existence of this stop band above the cutoff frequency.

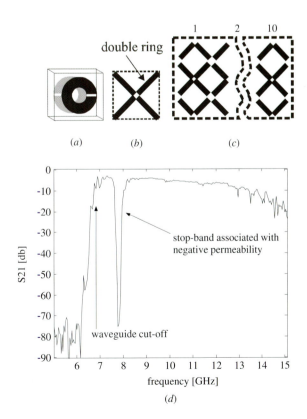

Figure 3.8 Waveguide filled with 2D isotropic MNG metamaterial: (a) double-ring inclusion; (b) arrangement of unit cell; (c) top view of waveguide; (d) measured transmission coefficient (S_{21}).

The effective permeability of an isotropic MNG metamaterial cannot be determined in a simple way, as it was done in the case of the isotropic ENG medium. The double-ring resonator operates in the vicinity of the resonant frequency, where the high currents flowing along the particle cause losses. The influence of these losses may be significant. Consequently, it would be highly desirable to measure both the real and imaginary parts of the effective permeability. The equivalent series impedance (Z in Fig. 3.6c) would then consist of an inductance (associated with μ') and a resistance (associated with μ''). Similarly, an equivalent shunt admittance, Y_2, is composed of both a susceptance and a conductance. The wave impedance in the presence of the magnetic losses and the new equivalent distributed elements are given by the expressions

$$Z = \omega\mu_t'' + j\omega\mu_t' \qquad Y_2 = \frac{\pi^2}{\omega d^2(\mu_l'^2 + \mu_l''^2)}\mu_l'' - j\frac{\pi^2}{\omega d^2(\mu_l'^2 + \mu_l''^2)}\mu_l'$$

$$\mu_t = \mu_l = \mu \qquad Y_1 = j\omega\varepsilon \qquad Z_w = \sqrt{\frac{Z}{Y_1 + Y_2}}$$

$$(3.9)$$

Unfortunately, one cannot extract μ' and μ'' directly from (3.9) since both the numerator and denominator of the wave impedance Z_w include an unknown complex permeability. One may try to use a standard transmission/reflection technique to determine the unknown permeability. However, the insertion loss in the stop band (Fig. 3.8d) is very high ($\sim$75 dB), primarily due to the negative real part of the permeability, that is, due to the evanescent nature of the field. The output signal is very weak and close to the noise floor of the network analyzer. It causes a high measurement uncertainty (particularly in the measurement of the phase of the transmission coefficient). One should use a waveguide whose length is short enough to meet the constraints associated with the network analyzer dynamics. On the other hand, the waveguide must contain a large enough number of inclusions to allow the use of effective media theory (i.e., to have homogenization be applicable). Additional difficulties are caused by the fact that one should use at least two different lengths of the waveguides (i.e., two different numbers of rows with unit cells in Fig. 3.8c) in order to resolve the ambiguity associated with the background mathematics. Hence, it appears that the determination of the relative permeability of an isotropic SRR-based MNG metamaterial is a quite complicated task; it was not attempted in this experimental investigation.

3.6 INVESTIGATION OF RECTANGULAR WAVEGUIDE FILLED WITH 2D UNIAXIAL MNG METAMATERIAL

The following experiments on a waveguide filled with a 2D uniaxial MNG metamaterial were carried out to verify the peculiar phenomenon of propagation below the cutoff frequency of the corresponding empty waveguide. The same double-ring inclusions used in the previous experiments were used for the construction of the uniaxial MNG metamaterial.

To verify the theoretical analysis, it was decided first to insert the inclusions into a waveguide that operates above the cutoff frequency. Nine double rings were placed along the line of symmetry of a 60-mm-long section of a standard J-band waveguide ($d = 35$ mm) with a lattice constant $a = 6$ mm (see Fig. 3.9a). Using the theory presented in Section 3.2.2, one concludes that the filling should act as an uniaxial MNG metamaterial described by a permeability tensor of the form given by Eq. (3.3). The real part of the longitudinal relative permeability (μ'_{lr}) is approximately equal to unity, whereas the real part of the transversal permeability (μ'_{tr}) may be negative within a finite frequency band.

The measured transmission coefficient (S_{21}) is depicted with the dashed curve in Figure 3.9b. Since the resonant frequency of the inclusions is higher than the cutoff frequency of the waveguide, the negative transverse permeability causes the stop band, as was predicted by the theoretical analysis in Section 3.3.

With these experiments completed, a new experimental waveguide was fabricated from copper, with a length of 60 mm and a square cross section of 12 mm $\times$ 12 mm ($f_{c0} = 12.5$ GHz). Nine double rings were again placed along the line of symmetry of the waveguide to form a metamaterial with a lattice constant $a = 6$ mm. The waveguide was directly interfaced with two standard J-band waveguides without tapers or any other matching element and the transmission coefficient (S_{21}) was measured using the network analyzer (the solid curve in Fig. 3.9b). One can clearly see the propagation passband located well below the cutoff frequency of the corresponding empty waveguide, a phenomenon that was reported for the first time in [10] and further investigated in [11, 12]. According to the theory presented in Section 3.3, the propagation of electromagnetic waves within the passband observed in Figure 3.9b takes place in the form of backward waves.

A very simple method of verification of the backward-wave propagation was proposed in [11]. It is based on the properties of the phase distribution along the transmission line (Fig. 3.9c). At a fixed frequency, the phase of a signal (argument of the transmission coefficient S_{21}) decreases along an ordinary (forward-wave) transmission line. Thus, a physically longer ordinary transmission line should exhibit a smaller argument of S_{21}. In contrast, the phase of a signal increases along the backward-wave transmission line since the direction of phase velocity is opposite to the energy flow.

The physically longer backward-wave transmission line exhibits a larger argument of S_{21}. Using standard microwave engineering terminology, one may say that the physically longer backward-wave transmission line appears to be electrically shorter. Following this principle, an additional longer waveguide (length of 66 mm) was manufactured and was filled in the same manner with 10 double rings. The phases of the transmission coefficients of both waveguides were measured across the passband; the results are shown in Figure 3.9d. It can be seen that the physically longer waveguide indeed appears to be electrically shorter. This is a direct proof of the backward-wave propagation and the theory presented in Section 3.3. One can also notice that an increase of the frequency causes a corresponding decrease of the phase of the transmission coefficient. Thus, a backward-wave transmission line does obey Foster's theorem [16].

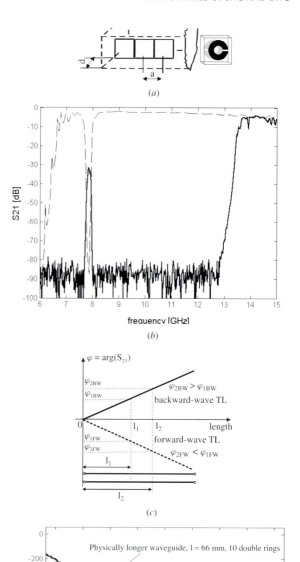

Figure 3.9 (*a*) Rectangular waveguide filled with 2D uniaxial MNG metamaterial. (*b*) Measured transmission coefficient (S_{21}): dashed line, resonance of inclusion located above cutoff; solid line, resonance of inclusion located below cutoff. (*c*) Phase along different kinds of transmission lines. (*d*) Measured phase of S_{21} parameter across passband of waveguides filled with 2D uniaxial metamaterial. From [11] Copyright © 2005 by the Institute of Electrical and Electronics Engineers, Inc.

Theoretically, the wave propagation is possible at an arbitrary frequency below the cutoff frequency of the corresponding empty waveguide, provided that the transverse permeability of the metamaterial filling is negative. Therefore, the classical constraint on the minimum transverse dimension of a waveguide does not hold in this case, that is, the waveguide width can be arbitrarily smaller than half of the wavelength in the filling material. This property may be used for waveguide miniaturization [11].

This waveguide miniaturization was tested with an experimental waveguide filled with the 2D uniaxial MNG metamaterial described above (Figs. 3.9a, b). It has a transverse width that is approximately 30 percent of the width of a standard J-band waveguide. In principle, the reduction in size can be arbitrarily large if it were possible to fabricate an appropriate metamaterial that has a negative transverse permeability at some low frequency below the cutoff frequency of the corresponding empty waveguide. In the case of a metamaterial with resonant inclusions, this can be accomplished by increasing the capacitive loading of the rings. To test the proposed approach, a laboratory model of this different type of metamaterial was fabricated. It was comprised of 12 solid copper rings (inner radius of 6 mm, outer radius of 8 mm, thickness of 1 mm, slit width of 1 mm; see inset in Fig. 3.10). Each ring was loaded with a 33-pF chip capacitor (2 mm × 2 mm × 1 mm in size). This inclusion behaves similarly to the SRR [8,9]. The rings were mounted on a foam with a lattice constant of 8 mm, and the assembly was inserted into a 115-mm-long section of waveguide (cross section of 16 mm × 16 mm, $f_{c0} = 9.4$ GHz) along its line of symmetry. Measurements of the transmission coefficient revealed a backward-wave passband located at a frequency of 350 MHz (Fig. 3.10). The transverse dimension of this waveguide was approximately only 3.7 percent of half of the wavelength.

The measured transmission coefficients of both of the waveguides filled with the uniaxial MNG metamaterial (Figs. 3.9d and 3.10) revealed very narrow operating bandwidths (5 to 10 percent). A narrow bandwidth is the direct consequence of the resonant nature of the metamaterial filling. This drawback might be

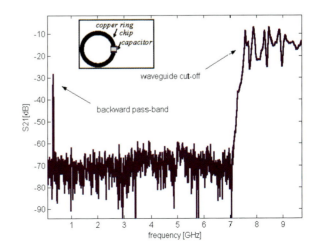

Figure 3.10 Measured S_{21} parameter of experimental waveguide (with cross section 16× 16 mm) filled with a uniaxial MNG material based on capacitively loaded rings. Inset: capacitively loaded ring. From [11] Copyright © 2005 by the Institute of Electrical and Electronics Engineers, Inc.

overcome in the future using some other types of metamaterial which would have a larger bandwidth. Another obvious drawback is the large insertion loss, almost 25 dB, in the measurement results given in Figure 3.9*b*. The main reason for this poor result is the big difference between the wave impedance of the experimental waveguide and the feeding J-band waveguide. Since the waveguides are directly interfaced, there is an abrupt change in the waveguide cross section that causes the large mismatch. Therefore, one-stub waveguide tuners were added at the input and output ends of the waveguide in an attempt to match the feeding and the experimental waveguides. It was possible to achieve an insertion loss of approximately 5 dB with a corresponding return loss of 15 dB (not shown in the figures). Thus, the measured insertion loss was approximately 0.8 dB/cm. This result was obtained without any optimization either of the metamaterial design parameters or the location of the metamaterial filling inside the waveguide. It is envisaged that it should be possible to further decrease this insertion loss by an optimization of the design parameters.

It would still be interesting to estimate the amount of losses associated with the double-ring inclusion itself, that is, to determine the associated complex permeability. In a previous section it was concluded that this determination is a difficult problem in the case of an isotropic MNG metamaterial. Fortunately, in the case of a uniaxial MNG metamaterial the losses are associated only with the series impedance Z in Eq. (3.9) (since $\mu_t = \mu_0$). In other words, the denominator of Z_w depends only on the geometry of the waveguide (dimension d) and the free-space permittivity and permeability (ε_0 and μ_0), as in the lossless case given in Eq. (3.8). Thus, one can simply measure the wave impedance Z_w and directly calculate the unknown complex permeability μ_t. The same standard

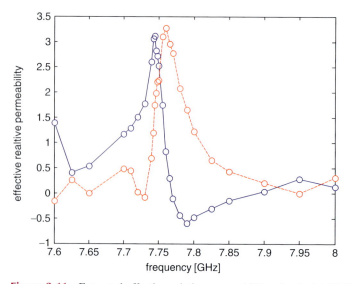

Figure 3.11 Extracted effective relative permeability of uniaxial MNG metamaterial based on double-ring inclusions: solid line, real part; dashed line, imaginary part.

J-band waveguide filled with one row of double-ring inclusions, which was used for the measurements of the stop band in Figure 3.9*b*, was terminated with a standard J-band waveguide matched load. The input impedance was measured at the frequency of the stop band (where the metamaterial filling has a negative transverse permeability). Assuming that the input impedance is equal to the wave impedance, the complex permeability was calculated from Eq. (3.9). The recovered values of the effective permeability are depicted in Figure 3.11.

It can be seen from Figure 3.11 that the curves for both μ' and μ'' qualitatively agree with the theoretical curves. The real part of the permeability becomes negative in a narrow frequency band (7.75 to 7.9 GHz) and reaches a minimal value of ~ -0.7. This relatively small value of negative permeability indicates that there was a significant influence on its value by the losses associated with the high current density along the rings in the vicinity of the resonant frequency.

3.7   INVESTIGATION OF RECTANGULAR WAVEGUIDE FILLED WITH 2D ISOTROPIC DNG METAMATERIAL

Following the basic ideas presented in [1, 2], the double rings were interleaved with wires (Fig. 3.12*a*) to achieve a DNG metamaterial. As was done in the

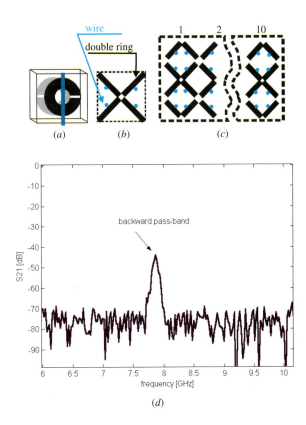

Figure 3.12 Waveguide filled with 2D isotropic DNG metamaterial: (*a*) inclusion based on double ring and thin wire; (*b*) arrangement of unit cell; (*c*) top view of waveguide; (*d*) measured transmission coefficient (S_{21}).

experiment with the isotropic MNG filling (Section 3.5), four double rings were used in the arrangement of the unit cell shown in Figure 3.12b. A standard X-band waveguide ($f_{c0} = 6.6$ GHz, length of 70 mm) was filled with 40 double rings (with properties identical to those used in the previous experiments with the MNG filling) and 40 copper wires (again with properties identical to those used in the previous experiments with the isotropic ENG filling). This waveguide contained 10×2 unit cells, each with dimensions 7×7 mm, as depicted in Figure 3.12c. It was expected that such a structure would behave as a waveguide filled with a DNG metamaterial and, thus, would exhibit propagation above the cutoff frequency of the corresponding empty waveguide in the form of backward waves (Fig. 3.5). The measured transmission coefficient (S_{21}) of this waveguide is shown in Figure 3.12d. It can be seen that a passband appeared in the vicinity of the particle's resonance, as expected. Again it was attempted to verify the backward-wave nature of the propagation. Therefore, an additional waveguide filled with 9×2 unit cells (overall length of 63 mm) was fabricated. Measurements of the phase of the transmission coefficient across the passband again revealed (not shown in figures) that the physically longer waveguide appeared to be electrically shorter; thus the propagation was indeed in a form of backward waves.

3.8   INVESTIGATION OF SUBWAVELENGTH RESONATOR

The very interesting idea of a thin, subwavelength resonator that consists of a combination of a DPS slab and a DNG slab sandwiched between two infinite conducting planes was introduced theoretically in [13]. The equivalent transmission line model is a short-circuited forward-wave stub connected in parallel with a short-circuited backward-wave stub (Fig. 3.13). As explained in [13], the phase delay introduced by the forward-wave line can be completely compensated by the phase advance of the backward-wave line for some special line lengths d_1 and d_2. In that case, the overall phase shift is equal to zero and the phase of the resultant electric (or magnetic) field is equal at any point along the structure; that is, a resonance condition is met. This mechanism is fundamentally different from the conventional case of an ordinary resonator with two forward-wave stubs where the overall phase shift must be a multiple of 2π due to the phase delays.

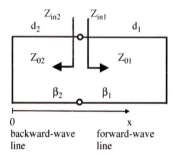

0
backward-wave line   forward-wave line

Figure 3.13   Transmission line model of subwavelength resonator. From [17] Copyright © 2004 the Institute of Electrical and Electronics Engineers, Inc.

Using conventional transmission line theory, one derives the input impedances of the stubs (Fig. 3.13) as

$$Z_{in1} = jZ_{01} \tan(\beta_1 d_1) \qquad Z_{01} = \left| \sqrt{\frac{\mu_1}{\varepsilon_1}} \right|$$
$$\beta_1 = +\omega|\sqrt{\mu_1 \varepsilon_1}| \qquad \beta_1 > 0 \tag{3.10}$$

$$Z_{in2} = jZ_{01} \tan(\beta_2 d_2) = -jZ_{02}\tan(|\beta_2|d_2) \qquad Z_{02} = \left| \sqrt{\frac{\mu_2}{\varepsilon_2}} \right|$$
$$\beta_2 = -\omega|\sqrt{\mu_2 \varepsilon_2}| \qquad \beta_2 < 0 \tag{3.11}$$

From (3.10) and (3.11) and bearing in mind that the sum of Z_{in1} and Z_{in2} must be zero, one derives the resonance condition for the structure [13]:

$$\frac{\tan(\omega|\sqrt{\mu_1\varepsilon_1}|d_1)}{\tan(\omega|\sqrt{\mu_2\varepsilon_2}|d_2)} = \frac{\sqrt{\mu_2/\varepsilon_2}}{\sqrt{\mu_1/\varepsilon_1}} \tag{3.12}$$

If the lengths d_1 and d_2 are much shorter than a wavelength, the expression (3.12) simplifies to [13]

$$\frac{d_1}{d_2} \approx \left| \frac{\mu_2}{\mu_1} \right| \tag{3.13}$$

Thus, the resonance condition can be met for the line lengths d_1 and d_2, which are arbitrarily shorter than the wavelength, provided that their ratio satisfies (3.13). It is interesting to notice that for short line lengths the resonance condition does not depend on the permittivity at all. This is explained by the equivalent circuit sketched in Figure 3.14. The differential length of the forward-wave line can be modeled with the well-known L-C equivalent circuit, whereas the backward line can be modeled as a C–L line, as was explained in Section 3.3. One can approximate a very short line (much shorter than a wavelength) with only one differential element. Since the lines are terminated with short circuits, the distributed admittances are short circuited, and, thus, one is left only with series impedances. Thus, the series inductance of the forward-wave line can come into resonance with the series capacitance of the backward-wave line (which actually occurs because the inductance is negative as a result of the negative permeability). This series resonance explains the condition given by (3.13).

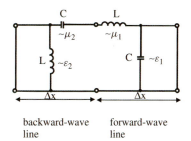

Figure 3.14 Equivalent circuit of resonator for short line lengths. From [17] Copyright © 2004 by the Institute of Electrical and Electronics Engineers, Inc.

To test this remarkable idea experimentally, it is not necessary to use an isotropic DNG metamaterial since the whole structure is essentially one dimensional. One actually only needs the 1D structure that supports backward-wave propagation. This fact initiated the idea of a possible design of a subwavelength resonator based on the miniaturized waveguide filled with the uniaxial MNG metamaterial described in Section 3.6 [17]. A 55-mm-long waveguide (cross section of 16 mm × 16 mm, $f_{c0} = 9.4$ GHz) was partially filled with an uniaxial MNG metamaterial based on the same capacitively loaded rings introduced in Section 3.6 (350 MHz resonant frequency). Four rings were mounted on a foam with a lattice constant of 8 mm and were then inserted along the line of symmetry of a waveguide spanning 35 mm in length (Fig. 3.15). The N connectors were

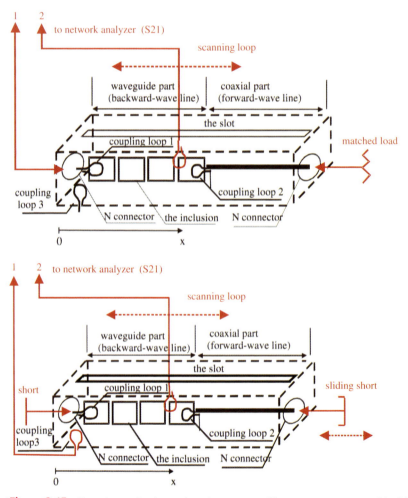

Figure 3.15 Experimental subwavelength resonator. Upper: measurement of incident magnetic field. Lower: measurement of standing-wave distribution. From [17] Copyright © 2004 by the Institute of Electrical and Electronics Engineers, Inc.

mounted at both ends of the waveguide. The first ring was (via a small loop) inductively coupled to the N connector mounted at the left end of the waveguide. The copper rod (3 mm diameter and 20 mm length) was inserted along the line of symmetry of the right-hand side of the waveguide. The rod was soldered to the central pin of the N connector mounted at the right end of the waveguide. The very left end of the rod was inductively coupled to the last (fourth) ring. The left-hand part of this fabricated structure behaved as a waveguide filled with a uniaxial MNG filling while the right-hand part behaved as an ordinary coaxial line (due to the inserted rod). In this way a hybrid waveguide–coaxial structure was fabricated which was essentially the cascade of a forward-wave transmission line and a backward-wave transmission line. A slit was machined on the top wall of the structure to enable the insertion of a small loop antenna. The forward-wave part of the structure was terminated with a coaxial matched load, and the structure was excited with a continuous-wave (CW) signal at a frequency of 350 MHz (frequency at which the metamaterial-filled waveguide part behaved as a backward-wave line). Since the structure was terminated with a matched load, only the presence of the incident wave was expected. The phase distribution of the incident magnetic field along the structure was measured by the help of a small loop and the network analyzer (schematic diagram in upper part of Fig. 3.15). The results of the measurements are given in Figure 3.16 (solid line). It can be seen that the phase increased along the backward-wave part and decreased along the forward-wave part of the structure. There are some irregularities in the phase distribution in the vicinity of the inductive loops due to mismatch. It can also be noticed that there is a steeper slope for the curve in the backward-wave part and, thus, $\beta_2 > \beta_1$.

In the next experiment, the backward-wave part of the structure was terminated with a short circuit and the forward-wave part was terminated with a sliding short. The structure was excited with an additional loop that was located

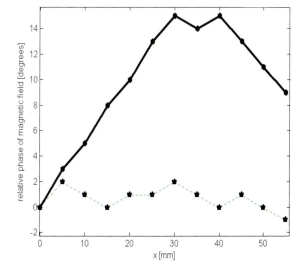

Figure 3.16 Measured phase distribution of *H* field: solid line, phase of incident wave; dashed line, phase of standing wave. From [17] Copyright © 2004 by the Institute of Electrical and Electronics Engineers, Inc.

near the fixed short circuit (schematic diagram in lower part of Fig. 3.15). The phase distribution was again scanned along the structure for different settings of the sliding short. It was possible to excite the structure at only one position of the sliding short ($d_1 = 35$ mm, $d_2 = 70$ mm). The measured distribution of the phase along the structure was approximately uniform (dashed line in Fig. 3.16), indicating the presence of the predicted resonance. Thus, it was indeed possible to meet the resonance conditions, although the overall length of the structure was much smaller than a wavelength (approximately $\lambda/10$). This proves the basic idea introduced theoretically in [16]. The resonance conditions were met for a ratio d_2/d_1 equal to 2; consequently, the equivalent permeability of the backward-wave part calculated with Eq. (3.13) was equal to -2.

3.9   CONCLUSIONS

The basic properties of SNG and DNG metamaterials have been experimentally investigated in a waveguide environment. The obtained experimental results were found to be in good agreement with the theoretical predictions. It was shown that a waveguide filled with an isotropic SNG (either ENG or MNG) metamaterial does not support the propagation of electromagnetic waves. The waveguide filled with an isotropic DNG material supports the propagation of backward waves above the cutoff frequency. The waveguide filled with a uniaxial MNG metamaterial supports propagation of backward waves at an arbitrary frequency below the cutoff frequency provided that the transverse permeability is negative. This peculiar phenomenon can be used for waveguide miniaturization. Such a miniaturized waveguide was used for the experimental verification of the previously introduced idea of a subwavelength resonator.

REFERENCES

1. D. R. Smith, W. J. Padilla, D. C. Vier, S. C. Nemat-Nasser and S. Schultz, "A composite medium with simultaneously negative permeability and permittivity," *Phys. Rev. Lett.*, vol. 84, no. 18, pp. 4184–4187, May 2000.
2. R. A. Shelby, D. R. Smith, S. C. Nemat-Nasser and S. Schultz, "Microwave transmission through a two-dimesonal, isotropic, left-handed metamaterial," *Appl. Phys. Lett.*, vol 78, no. 4, pp. 489–491, Jan. 2001.
3. W. Rotman, "Plasma simulation by artificial dielectrics and parallel-plate media," *IEEE Trans. Antennas Propag.*, vol. 10, no. 1, pp. 82–95, Jan. 1962.
4. J. B. Pendry, J. A. Holden, J. D. Robbins, and J. W. Stewart, "Low frequency plasmons in thin-wire structures," *J. Phys. Condensed Matter*, vol. 10, pp. 4785–4809, 1998.
5. S. Tretyakov, *Analytical Modeling in Applied Electromagnetics,* Artech House, Norwood, MA, 2003.
6. J. B. Pendry, A. Holden, J. D. Robbins, and J.W. Stewart, "Magnetism from conductors and enhanced nonlinear phenomena," *IEEE Trans. MTT*, vol. 47, no. 11, pp. 2075–2084, Nov. 1999.
7. R. Marques, F. Medina, and R. Rafii-El-Idrissi, "Role of anisotropy in negative permeability and left-handed metamaterials," *Phys. Rev. B*, vol. 65, pp. 1–6, 2002.

8. S. Hrabar and J. Bartolic, "Capacitively loaded loop as basic element of negative permeability metamaterial," in *Proc. European Microwave Conference, vol. 2, 2002*, Milan, 2002, pp. 357–361.

9. S. Hrabar and J. Bartolic, "Simplified analysis of split ring resonator used in backward meta-material," in *Proc. International Conference on Mathematical Methods in Electromagnetic Theory*, Kyev, vol. 2, 2002, pp. 560–562.

10. R. Marques, J. Martel, F. Mesa, and F. Medina, "Left-handed-media simulation and transmission of EM waves in subwavelength split-ring-resonator-loaded metallic waveguides," *Phys. Rev. Lett.*, vol. 89, 183901, Oct. 2002.

11. S. Hrabar, J. Bartolic, and Z. Sipus, "Waveguide miniaturization using uniaxial negative permeability meta-material," *IEEE Trans. Antennas Propagat.*, vol. 53, no. 1, pp. 110–119, Jan. 2005.

12. S. Hrabar and J. Bartolic, "Experimental investigation of backward metamaterials in waveguide environment," in *Proceedings ICEAA 2003*, Torino, 2003 pp. 451–454.

13. N. Engheta,"An idea for thin subwavelength cavity resonators using metamaterials with negative permittivity and permeability," *IEEE Antennas Wireless Propag. Lett.*, vol. 1, no. 1 pp. 10–13, 2002.

14. C. R. Simovski, P. A. Belov, and H. Sailing, "Backward wave region and negative material parameters of a structure formed by lattices of wires and split-ring resonators," *IEEE Trans. Antennas Propag.*, vol. 51, no. 10, pp. 2582–2591, Oct. 2003.

15. L. D. Landau and E. M. Lifshitz, *Electrodynamics of Continuous Media*, Butterworth Heinemann, Oxford, 2002.

16. N. Engheta, "Is Foster's reactance theorem satisfied in double-negative and single-negative media?," *Microwave & Optical Technology Letters*, vol. 39, pp. 11–14, Oct. 2003.

17. S. Hrabar, J. Bartolic, and Z. Sipus, "Experimental investigation of subwavelength resonator based on backward-wave meta-material," in *Proceedings IEEE AP-S Symposium 2004*, Monterey, 2004, pp. 2568–2571.

REFRACTION EXPERIMENTS IN WAVEGUIDE ENVIRONMENTS

Tomasz M. Grzegorczyk, Jin Au Kong, and Ran Lixin

4.1 INTRODUCTION

In 1968, the existence of *"electromagnetics of substances with simultaneously negative values of ε and μ"* was postulated theoretically [1], and a description of some of the properties related to these media was presented, such as negative index of refraction, backward phase, left-handed triad, reversed Vavilov–Čerenkov effect, reversed Doppler effect, flat lens, anisotropy frequency dispersions, and so on. In the late 1990s, the realization of these substances at microwave frequencies was studied in two steps: first, it was shown that negative values of ε (the permittivity) could be achieved at microwave frequencies [2] and, second, that negative values of μ (the permeability) could also be achieved at similar frequencies [3]. These two respective effects, based on particular shapes of metallizations, were further studied in [4] and the first experimental verification of a negative index of refraction was reported in [5].

This chapter reviews two main topics related to the experimental study of these new media, also referred to as metamaterials: the refraction of waves at their boundaries and their experimental implementation and verification in waveguide environments. The first topic is briefly reviewed theoretically based on the dispersion relations exhibited by the media, from which the refraction of the wave vectors and the Poynting power can be obtained. Few instances of negative refraction are identified, which can all be measured experimentally. The second topic deals with the experimental aspect of the work, which aims at verifying some of the properties outlined theoretically. Toward this purpose, we first review a few possible implementations of these new substances and then show how they are used in experimental setups. Most of the measurements performed to characterize these substances have been carried out in a parallel-plate waveguide, essentially because it efficiently shields the sample from the perturbations of the external environment and because the boundary conditions allow the samples to remain small yet of seemingly infinite extent. However, it turned out that some structures are not well suited for waveguide measurements, essentially because of contacting issues with the boundaries of the waveguide.

Metamaterials: Physics and Engineering Explorations, Edited by N. Engheta and R. W. Ziolkowski
Copyright © 2006 the Institute of Electrical and Electronics Engineers, Inc.

Therefore, some groups have resorted to open-space measurements, for which successful results have been reported at the cost of much larger samples [6].

4.2 MICROSCOPIC AND MACROSCOPIC VIEWS OF METAMATERIALS

The substances postulated in [1] have not yet been found in nature and need to be fabricated in the laboratory. Currently, they are realized as an arrangement of metallizations properly oriented in space, yielding metamaterials that are therefore intrinsically inhomogeneous and *microscopic*. On the other hand, the metallizations themselves as well as their separations are very small compared to the operating wavelength. Calling upon the effective medium theory, it is therefore legitimate to look for bulk properties, in this case a bulk permittivity and permeability, that govern the *macroscopic* behavior of the medium.

These two views, the microscopic view on one hand and the macroscopic view on the other, are two aspects of the same problem that are connected by *retrieval algorithms* which, from a set of parameters measured on the microscopic metamaterials, yield the bulk properties of the macroscopic metamaterials. Various retrieval algorithms have been published in the literature [7–9], all with the same purpose of establishing the connection between the metallizations and the constitutive parameters of the effective medium. In the next section, we shall briefly describe both points of view to ascertain how these metamaterials are *realized* and how they are *modeled*.

4.2.1 Microscopic View: Rods and Rings as Building Blocks of Metamaterials

Current implementations of metamaterials rely on "infinite" rods and split-ring resonators (SRRs) to achieve a negative permittivity and a negative permeability, respectively. The rings can take various shapes, some of which will be detailed in the forthcoming sections. The rings are the building blocks to achieve an effective frequency-dispersive permeability, which has been shown to obey the frequency-dispersive Lorentz model [3] illustrated in Figure 4.1*a*:

$$\mu_r = 1 - \frac{f_{mp}^2 - f_{mo}^2}{f^2 - f_{mo}^2 + \mathrm{i}\gamma_m f/2\pi}, \tag{4.1}$$

where the subscript *r* refers to relative values, γ_m is the magnetic damping factor, and f_{mo} and f_{mp} are the magnetic resonant and plasma frequencies, respectively, with $f_{mo} < f_{mp}$. It is seen that for frequencies between f_{mo} and f_{mp} the permeability assumes negative values. The shape of the rings, their effective radii, the width of their metallizations, and many other factors directly translate into their properties and govern their resonant and plasma frequencies, which are directly related to the bandwidth where negative values occur. Hence, it is of little surprise that the design and optimization of the geometry of the rings have been an active area of research.

The rods, on the other hand, require little optimization as such but impose a tremendous limitation on the experimental realization in that they need to be infinite or very long to operate as an effective plasma medium described by the

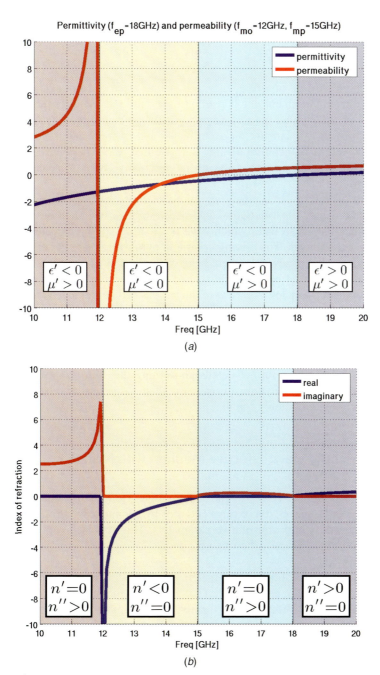

Figure 4.1 Permittivity, permeability, and associated isotropic index of refraction for a medium characterized by a Drude model and a Lorentz model. For the sake of illustration, the characteristic frequencies have been chosen as $f_{ep} = 18$ GHz, $f_{mo} = 12$ GHz, $f_{mp} = 15$ GHz. (*a*) Real part of relative permittivity and relative permeability. (*b*) Real and imaginary part of index of refraction.

frequency-dispersive Drude model illustrated in Figure 4.1a:

$$\epsilon_r = 1 - \frac{f_{ep}^2}{f^2 + i\gamma_e f/2\pi} \tag{4.2}$$

where f_{ep} is the electric plasma frequency and γ_e the electric damping factor. Within a waveguide environment, the constraint of infinite rods requires them to touch the plates of the parallel-plate waveguide (PPW), which turned out to be an effect very difficult to guarantee in experiments and of tremendous impact. In this regard as well, it is therefore of little surprise that researchers have tried to bypass the infinite-rod requirement and to find alternative ways to achieve a negative permittivity.

As a result of this research, many variations of rings and rods have been devised to achieve negative permittivities and permeabilities. We shall describe here the main ones known to date, which have been successfully characterized and measured in experiments. The driving criteria in optimizing these geometries have been to increase the bandwidth where negative properties can be measured, to reduce the losses of the effective medium, and to yield stable and repeatable results in measurements. A quick overview of the various geometries can be found in Table 4.1, where five designs are depicted: the edge-coupled SRR [3, 4], the broadside SRR [10], the axially symmetric SRR [11], the omega SRR [12, 13], and the S ring [14]. The first ring has been often used in the literature and a variety of theoretical and experimental papers have studied its properties. For this reason, we shall not give further details concerning this ring and refer the reader directly to the open literature. Instead, we shall focus on the other ring designs, which are less commonly found.

4.2.2   Macroscopic View: Effective Medium with Negative Constitutive Parameters

4.2.2.1   *Modeling Metamaterials*   The macroscopic view of metamaterials consists in replacing the succession of rings and rods by a homogeneous effective medium characterized by bulk constitutive parameters. This approach is possible since the rings and the separations between the rods are very small compared to the operating wavelength.

At first, isotropic constitutive parameters were sought, namely a scalar permittivity (ϵ) and a scalar permeability (μ). From the measurements of the reflection and transmission coefficients, the retrieved permittivity and permeability were shown to obey a Drude model and a Lorentz model, respectively, described by Eqs. (4.1) and (4.2) and shown in Figure 4.1a.

The isotropic approximation, however, is valid for a single polarization impinging onto a metamaterial containing rings in two dimensions, such as in the original experiment [5]. Later versions of metamaterials were constructed using rings in one direction only, thus breaking the isotropy and creating intrinsically

TABLE 4.1 Overview of Geometries and Features of Main Rings Known to Date

Nomenclature	Geometry	Features
Edge-coupled SRR		• Presents an asymmetry which seems detrimental to good transmission in waveguides (seen in simulations) • Contact issue of the rods
Broadside SRR		• Similar to the previous ring but cancels bi-anisotropy • Contact issue of the rods
Axially symmetric SRR		• Numerical simulations show a very good transmission and a good field symmetry • Contact issue of the rods
Omega SRR		• Single structure combining a ring and a rod • Two rings back to back cancel bi-anisotropy • Contact issue of the rods
S ring		• Single structure producing a permittivity and a permeability effect • No rod issue, which is a very significant advantage in waveguide measurements • Wide bandwidth

anisotropic metamaterials. The relative constitutive parameters had therefore to be considered in tensor form, and biaxial media became the accepted model, described by

$$\overline{\overline{\epsilon}} = \text{diag}[\epsilon_x, \epsilon_y, \epsilon_z] \qquad \overline{\overline{\mu}} = \text{diag}[\mu_x, \mu_y, \mu_z] \qquad (4.3)$$

Obviously, depending on the incident polarization, only some of these parameters are relevant. In this chapter, we consider transverse electric (TE) incident waves polarized in the $\hat{y}$ direction, so that the relevant parameters are $(\epsilon_y, \mu_x, \mu_z)$ in the dispersion relation:

$$\frac{k_x^2}{\epsilon_y \mu_z} + \frac{k_z^2}{\epsilon_y \mu_x} = k_0^2 \tag{4.4}$$

Therefore, depending on the presence or not of rods in the $\hat{y}$ direction, ϵ_y is either described by a Drude model or by a constant which we take equal to 1 since the background medium in the metamaterial is essentially free space. Similarly, depending on the orientation of the ring, either μ_x or μ_z (or both) obey the Lorentz model and are equal to 1 otherwise.

As we shall see in the next section, most of the experiments performed on metamaterials are based on transmission and refraction at boundaries. These concepts are straightforward in the isotropic case: Snell's law provides the refracted angles while the Fresnel coefficients provide the transmission levels. For anisotropic media, where an index of refraction cannot be uniquely defined, the refraction has to be obtained from the dispersion relations and the fact that the direction of the Poynting vector coincides with $\nabla_{\bar{k}} \omega$ [15, 16]. Note that after some calculations, a generalization of Snell's law can still be obtained for the refraction of both the wave vector and the Poynting vector, as has been shown in [17].

4.2.2.2 *Properties of Metamaterials*

The fundamental properties that can be measured in an experimental configuration are the level of transmission and the refraction of an incident wave. These features can be measured in a waveguide environment on both isotropic and anisotropic metamaterials: Refraction phenomena are governed by the laws derived in [17], while the levels can be obtained by a direct generalization of Fresnel's reflection and transmission coefficients [18]. Depending on the signs of ϵ_y, μ_x, and μ_z, the metamaterial may or may not refract negatively an incident plane wave from free space. The possible combinations are summarized in Table 4.2 [19], and an illustration of each case is shown in Table 4.3. All the power refraction cases listed in Table 4.2 can be directly measured experimentally, while it is much more difficult to measure the refraction angles of the phase.

The measurements are most commonly performed using either a slab or a prism geometry. The slab configuration is very versatile and can be used to measure the transmission levels of plane waves at normal as well as at oblique incidences, to measure the deflection of a Gaussian beam [20–23], or to measure the focusing properties of metamaterials when a point source or a line source is used. The prism geometry is essentially used to measure the refraction properties and has been the setup most commonly used to identify positive or negative refraction. In addition, the slab configuration could also be used to measure the backward phase propagating inside an isotropic metamaterial. This measurement is directly feasible in numerical simulations [24] but is very delicate to perform in measurements, and for this reason, we shall omit it from our discussion.

Transmission Through a Slab   The transmission of a plane wave normally incident onto a metamaterial slab has been one of the first experiments performed

TABLE 4.2 Nature of Dispersion Relation for Various Cases of Signs for ϵ_y, μ_x, and μ_z and Angles of Transmission of Wave Vector θ_k and Transmitted Poynting Vector θ_s

Case Number	ϵ_y	μ_x	μ_z	Dispersion Relation Shape	θ_k	θ_s
1	+	+	+	Ellipse	+	+
2	−	−	−	Ellipse	−	−
3	−	−	+	Hyperbola 1	−	+
4	+	+	−	Hyperbola 1	+	−
5	+	−	+	Hyperbola 2	−	+
6	−	+	−	Hyperbola 2	+	−

Note: The signs + or − refer to positive or negative values, respectively, and phase matching is taken along the $\hat{x}$ direction. Circles are degenerate cases of ellipses. An illustration of hyperbola 1 and 2 can be seen in Table 4.3. Cases 5 and 6 are valid only when phase matching exists.

to indicate a frequency region where both the permittivity and the permeability are negative. The concept is directly related to the index of refraction n shown in Figure 4.1*b*. Neglecting the losses, we see that when n is purely imaginary [in the frequency regions (10 GHz, 12 GHz) and (15 GHz, 18 GHz) in Fig. 4.1*b*], the wave vector is also purely imaginary and the wave is strongly attenuated inside the slab. When n is real on the other hand, either positive or negative, the wave vector is also real and propagation occurs. In the example of Figure 4.1*b*, transmission would therefore occur in the frequency bands of (12 GHz, 15 GHz) and (18 GHz, 20 GHz) and would be surrounded by stop bands (note that having a real index of refraction does not necessarily yield a good transmission level since the slab can still be mismatched to air). Examples of transmission will be given subsequently with various ring geometries to show that the passband effect is indeed very clearly observed in measurements, independently of the ring geometry used.

However, the logical implication of this experiment should be well understood: It is because the retrieved parameters are negative between 12 and 15 GHz in the example above that the transmission band witnessed in the measurements can be identified as LH (where LH stands for *left handed*, a terminology borrowed from [1]) and not because there is a transmission band that the constitutive parameters can be concluded to be negative. Therefore, the transmission experiment alone is not a verification of an LH behavior since complex coupling mechanisms within the metamaterial itself can perturb the effective permittivity and permeability enough to make them become positive, in which case a transmission would also be witnessed. Consequently, the transmission experiment is only an indication of the frequency band in which the metamaterial *might* exhibit LH properties and by no means constitutes a proof. Further experiments are necessary to ascertain the properties of the medium.

Refraction Negative refraction is sometimes viewed as the key property that justifies the interest in these metamaterials. In the previous setup, a plane wave is made normally incident onto a metamaterial, where "normally" refers

TABLE 4.3 Illustration in Spectral and Spatial Domains of Transmitted Waves in Cases 2, 3, and 6 of Table 4.2

ϵ_y	μ_x	μ_z	Dispersion	Illustration
-0.8	-1.9	-1.9		
-0.8	1	-1.9		
-0.8	-1.9	1		
-0.6	1	-0.4		
-0.6	-0.4	1		

Note: The blue circle represents the free-space dispersion relation in the (k_x, k_z) plane which supports an incident wave shown by the thin blue arrow. The red circle or hyperbola represents the dispersion relation of the medium. The dashed curves correspond to the dispersion relation of the corresponding media at a slightly higher frequency. When a transmitted wave is supported, its wave vector and power direction are shown by a thin red arrow and a thick red arrow, respectively. The illustrations on the right have been calculated analytically for a Gaussian beam incident from the top-left corner onto a flat boundary.

to a zero incidence angle when the boundary of the medium and the principal axes of the biaxial medium are aligned. Under these circumstances, no refraction is witnessed since the dispersion relations are not rotated. Refraction can therefore be obtained either by rotating the axes of the metamaterial or by having the plane wave impinging at oblique incidence onto the slab. The latter setup is easier to realize in experiments since the same slab as in the transmission measurements can be used. The former, although more difficult, has also yielded good experimental results.

The immediate generalization of the transmission experiment mentioned above is therefore to create a beam obliquely incident onto a negative metamaterial. The isotropic version of this setup has been theoretically studied in [20–22], then experimentally realized in [23], while the anisotropic version can directly be generalized knowing the refraction laws presented in [17]. The deflection of the beam can therefore be predicted theoretically and measured experimentally.

Another refraction experiment, made popular by the setup used in [5], is to use a prism-shaped metamaterial: The wave impinges at normal incidence on the first boundary without experiencing refraction, propagates through the metamaterial, and impinges onto the second boundary at an incident angle determined by the angle of the prism. This setup is not the most general, though, since the incident angle at the second boundary is exactly opposite to the rotation angle of the dispersion relation. Yet, for this particular point on the dispersion relation, refraction occurs which can be measured to be either positive or negative.

Focusing   The experiments listed above make all use of a plane-wave or Gaussian source to create an almost uniform phase front impinging onto a metamaterial. Yet, another source can be used which takes advantage of the negative-refraction property of certain metamaterials, namely a point source or a line source in two dimensions. Theoretically, a line source can be decomposed into a spectrum of plane waves where all propagating and all evanescent waves are included (in other words, the Hankel function can be represented as an integral of plane waves). Hence, propagating waves from a line source can all be treated separately with the plane-wave refraction laws derived in [17], where one must still make the distinction between isotropic and anisotropic metamaterials.

This ray diagram for the propagating waves in the isotropic case, already presented in [1], reveals an interesting focusing pattern: For a slab thick enough, the propagating waves emanating from a point source are bent negatively at the first and second interfaces and yield an image inside and an image outside the slab. This focusing of a point source is therefore achieved using a slab, yielding the "flat-lens" effect illustrated in Figure 4.2.

This ray diagram is only relevant for propagating waves, which form only a portion of the spectrum of a line source. In general, evanescent waves are not considered because of their exponential decay, which makes their contribution in the image plane negligible. However, isotropic metamaterials exhibit the remarkable and unique property of amplifying the evanescent waves, allowing them to contribute in the image plane in addition to the propagating waves. This truly unique property has been pointed out in [25], where it is shown that since

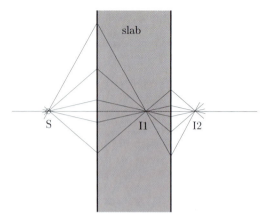

Figure 4.2 Concept of flat lens with anisotropic metamaterial. The rays emanating from the source (S) are refocused by a slab of proper thickness to form an image both inside (I1) and outside (I2) the slab. Note that this concept was already mentioned in [1].

the evanescent waves are now contributing to the image, the diffraction limit of standard lenses can be overcome and a *perfect* lens can be realized (where "perfect" means with a perfect resolution). Of course, this would happen in ideal conditions, when the medium is lossless and made such that $(\epsilon, \mu) = (-\epsilon_0, -\mu_0)$, which are very stringent constraints. However, even if a perfect case cannot be achieved making perfect resolution unachievable so far, subwavelength focusing is still possible.

The amplification of evanescent waves has already been proven theoretically [25] and verified numerically using methods such as the finite-difference time-domain method [26]. However, no conclusive three-dimensional experiments have been provided yet. It should be mentioned, however, that the amplification of evanescent waves has been reported for planar transmission line–based structures [27, 28] and that subwavelength experiments have been reported already in [29] with silver plates, namely using a medium where the permittivity only is negative, as had been suggested in [25].

Figure 4.2 clearly shows that refocusing a line source by a slab is closely related to a negative-refraction phenomenon. From Table 4.2, it is seen that two instances of anisotropic metamaterial can also achieve negative refraction of the power: cases 4 and 6. The latter is less interesting since part or all the rays are totally reflected but *would* be negatively refracted if they *were* transmitted. Case 4 is more interesting, though, since all the spectrum is transmitted and all the propagating rays experience a negative refraction of the power. Contrary to the isotropic case of Figure 4.2, however, the refraction angles are not such that the negatively refracted rays all converge to a single point. Instead, the crossing region is smeared in an area around the perfect lens image.

The experimental verifications of focusing by both an isotropic slab and an anisotropic slab have been presented in the literature [30–32]. However, more experimental verifications are still needed before we can claim that this phenomenon is well understood.

4.3   MEASUREMENT TECHNIQUES

Measuring the properties of metamaterials proves to be an essential complement to the theoretical works developed in parallel. The transmission levels as well as the positive or negative refractions described in the previous section can all be measured experimentally without many difficulties, providing strong verifications of the new properties postulated. In the following sections, we shall illustrate some of these experiments carried out on metamaterials based on various ring designs. Most of the measurements reported in this chapter have been performed with an HP8350B source and an HP82592A plug-in option, which can output 15 dBm in frequency-scanning mode between 20 MHz and 20 GHz, or up to 17 dBm at a fixed frequency. The source was modulated by a 28.8-KHz square wave in order to be plugged into an HP8756A scalar network analyzer and the detector used was an HP11664A.

4.3.1   Experimental Constraints

4.3.1.1   Obtaining a Plane-Wave Incidence   The first step to perform transmission measurements on metamaterials is to obtain a proper incident beam. In most cases, a plane wave is the ideal input, but it cannot be obtained experimentally. However, an approximated plane wave can be created by eliminating the interference from the external environment as much as possible. This can be achieved using a PPW configuration with microwave absorbers on the two lateral sides [5], which yields a nonnormalized plane wave. Figure 4.3 shows the coupling method of the source, such that a TE_{01} mode is fed into the PPW chamber and a similar coupler is used for the reception. Figure 4.4 shows the amplitude and phase of the output electromagnetic wave, indicating that at each frequency the output beam is indeed a nonnormalized plane wave with a Gaussian-like amplitude distribution.

For the experiments, a sample of metamaterial is placed inside the PPW chamber. The position of the sample (the sample being a slab or a prism in our cases, depending on the properties measured) is a result of a trade-off between incidence and reception: It needs to be far enough from the source so that the wave front exiting the waveguide coupler has enough space to flatten out and to approach a plane-wave front while it has to be far enough from the reception to minimize all near-field effects. In addition, the plane-wave assumption is only valid within a region around the symmetry axis of the propagating channel in Figure 4.3, since the tapering effect of the absorbers cannot be avoided, and neither can the Gaussian far-field distribution due to the aperture source. For this reason, working with wide samples in the transverse direction guarantees more controlled experimental conditions.

The first experiments with this setup were performed with a slab aperture $D = 5$ cm at frequencies around 10 GHz, yielding an approximate far-field limit of $2D^2/\lambda \approx 17$ cm [33]. The distance between the source and the sample was 1 m, allowing the phase front of the incident beam to flatten out, while the

Figure 4.3 Picture of waveguide coupler being fed into PPW (in which top plate has been removed).

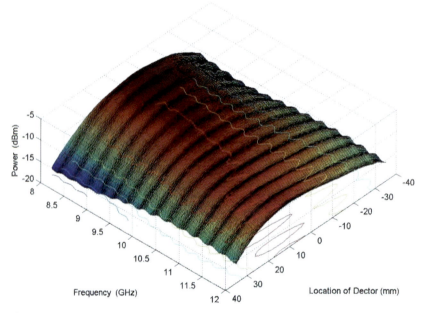

Figure 4.4 Amplitude of incident beam as it leaves the waveguide coupler shown in Figure 4.3. It is seen that a high amplitude is kept close to the center of the propagating channel and tapers off toward the absorbers.

distance from the sample to the receiver was about 30 cm, well beyond the far-field limit. Later, the experiment was repeated by progressively shrinking the dimensions of the propagating channel, and it has been found that a distance between source and reception of about 40 cm is enough to yield the expected results.

4.3.1.2   Contacting Issue with Waveguide Walls   Another stringent experimental constraint relates to the infinite-rod requirement. In theory, a plasma medium is obtained for the permittivity when infinite aligned rods are closely spaced together. The effective medium thus created exhibits a uniaxial permittivity tensor, where the permittivity component aligned with the rods obeys a Drude model. Most of the designs of metamaterials (yet not all, as will be shown subsequently) rely on this effect to produce a negative permittivity.

Invoking the boundary conditions of the electric field and the image theory, researchers have immediately realized that rods oriented perpendicularly to the top and bottom plates of a PPW behave as if they were infinite. Experimentally, this requires the rods to be touching the top and bottom plates of the PPW, which turned out to be very challenging since all the boards on which the rods are printed need to be cut of the exact same height, and the metallizations of the rods need to extend until the very edge of the boards. Guaranteeing these properties throughout a whole set of rods turns out to be very difficult, yet crucial.

4.3.2   Measurements of Various Rings

Negative metamaterials are realized from the arrangement of rings and rods or rings alone and exhibit specific transmission and refraction properties which have been outlined in the previous sections. In this section, we present how these two properties are measured within an experimental framework when the metamaterials are realized based on the ring geometries shown in Table 4.1. Since the first geometry in the table (the edge-coupled SRR) and its immediate variation (the broadside SRR) have been extensively studied in the literature, we shall concentrate here on the other geometries solely: namely the axially symmetric SRR, the omega SRR, and the 'S' ring.

4.3.2.1   Axially Symmetric SRR   The geometry of the axially symmetric ring is shown in Figure 4.5. A set of specific dimensions is proposed in the caption, which corresponds to resonant and plasma frequencies of $f_{mo} \approx 8$ GHz, $f_{mp} \approx 9$ GHz, and $f_{ep} \approx 11$ GHz. Obviously, as has been mentioned before, these frequencies are directly related to the thicknesses of the metallizations, the gaps, and other geometric parameters in the design of the SRR itself which can be designed to achieve other frequency values than the ones proposed above.

Measuring the transmission level through a slab of metamaterial within a waveguide environment is fairly straightforward: Two waveguide ports are used as transmitter and receiver and are located on opposite sides of an aluminum plate. A slab of metamaterial is located in between the two waveguide ports, and the

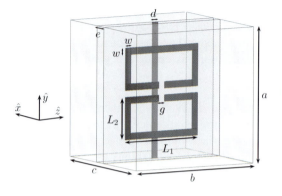

Figure 4.5 Unit cell of axially symmetric ring: $a = 12$ mm, $b = 5.04$ mm, $c = 2.5$ mm, $d = 0.6$ mm, $e = 1$ mm, $w = g = 0.24$ mm, $L_1 = 3.13$ mm, $L_2 = 2.2$ mm, $\epsilon_r = 4.6$.

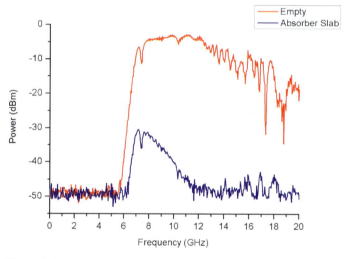

Figure 4.6 Transmission level through empty waveguide and 6-cm-thick slab of absorbers. The attenuation is seen to be less than -30 dB throughout the frequency range, which is acceptable. This type of absorber has been used for all the experiments reported in this chapter.

sides are padded with absorbers to reduce reflections and noise measured by the transmission port. The characteristics of the absorbers used in the experiments reported in this chapter are shown in Figure 4.6. The absorption level is seen to be less than -30 dB, an acceptable level for these types of measurements. Depending on the orientations of the rings and rods within the slab, transmission or no transmission within specific frequency bands is expected.

The transmission experiment was performed on a slab sample of negative metamaterial, as shown in Figure 4.7, composed of rings-only building blocks, as shown in Figure 4.5. The measurements were performed on a structure of 10×60 unit cells while the numerical simulations, because of limitation in computer memory, were performed on a structure of 8×18 unit cells. The respective

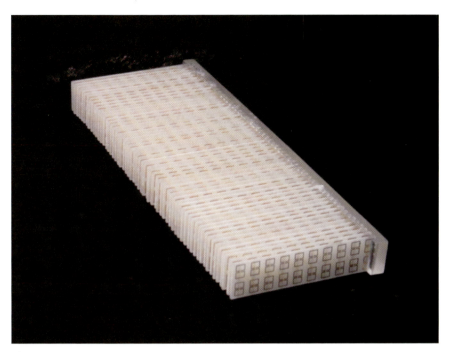

Figure 4.7 Slab of metamaterial realized based on unit cell of Figure 4.5. All PC boards are pasted on a frame made of organic plastic in which multiple grooves have been etched.

results are shown in Figure 4.8*a* and *b*. Since no rods are present in the structure, the permittivity is positive and equal to 1, and only the permeability can assume negative values. Therefore, as expected, the simulation results show a stop band between 8 and 9 GHz, in between the plasma and resonant frequencies of the permeability. The measurement results of Figure 4.8*a* show a similar stop band at similar frequencies, indicating that the rings indeed operate as expected. The low transmission witnessed at about 6 GHz corresponds to the cutoff frequency of the 3-cm waveguide coupler and is therefore not an effect due to the metamaterial.

Following the transmission experiment, a refraction experiment can be performed to look for frequency regions where negative refraction occurs. In this case, the unit cell of the metamaterial contains both rings and rods and is shaped into a prism structure of the same dimensions as the one described in [5]. The angle and levels of transmission are recorded as functions of frequency, and the results are summarized in Figure 4.9. It is clearly seen that there exists a stand alone peak in the frequency region between 8.2 and 8.7 GHz at a refraction angle of about $-30°$. Consequently, within this frequency band where both the permittivity and the permeability are negative, an effective index of refraction assuming negative values is created, although the power associated with this frequency band is low.

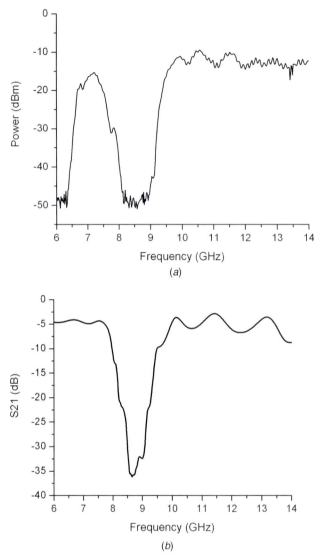

Figure 4.8 Transmission level through 5-cm-thick slab of negative metamaterial based on unit cell shown in Figure 4.5 where rod has been removed: (*a*) experiments; (*b*) simulations. The stop band between 8 and 9 GHz corresponds to the frequency band where $\mu < 0$. The lower cutoff corresponds to the cutoff of the waveguide coupler used in the experiment.

4.3.2.2 *Omega (Ω) SRR*

The Ω ring is a geometry that has already been studied in another context. In fact, in 1992 already, Ω rings were introduced as a way to achieve pseudochirality or bi-anisotropy [34, 35]. More recently, these geometries have been shown to exhibit frequency-dependent responses associated

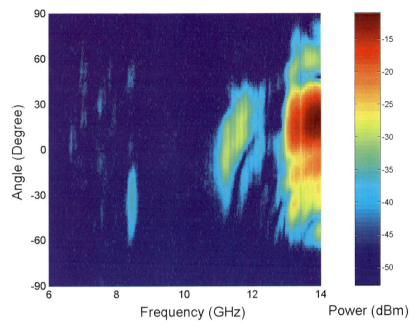

Figure 4.9 Transmission level through prism of metamaterial based on unit cell shown in Figure 4.5.

with negative values of permittivities and permeabilities, and the inherent bi-anisotropic character of these rings has been canceled by printing two reversed Ω patterns back to back on each side of a substrate board [12].

The current realizations of metamaterials based on Ω structures use this configuration, as can be seen in Figure 4.10. For the purpose of realization, three Ω's are stacked in series to form the unit cell in the vertical direction. The FR4 substrate on which the metallizations are printed has a thickness of 0.4 mm, which is small in order to increase the coupling between the back-to-back Ω rings and to cancel their respective chiral effects. Figure 4.10a shows the basic unit cells of the Ω-like structure while Figure 4.10b shows various dielectric cards repeated with a periodicity of 2.5 mm. This metamaterial is then used for the transmission and refraction experiments.

The power transmission experiment and simulations have been performed on a metamaterial slab consisting of 10×80 unit cells and 10×40 unit cells, respectively, along the $\hat{y}$ and $\hat{z}$ directions (for a beam incident in the $\hat{z}$ direction). The simulation results are shown in Figure 4.11a, where a clear transmission band is seen between 12 GHz and 13 GHz. The corresponding measured results are shown in Figure 4.11b, where a similar transmission band is seen between 12 GHz and 13.2 GHz. The frequency corresponding to the peak value of the power is 12.6 GHz, corresponding to a wavelength in air of about 24 mm. Therefore, the 4-mm repetition of the unit cell in the $\hat{z}$ direction and the 3.4-mm

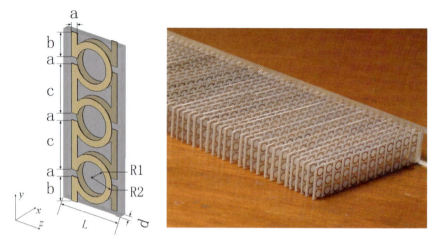

Figure 4.10 Illustration of (*a*) unit cell and (*b*) final metamaterial based on Ω structure. The unit cell measures 2.5 mm, 10 mm, and 4 mm for the $\hat{x}$, $\hat{y}$, and $\hat{z}$ directions, respectively. Other parameters are $a = 0.4$ mm, $b = 1.5$ mm, $c = 2.9$ mm, $R_1 = 1$ mm, and $R_2 = 1.4$ mm.

repetition of the unit cell in the $\hat{y}$ direction correspond to approximately one-sixth or less of the wavelength so that the slab can be regarded as a homogeneous material at these frequencies.

Figure 4.12 shows the result of the refraction experiment in which the setup is similar to the one described in Figure 2 in [5]. Since the size of the basic unit cell in this metamaterial is smaller than that in [5] (4 mm × 2.5 mm vs. 5 mm × 5 mm), the angle of the hypotenuse and the longer side of the triangular prism is about $15.46°$ instead of $18.43°$. The experimental results are shown in Figure 4.12*a*, where a peak at negatively transmitted angles can be seen at around 12.6 GHz. The band that would correspond to an LH behavior of the metamaterial extends over 1.2 GHz, which agrees very well with the passband witnessed in the power transmission experiment. It should also be mentioned that the losses of the prism at 12.6 GHz are smaller than 14 dB, which is acceptable. Figure 4.12*b* shows a 2D power–angle plane extracted from Figure 4.12*a* at 12.6 GHz, from which the refracted beam is seen to bend at an angle of about $27°$, corresponding to an effective index of refraction of about -1.7.

Finally, it is important to mention the sensitivity of the experiment to the electrical contact of the arms of the Ω's with the top and bottom plates of the PPW. The trend that has been witnessed, confirmed by numerical simulations, is to see a wider and less lossy LH frequency band when the electric contact is improved. This suggests that within an experimental measurement a pressure has to be applied between the two plates of the PPW or the arms of the Ω's need to be separately treated to ensure contact. This issue, although rarely mentioned in the literature, has proven to be of tremendous importance in designs which rely on a rod-like structure for the realization of the negative permittivity.

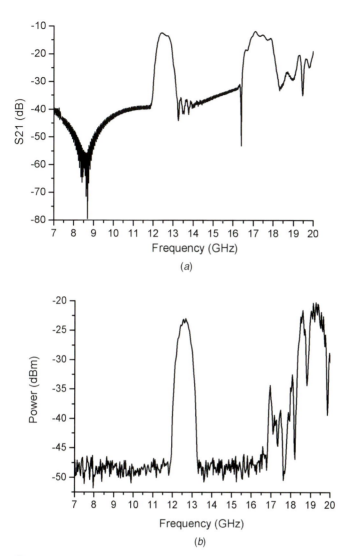

Figure 4.11 Transmission results for material shown in Figure 4.10: (*a*) simulations; (*b*) experiments.

4.3.2.3 *Solid-State Structure* A major drawback in most metamaterials realized to date is the losses that they exhibit. Although the LH properties are undoubtedly verified experimentally, the level of transmission exhibited by these metamaterials is still too low to make them useful in industrial applications. Various causes have been identified as a possible origin of losses, one of them being mismatch. Another major drawback of most of the current implementations of negative metamaterials is their mechanical fragility: Being essentially constituted of dielectric boards separated by air, they are very delicate to manipulate.

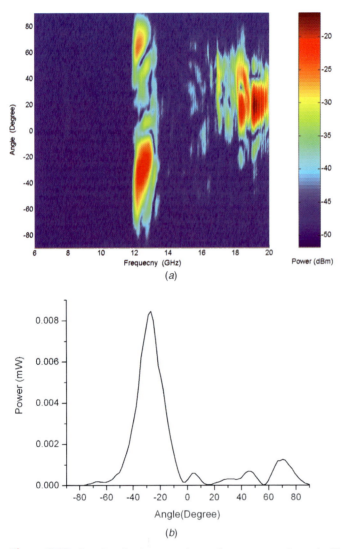

Figure 4.12 Results of prism experiment for geometry shown in Figure 4.10: (*a*) transmission level as function of frequency and angle; (*b*) transmission level at 12.6 GHz.

A way that appears to avoid both drawbacks is to realize a *solid-state* metamaterial. By solid-state metamaterial, we mean here a rigid structure composed of a periodic arrangement of rings small compared to the wavelength and which exhibits LH properties at certain frequencies. Such a solid-state metamaterial can be easily obtained with standard hot-press techniques used in the manufacture process of multilayered printed circuit boards obtained by the compression of multiple PC boards at high temperature [36]. Note that this concept was also postulated in [8] and realized by using high-dielectric spacers between the rings, although not compressed together.

(a)

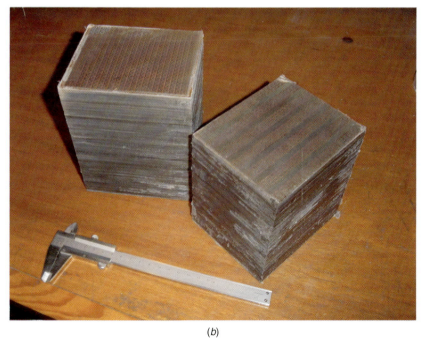

(b)

Figure 4.13 Toward a solid-state metamaterial: (a) machine used to compress samples; (b) solid-state metamaterial.

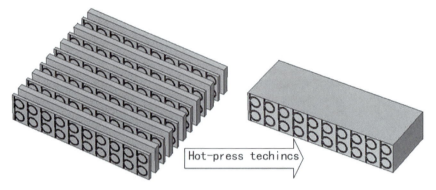

Figure 4.14 Concept of solid-state metamaterial.

Figure 4.13 shows the machine where the hot compression is performed and the photograph of the original compressed sample. The material thus produced has more stable and better characteristics, both from a mechanical standpoint (it is less fragile) and an electromagnetic standpoint (it exhibits less mismatched boundaries).

The fabrication process is illustrated here on a metamaterial based on the Ω ring described in the previous section. Figure 4.14 shows the concept underlying the production of solid-state metamaterial samples from a succession of dielectric boards.

In this case, two types of dielectric boards are alternated: The first type contains two stacked Ω rings with their corresponding inverted images and the second one does not contain any metallization. The geometry of the Ω rings is different from the previous ones: The inner and outer radii of the Ω pattern are 1.5 and 1.9 mm, respectively; the length of the arm is 2.3 mm; the gap between the two arms is 0.4 mm; the width of the printed track is 0.4 mm; and each unit cell occupies a 5-mm space. The boards without Ω patterns have the same permittivity as those on which the Ω patterns are printed. The final product of this process is shown in Figure 4.15, where the solid-state metamaterial measures 5 cm $\times$ 1 cm $\times$ 9 cm.

The transmission level as function of frequency is shown in Figure 4.16, where a very clear passband can be identified around 8.8 GHz. A closer examination reveals that the bandwidth where high transmission occurs is about 1 GHz, which is very significant, and the level at 8.85 GHz (the exact central frequency of the passband) is -14.8 dBm. To estimate the insertion losses of the metamaterial slab, we have measured the transmission power with and without the metamaterial sample. The corresponding value at 8.85 GHz is -9.8 dBm, which can be approximated as the power incident upon the interface of the slab and air. Thus, with a return loss at the incident interface, the maximum insertion loss of the slab composed of 10 unit cells is less than 5 dB, which corresponds to less than 0.5 dB/unit cell. Such a level is comparable with the one exhibited currently by microwave devices and indicates that such a solid-state metamaterial could be used in industrial applications.

Figure 4.15   Photograph of solid-state material.

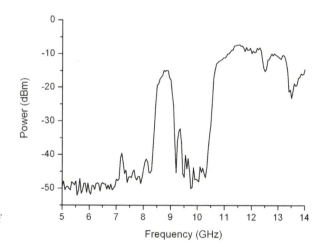

Figure 4.16   Transmission level through solid-state metamaterial of Figure 4.15.

The results obtained from the refraction by a prism are illustrated in Figure 4.17. A clear deflection toward negative angles can be seen between 8.3 and 9.3 GHz, again with small losses.

4.3.2.4 S Ring   The last structure we shall review here is the S structure depicted in Figure 4.18. This ring has been extensively studied in [14, 37] both theoretically and numerically, and we shall only review two of its interesting properties here.

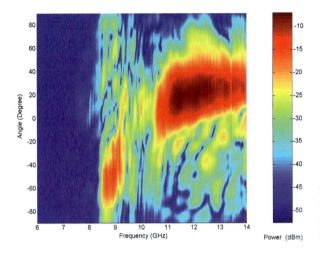

Figure 4.17 Transmission result through solid-state prism shown in Figure 4.15.

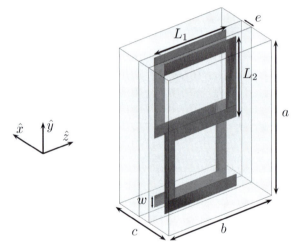

Figure 4.18 Unit cell of the S-ring geometry: $e = 0.5$ mm, $L_1 = L_2 = 2.8$ mm, $w = 0.4$ mm, $a = 5.4$ mm, $b = 4$ mm, $c = 2.5$ mm.

The first property of foremost importance is that the S ring does not require the addition of a rod to exhibit a negative permittivity at similar frequencies to where it exhibits a negative permeability. In fact, all the rings exhibit a frequency-dispersive permittivity response, in addition to the required frequency-dispersive permeability response (see, e.g., [38] for the broadside-coupled SRR). However, the interesting region of permittivity response where negative values are achieved is usually much higher in frequency than the region where the permeability is negative, making this effect not usable. The S ring, on the contrary, exhibit a negative-permittivity response at *similar* frequencies as the negative-permeability response. In addition to being a design advantage since a single entity can now control both parameters, it is mostly an experimental advantage since it avoids

Figure 4.19   Photograph of S-based solid-state metamaterial.

the necessity of ensuring the electrical contact between the plates of the PPW and the rodlike structure present in other designs.

A realization of a metamaterial based on this structure is shown in Figure 4.19. As can be seen, the top and bottom arms of the S shape are not reaching the edges of the dielectric boards, which indicates that they need not be in contact with the PPW. A prism experiment based on the metamaterial of Figure 4.19 has been performed, yielding the results shown in Figure 4.20. A high transmission peak corresponding to negatively refracted angles can be seen between 10.9 and 13.5 GHz, indicating a bandwidth of operation of 2.6 GHz. In addition, the insertion losses were estimated in the same way as in the solid-state case, revealing losses of about 0.7 dB per unit cell. The S ring therefore yields a low-loss metamaterial where LH properties are obtained over a large bandwidth.

The second important feature of this ring is that its shape can be easily modified to achieve desired frequency responses: The two loops in the S pattern need not necessarily be of the same size or, if needed, additional loops can be added. The whole frequency response of the S ring can be directly predicted from its circuit model, as has been shown in [37]. The various capacitances and inductances are directly related to the geometry of the ring and can be modified or new elements can be added to achieve new properties. This flexibility has been illustrated in [39], where a modified S-ring design has been proposed to achieve two frequency bands where the effective index of refraction is negative, while the overall transmission exhibits small losses over a large frequency band. Such flexibility is of the foremost importance for the use of negative metamaterial in industrial applications.

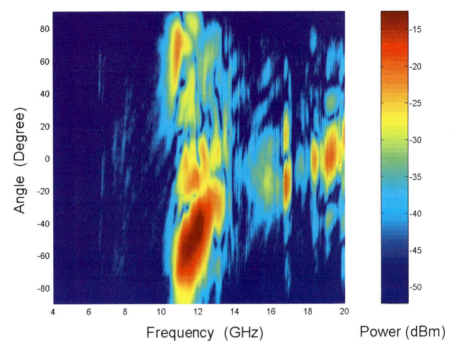

Figure 4.20   Transmission through prism of metamaterial as shown in Figure 4.19.

4.4   CONCLUSION

This chapter has presented various experiments on metamaterials, emphasizing properties such as transmission and negative refraction. It should be noted, however, that negative refraction can also be achieved with traditional anisotropic media, while some anisotropic metamaterials do not exhibit it. Therefore, the negative or positive refraction should be merely viewed as a consequence of the deeper properties of metamaterials. Among some other consequences, we can also list focusing, backward phase, reversed Čerenkov radiation, reversed Doppler shift, and so on. In this regard, all the experimental works on metamaterials are fundamentally important. By offering repeatable experimental verifications, they help us to understand this class of media and lead us to the exploration of their designs and applications.

ACKNOWLEDGMENTS

We would like to acknowledge all the people who have been involved directly or indirectly in the elaboration of this chapter: Hongsheng Chen, Jianbing Chen, Kangsheng Chen, Xudong Chen, Jiangtao Huangfu, Xianming Zhang, Jie Lu, Zachary Thomas and Bae-Ian Wu.

REFERENCES

1. V. Veselago, "The electrodynamics of substances with simultaneously negative values of ε and μ," *Sov. Phys. USPEKHI*, vol. 10, pp. 509–514, Jan./Feb. 1968.
2. J. Pendry, A. Holden, D. Robbins, and W. Stewart, "Low frequency plasmons in thin-wire structures," *J. Phys Condens. Matter*, vol. 10, pp. 4785–4809, 1998.
3. J. Pendry, A. J. Holden, D. Robbins, and W. J. Stewart, "Magnetism from conductors and enhanced nonlinear phenomena," *IEEE Trans. Microwave Theory Tech.*, vol. 47, pp. 2075–2084, Nov. 1999.
4. D. R. Smith, W. J. Padilla, D. Vier, S. Nemat-Nasser, and S. Schultz, "Composite medium with simultaneously negative permeability and permittivity," *Phys. Rev. Lett.*, vol. 84, pp. 4184–4187, May 2000.
5. R. Shelby, D. Smith, and S. Schultz, "Experimental verification of a negative index of refraction," *Science*, vol. 292, pp. 77–79, Apr. 2001.
6. K. Li, S. J. McLean, R. B. Greegor, C. G. Parazzoli, and M. H. Tanielian, "Free-space focused-beam characterization of left-handed materials," *Appl. Phys. Lett.*, vol. 82, no. 15, pp. 2535–2537, Apr. 2003.
7. D. R. Smith, S. Shultz, P. Markoš, and C. M. Soukoulis, "Determination of effective permittivity and permeability of metamaterials from reflection and transmission coefficients," *Phys. Rev. B*, vol. 65, pp. 195104, 2002.
8. R. W. Ziolkowski, "Design, fabrication, and testing of double negative metamaterials," *IEEE Trans. Antennas Propag.*, vol. 51, pp. 1516–1529, July 2003.
9. X. Chen, T. M. Grzegorczyk, B.-I. Wu, J. Pacheco Jr., and J. A. Kong, "Robust method to retrieve the constitutive effective parameters of metamaterials," *Phys. Rev. E*, vol. 70, 016608, 2004.
10. R. Marqués, F. Medina, and R. Rafii-El-Idrissi, "Role of bianisotropy in negative permeability and left-handed metamaterials," *Phys. Rev. B*, vol. 65, 144440, 2002.
11. S. O'Brien and J. Pendry, "Magnetic activity at infrared frequencies in structured metallic photonic crystals," *J. Phys. Condens. Matter*, vol. 14, pp. 6383–6394, 2002.
12. C. R. Simovski and S. He, "Frequency range and explicit expressions for negative permittivity and permeability for an isotropic medium formed by a lattice of perfectly conducting Ω particles based on the quasi-static Lorentz theory," *Phys. Lett. A*, vol. 311, pp. 254–263, 2003.
13. J. Huangfu, L. Ran, H. Chen, X. Zhang, K. Chen, T. M. Grzegorczyk, and J. A. Kong, "Experimental confirmation of negative refractive index of a metamaterial composed of Ω-like metallic patterns," *Appl. Phys. Lett.*, vol. 84, pp. 1537–1539, Mar. 2004.
14. H. Chen, L. Ran, J. Huangfu, X. Zhang, K. Chen, T. M. Grzegorczyk, and J. A. Kong, "Left-handed metamaterials composed of only S-shaped resonators," *Phys. Rev. E*, vol. 70, 057605, 2004.
15. L. Landau and E. Lifshitz, *The Classical Theory of Fields*, Vol. 2, Pergamon, 1975.
16. J. A. Kong, *Electromagnetic Wave Theory*, EMW, 2000.
17. T. M. Grzegorczyk, M. Nikku, X. Chen, B.-I. Wu, and J. A. Kong, "Refraction laws for anisotropic media and their application to left-handed metamaterials," *IEEE Trans. Microwave Theory Tech.*, vol. 53, no. 4, pp. 1443–1450, Apr. 2005.
18. T. M. Grzegorczyk, X. Chen, J. Pacheco Jr., J. Chen, B.-I. Wu, and J. A. Kong, "Reflection coefficients and Goos Hänchen shifts in anisotropic and bianisotropic left-handed metamaterials," *Prog. Electromagn. Res.*, Special Issue on Left-Handed Metamaterials, vol. 51, pp. 83–113, 2005.
19. D. R. Smith and D. Schurig, "Electromagnetic wave propagation in media with indefinite permittivity and permeability tensors, " *Phys. Rev. Lett.*, vol. 90, 077405, 21 Feb. 2003.
20. J. A. Kong, "Electromagnetic wave interaction with stratified negative isotropic media," *Prog. Electromagn. Res.*, vol. 35, pp. 1–52, 2002.

21. J. A. Kong, B.-I. Wu, and Y. Zhang, "Lateral displacement of a Gaussian beam reflected from a grounded slab with negative permittivity and permeability," *Appl. Phys. Lett.*, vol. 80, pp. 2084–2086, Mar. 2002.

22. J. A. Kong, B.-I. Wu, and Y. Zhang, "A unique lateral displacement of a Gaussian beam transmitted through a slab with negative permittivity and permeability," *Microwave Opt. Tech. Lett.*, vol. 33, pp. 136–139, Apr. 2002.

23. L. Ran, J. Huangfu, H. Chen, X. Zhang, K. Chen, T. M. Grzegorczyk, and J. A. Kong, "Beam shifting experiment for the characterization of left-handed properties," *J. Appl. Phys.*, vol. 95, pp. 2238–2241, Mar. 2004.

24. C. Moss, T. M. Grzegorczyk, Y. Zhang, and J. A. Kong, "Numerical studies of left-handed metamaterials," *Prog. in Electromagn. Res.*, vol. 35, pp. 315–334, 2002.

25. J. Pendry, "Negative refraction makes a perfect lens," *Phys. Rev. Lett.*, vol. 85, pp. 3966–3969, Oct. 2000.

26. X. S. Rao and C. K. Ong, "Amplification of evanescent waves in a lossy left-handed material slab," *Phys. Rev. B*, vol. 68, 113103, 2003.

27. A. Grbic and G. V. Eleftheriades, "Negative refraction, growing evanescent waves, and sub-diffraction imaging in loaded transmission-line metamaterials," *IEEE Trans. Microwave Theory Tech.*, vol. 51, pp. 2297–2305, Dec. 2003.

28. T. Andrade, A. Grbic, and G. V. Eleftheriades, "Growing evanescent waves in continuous transmission-line grid media," *IEEE Microwave Wireless Components*, vol. 15, pp. 131–133, Feb. 2005.

29. M. M. Alkaisi, R. J. Blaikie, S. J. McNab, R. Cheung, and D. R. S. Cumming, "Sub-micron imaging with a planar silver lens," *Appl. Phys. Lett.*, vol. 84, pp. 4403–4405, May 2004.

30. A. A. Houck, J. B. Brock, and I. L. Chuang, "Experimental confirmation of a left-handed material that obeys Snell's law," *Phys. Rev. Lett.*, vol. 90, no. 13, 137401, 2003.

31. D. R. Smith, D. Shurig, J. Mock, P. Kolinko, and P. Rye, "Partial focusing of radiation by a slab of indefinite media," *Appl. Phys. Lett.*, vol. 84, pp. 2244–2246, Mar. 2004.

32. J. B. Brock and A. A. H. I. L. Chuang, "Focusing inside negative index materials," *Appl. Phys. Lett.*, vol. 85, no. 13, pp. 2472–2474, 2004.

33. C. Balanis, *Antenna Theory: Analysis and Design*, 2nd ed., Wiley, New York, 1997.

34. J. A. Kong, "Theorems of bianisotropic media," *Proc. IEEE*, vol. 60, no. 9, pp. 1036–1046, Sept. 1972.

35. M. M. I. Saadoum and N. Engheta, "A reciprocal phase shifter using novel pseudochiral or Ω medium," *Microwave Opt. Technol. Lett.*, vol. 5, pp. 184–188, Apr. 1992.

36. L. Ran, J. Huangfu, H. Chen, Y. Li, X. Zhang, K. Chen, and J. A. Kong, "Microwave solid-state left-handed material with a broad bandwidth and a ultra-low loss," *Phys. Rev. B*, vol. 70, 073102, 2004.

37. H. Chen, L. Ran, J. Huangfu, X. Zhang, K. Chen, T. M. Grzegorczyk, and J. A. Kong, "Magnetic properties of S-shaped split-ring resonators," *Prog. Electromagn. Res.*, Special Issue on Left-Handed Metamaterials, vol. 51, pp. 231–247, 2005.

38. A. Ishimaru, S.-W. Lee, Y. Kuga, and V. Jandhyala, "Generalized constitutive relations for metamaterials based on the quasi-static Lorentz theory," *IEEE Trans. Antennas Propag.*, vol. 51, pp. 2550–2557, Oct. 2003.

39. H. Chen, L. Ran, X. Z. Jiangtao Huangfu, K. Chen, T. M. Grzegorczyk, and J. A. Kong, "Metamaterial exhibiting left-handed properties over multiple frequency bands," *J. Appl. Phys.*, vol. 96, no. 9, pp. 5338–5340, 1 Nov. 2004.

TWO-DIMENSIONAL PLANAR NEGATIVE-INDEX STRUCTURES

CHAPTER 5

ANTENNA APPLICATIONS AND SUBWAVELENGTH FOCUSING USING NEGATIVE-REFRACTIVE-INDEX TRANSMISSION LINE STRUCTURES

George V. Eleftheriades

5.1 INTRODUCTION

In the 1960s Victor Veselago asked whether Maxwell's equations permit materials with simultaneously negative permittivity and permeability [1]. His conclusion was positive and he predicted a number of unusual electromagnetic phenomena associated with such hypothetical media. A characteristic property of these materials is that plane waves propagating in them would have their phase velocity antiparallel to the group velocity; hence these media would support backward waves. Likewise the vectors describing the electric field, the magnetic field, and the propagation direction would follow the left-handed rule; hence he coined the term "left handed" to describe these hypothetical media. Moreover, Veselago associated this backward-wave (BW) property with the notion of negative refraction and he described several unusual focusing devices (e.g., lenses) that operate based on negative refraction.

It was only recently, though, that people understood how to implement these left-handed or negative-refractive-index (NRI) media. The first such implementation was produced at the University of California at San Diego and comprised a volumetric periodic array of straight metallic wires and split-ring resonators to synthesize negative effective permittivity and negative permeability, respectively [2].

Portions of this material also appears in *Negative-Refraction Metamaterials: Fundamental Principles and Applications*, G.V. Eleftheriades and K.G. Balmain, Eds., IEEE PRESS/Wiley, Copyright © 2005, John Wiley & Sons.

Metamaterials: Physics and Engineering Explorations, Edited by N. Engheta and R. W. Ziolkowski
Copyright © 2006 the Institute of Electrical and Electronics Engineers, Inc.

Another way to implement materials that support the phenomenon of negative refraction was subsequently proposed based on the concept of loading planar transmission line (TL) grids with reactive elements (see Section 5.2). In this chapter we present a number of radio frequency (RF)/microwave passive devices that have been developed at the University of Toronto based on the concept of TL NRI metamaterials. The emphasis is placed on focusing and antenna applications (including their feed networks).

5.2 PLANAR TRANSMISSION LINE MEDIA WITH NEGATIVE REFRACTIVE INDEX

A 2D NRI metamaterial (MTM) can be physically implemented by reactively loading a host TL grid. A representative unit cell of such a periodic NRI-TL medium is depicted in Figure 5.1.

Specifically, a host TL medium (e.g., microstrip) is periodically loaded with discrete series capacitors and shunt inductors [3,4]. From the onset, the key observation is that there is a correspondence between negative permittivity and a shunt inductance (L) as well as between negative permeability and a series capacitance (C). This allows the synthesis of artificial media (MTMs) with a negative permittivity and a negative permeability and hence a negative refractive index. When the unit-cell dimension d is much smaller than a guided wavelength, the array can be regarded as a homogeneous effective medium and as such can be described by effective constitutive parameters $\mu_N(\omega)$ and $\varepsilon_N(\omega)$, which are determined through a rigorous periodic analysis to be of the form shown in Eq. (5.1) (assuming 2D transverse magnetic (TM_y) wave propagation in Fig. 5.1)

$$\varepsilon_N(\omega) = 2\varepsilon_p - \frac{g}{\omega^2 L_0 d} \qquad \mu_N(\omega) = \mu_p - \frac{1/g}{\omega^2 C_0 d} \qquad (5.1)$$

where ε_p and μ_p are positive constants describing the host TL medium and they are proportional to the per-unit-length capacitance and inductance of this host TL medium, respectively. On the other hand, the geometric factor g relates the characteristic impedance of the TL network to the wave impedance of the effective medium. Moreover, the factor of 2 in front of the effective permittivity of the 2D medium is necessary to properly account for scattering at the edges of the unit cell (this factor becomes 1 for 1D media). These TL media support backward

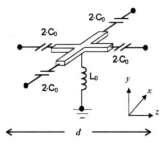

Figure 5.1 Unit cell for 2D NRI-TL metamaterial. A host TL is loaded periodically with series capacitors and shunt inductors in a dual (high-pass) configuration. From [7], copyright © 2003 by the Optical Society of America.

waves in which the phase and group velocities are antiparallel; hence, they implement the left-handed or NRI media envisioned by Veselago, as was pointed out in [4,5]. The relationship between 1D BW lines (but in the ideal case without any host medium) and left-handed lines was also pointed out in [6]. Naturally, due to the host microstrip medium, the practically realizable unit cell of Figure 5.1 contains both positive- and negative-refractive-index responses, as implied by Eq. (5.1), which was originally stipulated in [4,5,7] and in [8]. This particular arrangement of the inclusions L_0 and C_0 provides the desired negative material contribution that diminishes with frequency ω and ensures compatibility with the Poynting theorem for dispersive media [1]. When the parameters are simultaneously negative, these structures exhibit a negative effective refractive index and have experimentally demonstrated the predicted associated phenomena, including negative refraction and focusing [3,4,7,9] and focusing with subwavelength resolution [10,11].

In practical realizations, the subwavelength unit cell of Figure 5.1 is repeated to synthesize artificial 2D materials with overall dimensions that are larger than the incident electromagnetic wavelength. Therefore, the resulting structures are by definition distributed. However, the loading lumped elements could be realized either in chip [3,4] or in printed form [5,12,13].

5.3 ZERO-DEGREE PHASE-SHIFTING LINES AND APPLICATIONS

In conventional positive-refractive-index (PRI) TLs, the phase lags in the direction of positive group velocity, thus incurring a negative phase. It therefore follows that phase compensation can be achieved at a given frequency by cascading a section of a NRI line (e.g., BW line) with a section of a PRI line to synthesize positive, negative, or zero transmission phase over a short physical length (see Fig. 5.2) [14]. This idea of phase compensation is inherent in Veselago's flat-lens idea and was also proposed for implementing thin subwavelength resonators [15].

A physical implementation of this concept using TLs is shown in Figure 5.3. The structure of Figure 5.3 can be rearranged to form a series of symmetric MTM unit cells as proposed in [4,14]. Such a unit cell is shown in Figure 5.4, and it is nothing but a TL of characteristic impedance Z_0 periodically loaded with series capacitance C_0 and shunt inductance L_0. A representative dispersion diagram for typical host TL and loading parameters is shown in Figure 5.5.

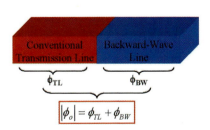

Figure 5.2 Method of phase compensation using conventional TL and BW lines.

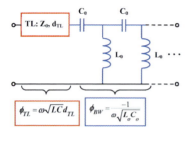

Figure 5.3 Phase-compensating structure based on conventional TL and NRI (BW) lines.

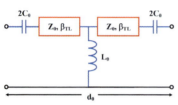

Figure 5.4 Unit cell of MTM phase-shifting line comprising host TL periodically loaded with series capacitors and shunt inductors. From [34], copyright © 2005 by the Institute of Electrical and Electronics Engineers, Inc.

The MTM phase-shifting lines can then be constructed by cascading a series of these unit cells. The edges of the stop band, f_{c1} and f_{c2}, in Figure 5.5 are determined at the series resonance between the inductance of the TL section and the loading capacitance C_0 and at the shunt resonance between the capacitance of the TL section and the loading inductance L_0, respectively. Alternatively these are the frequencies at which the effective permeability $\mu_N(\omega)$ and effective permittivity $\varepsilon_N(\omega)$ vanish, $\varepsilon_N(\omega) = 0$, $\mu_N(\omega) = 0$. Hence, by setting the effective material parameters of Eq. (5.1) to zero, these cutoff frequencies are readily determined to be

$$f_{c1} = \frac{1}{2\pi}\sqrt{\frac{1/g}{\mu_p C_0 d}} \qquad (5.2)$$

$$f_{c2} = \frac{1}{2\pi}\sqrt{\frac{g}{\varepsilon_p L_0 d}} \qquad (5.3)$$

where the characteristic impedance of the host transmission line is $Z_0 = g\sqrt{\mu_p/\varepsilon_p} = \sqrt{L/C}$. By equating f_{c1} and f_{c2}, the stop band in Figure 5.5 can be closed, thus allowing access to phase shifts around the zero mark. The condition for a closed stop band is therefore determined to be

$$Z_0 = \sqrt{\frac{L_0}{C_0}} \qquad (5.4)$$

This condition also implies that the PRI TL of Figure 5.2 is matched to the NRI line. The closed stop-band condition (5.4) was originally derived in [4] Eq. (29) and subsequently also reported in [8]. Under this condition, it has been shown in [14] that the total phase shift per unit cell is

$$\beta_{\text{eff}} \approx \omega\sqrt{LC} + \frac{-1}{\omega\sqrt{L_0 C_0}} \qquad (5.5)$$

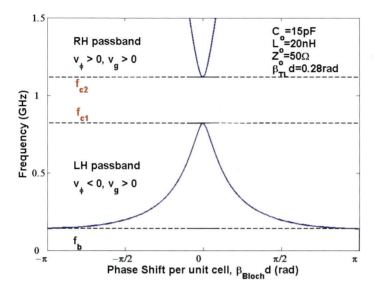

Figure 5.5 Dispersion diagram for periodic structure of Figure 5.4 with typical line and loading parameters. The edges of the stop band are designated by f_{c1} and f_{c2}. From [14], copyright © 2003 by the Institute of Electrical and Electronics Engineers, Inc.

This expression can be interpreted as the sum of the phase incurred by the host $C-L$ TL and a uniform BW $L-C$, line as shown in Figures 5.2 and 5.3.

Various 1D phase-shifting lines were constructed in coplanar waveguide (CPW) technology at 0.9 GHz, as shown in Figure 5.6. The simulated and measured phase responses for two- and four-stage zero-degree phase-shifting lines are shown in Figure 5.7, compared to the phase response of a conventional $-360°$ TL. Also shown is the magnitude response of the two- and four-stage $0°$ lines.

It can be observed that the experimental results correspond very closely to the simulated results, highlighting the broadband nature of the phase-shifting lines and their small losses.

It can be concluded that the MTM phase-shifting lines offer some significant advantages when compared to conventional delay lines. They are compact in size, can be easily fabricated using standard etching techniques, and exhibit a linear phase response around the design frequency. They can incur *either* a negative *or* a positive phase as well as a zero-degree phase depending on the values of the loading elements while maintaining a short physical length. In addition, the phase incurred is independent of the length of the structure. Due to their compact, planar design, they lend themselves easily to integration with other microwave components and devices. The MTM phase-shifting lines are therefore well suited for broadband applications requiring small, versatile, linear devices.

It should be pointed out that these phase-shifting lines offer an advantage in terms of size and bandwidth when phase shifts about the zero-degree mark are needed. In this case, the proposed devices have a clear advantage when compared to a corresponding delay line about the one-wavelength mark (see Fig. 5.7). The

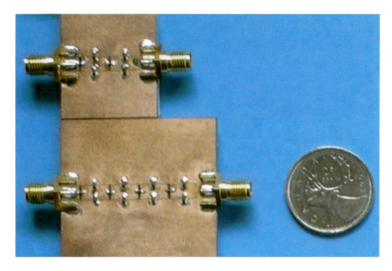

Figure 5.6 Top: Two-stage phase-shifting line (16 mm). Bottom: Four-stage phase-shifting line (32 mm) at 0.9 GHz. Note: Reference $-360°$ TL line, 283.5 mm (not shown).

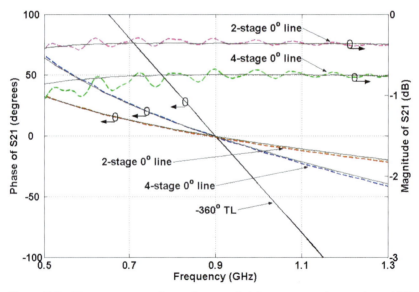

Figure 5.7 Phase responses of one-, four-, and eight-stage zero-degree phase-shifting lines compared to conventional $-360°$ TL at 0.9 GHz: $(---)$ measured; (---) Agilent-ADS Simulation. Also shown is the amplitude response for the two- and four-stage devices.

significant advantage arises from the short electrical length of these zero phase shift lines, which also implies a broadband response (always when comparing to a one-wavelength delay line). Examples of harnessing these advantages in practical applications are discussed below. It should be pointed out that for electrically long PRI/NRI phase-shifting lines their broadband nature could be retained if the constituent NRI section is also designed to exhibit a negative group velocity, as was done in [16]. In this case not only the signs but also the slopes of the propagation constants (vs. frequency) of the NRI and PRI lines compensate, thus leading to inherently broadband response. Of course the difficulty now becomes the issue of how to synthesize a negative group velocity over a broad bandwidth. Moreover the NRI lines of [16] are lossy and, hence, restoring amplifiers would need to be included for acceptable performance.

5.3.1 Nonradiating Metamaterial Phase-Shifting Lines

Any artificial TL that supports fast waves, that is, waves whose phase velocity is greater than the speed of light, will tend to radiate into free space if its electrical length is sufficiently long. When the MTM lines presented in the previous section are used to create zero-degree phase-shifting lines, the phase incurred by each unit cell and therefore the propagation constant are equal to zero at the design frequency. Since the phase velocity is defined as $v_\phi = \omega/\beta$, its value will be infinite at the design frequency. Thus the lines will support fast waves that will tend to radiate when they are long enough. A typical dispersion diagram with a closed stop band [see Eq. (5.4)] for a zero-degree MTM phase-shifting line of the type shown in Figure 5.4 is depicted in Figure 5.8. If the MTM lines are designed to operate anywhere within the radiation cone of the Brillouin diagram, the lines will be prone to radiation.

To ensure that the MTM phase-shifting lines do not radiate, they can be operated in the NRI BW region, while simultaneously ensuring that the propagation constant of the line exceeds that of free space. This will effectively produce a slow-wave structure with a *positive* insertion phase, Φ_{MM}. The Brillouin diagram

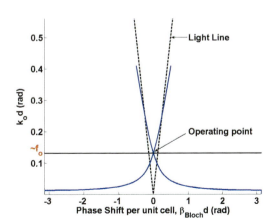

Figure 5.8 Dispersion diagram for zero-degree MTM unit cell; the horizontal line designates the operating frequency.

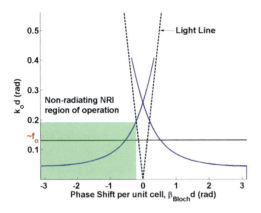

Figure 5.9 Dispersion diagram indicating regions of propagation outside radiation cone; the horizontal line designates the operating frequency.

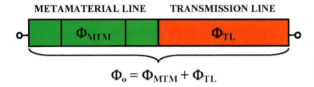

Figure 5.10 Nonradiating (slow-wave) MTM phase-shifting line.

for this scenario is shown in Figure 5.9. It can be observed that at the design frequency, propagation occurs outside of the radiation cone.

By cascading the MTM line of Figure 5.9 with a conventional (PRI) TL that inherently incurs a *negative* insertion phase, Φ_{TL}, a composite slow-wave MTM phase-shifting line is obtained that does not radiate, as shown in Figure 5.10. Furthermore, If Φ_{MTM} and Φ_{TL} are equal but opposite in sign, then the structure will incur a zero insertion phase, given by

$$\Phi_0 = \Phi_{MTM} + \Phi_{TL} = 0 \qquad (5.6)$$

Thus, it has been shown that it is possible to construct MTM phase-shifting lines that do not radiate, which can then be used for the design of antenna feed networks, without affecting the radiation patterns.

5.3.2 Series-Fed Antenna Arrays with Reduced Beam Squinting

An example of utilizing the slow-wave phase-shifting lines of the previous section to feed a set of in series printed dipole antennas has been reported in [17]. The main idea is to use zero-degree phase-shifting lines to feed the dipoles of a series-fed array in phase. Due to the broadband nature of these lines, the resulting array patterns squint much less with frequency when compared to their conventional series-fed counterparts using meandered one-wavelength lines.

In a typical series-fed linear array designed to radiate at broadside, the antenna elements must be fed in phase. In addition, an interelement spacing d_E of less than a half a free-space wavelength ($d_E < \lambda_0/2$) is necessary to avoid capturing grating lobes in the visible region of the array pattern. To achieve

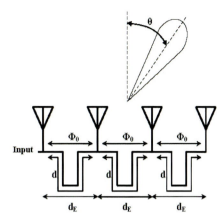

Figure 5.11   Series-fed linear array using conventional -2π TL meandered feed lines.

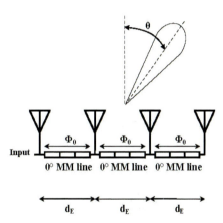

Figure 5.12   Series-fed linear array using Zero-degree MTM feed lines.

these design constraints, traditional designs employing TL-based feed networks have resorted to a meander line approach, as shown in Figure 5.11. This allows the antenna elements to be physically separated by a distance $d_E = \lambda_0/2$ while still being fed in phase with a one guided-wavelength λ_g long meander line that incurs a phase of -2π rad. Because the phase incurred by the TLs is frequency dependent, a change in the operating frequency will cause the emerging beam to squint from broadside, which is generally an undesirable phenomenon. In addition, the fact that the lines are meandered causes the radiation pattern to experience high cross-polarization levels, particularly in CPW implementations as a result of parasitic radiation due to scattering from the corners of the meander lines. The proposed feed networks employ nonradiating MTM phase-shifting lines within a series-fed linear array (see Fig. 5.12) to mitigate some of the problems encountered with conventional TL-based feed networks.

Assuming that the same type of TL sections are used for TL1 and TL2, then Z_0, L, and C will be the same for both lines. Therefore, Φ_{TL2} is given by $\Phi_{\mathrm{TL2}} = \omega\sqrt{LC}d_{\mathrm{TL2}}$. Correspondingly, for a transmission line of length λ_g, the phase as a function of frequency is given by $\Phi_{\lambda_g} = \omega\sqrt{LC}\lambda_g$. The scan angle

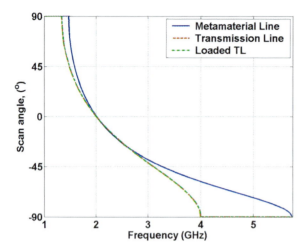

Figure 5.13 Scan angle performance of series-fed linear array with $d_E = \lambda_0/2$ using different feeding techniques.

for each of the MTM- and TL-based linear arrays with an interelement phase shift Φ_0 can therefore be written as

$$\theta_{\text{SCAN,MTM}} = \sin^{-1}\left(-\frac{\Phi_0}{k_0 d_E}\right) = \sin^{-1}\left(-\frac{\Phi_{\text{MTM}} + \Phi_{\text{TL2}}}{k_0 d_E}\right) \quad (5.7)$$

$$\theta_{\text{SCAN,TL}} = \sin^{-1}\left(-\frac{\Phi_0}{k_0 d_E}\right) = \sin^{-1}\left(-\frac{\Phi_{\lambda_g}}{k_0 d_E}\right) \quad (5.8)$$

The MTM- and TL-based feed networks were evaluated in CPW technology at a design frequency of 2 GHz. Two designs were considered: an array with an inter-element spacing of $d_E = \lambda_0/2$ and an array with a spacing $d_E = \lambda_0/4$. The scan angle characteristics for the MTM- and TL-based linear arrays with $d_E = \lambda_0/2$ are shown in Figure 5.13. It can be observed that the scan angle for the TL-fed array exhibits its full scanning range from $+90°$ to $-90°$ within a bandwidth of 2.67 GHz, while the corresponding scanning bandwidth for the MTM-fed array is 4.27 GHz. Thus, the MTM-fed array offers a more broadband scan angle characteristic while simultaneously eliminating the need for meander lines. Also shown in Figure 5.13 is the scan angle characteristic for a low-pass loaded TL also of length $\lambda_0/2$. It can be observed that the performance of this line is identical to that of the TL feed line. Thus, although the loaded line can eliminate the need for meander lines, it does not provide the advantage of an increased scan angle bandwidth that the MTM feed lines offer.

The scan angle characteristics for the $\lambda_0/4$ feed network are shown in Figure 5.14. It can be observed that the bandwidth of the scanning angle for the TL-fed array and the loaded TL array decreases to 1.07 GHz, while the corresponding scanning bandwidth for the MTM-fed array remains at 4.27 GHz. Thus, as the spacing decreases between the antenna elements, it can be seen that the scan angle characteristic for a MTM-fed array remains unchanged, while the corresponding scan angle characteristic for the TL-fed array becomes more narrow band. This is consistent with the previous observation that electrically short phase-shifting lines exhibit a broader bandwidth.

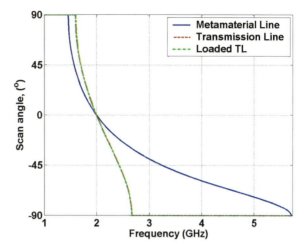

Figure 5.14 Scan angle performance of series-fed linear array with $d_E = \lambda_0/4$ using different feeding techniques.

5.3.3 Broadband Wilkinson Balun Using Microstrip Metamaterial Lines

Baluns are particularly useful for feeding two-wire antennas, where balanced currents on each branch are necessary to maintain symmetric radiation patterns with a given polarization. Two-wire antennas have input ports that are closely spaced; therefore their feeding structures should be chosen to accommodate for this requirement. This precludes the use of certain balun designs whose output ports are spaced far apart [18].

Printed balun designs can generally be classified as distributed-TL or lumped-element type. Distributed-TL designs are inherently narrow band due to the frequency dependence of the TLs used. These can be made broadband; however, they usually require TLs that are at least several wavelengths long and are therefore not very compact [19]. Lumped-element designs, albeit compact, can suffer from a relatively narrow-band differential output phase resulting from the inherent mismatch between the phase response of the low-pass/high-pass output lines that they employ [20].

The proposed MTM balun, shown in Figures 5.15 and 5.16, consists of a Wilkinson power divider followed by a $+90°$ MTM phase-shifting line along the top branch and a $-90°$ MTM phase-shifting line along the bottom branch. The design of the balun was based on the MTM unit cell shown in Figure 5.10 and was carried out by first selecting appropriate values for the loading elements of the $+90°$ MTM line to produce a $+90°$ phase shift at the design frequency f_0 while maintaining a short overall length. Then, the pertinent parameters for the $-90°$ MTM line were calculated such that the shape of the phase responses of the $+90°$ and $-90°$ MTM lines matched, thus maintaining a $180°$ phase difference over a large bandwidth.

To match the phase response of the $-90°$ MTM line with that of the $+90°$ MTM line and therefore create a broadband differential output phase, the slopes of

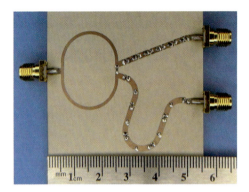

Figure 5.15 Photograph of fabricated MTM balun. From [34], copyright © 2005 by the Institute of Electrical and Electronics Engineers, Inc.

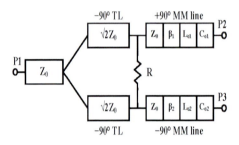

Figure 5.16 Proposed architecture of TMM balun. From [34], copyright © 2005 by the Institute of Electrical and Electronics Engineers, Inc.

their phase characteristics must be equal at the design frequency, thus satisfying

$$\frac{d\Phi_{\mathrm{MTM+}}}{d\omega}\bigg|_{\omega_0} = \frac{d\Phi_{\mathrm{MTM-}}}{d\omega}\bigg|_{\omega_0} \tag{5.9}$$

Moreover, to ensure that the MTM phase-shifting lines do not radiate, each unit cell must be operated in the region outside the light cone on the Brillouin diagram. Thus, the $+90°$ MTM phase-shifting lines must be operated in the NRI backward-wave region, while simultaneously ensuring that the propagation constant of the line exceeds that of free space, resulting in a slow-wave structure with a positive insertion phase. Correspondingly, the $-90°$ MTM phase-shifting lines must be operated in the PRI forward-wave region, while simultaneously ensuring that the propagation constant of the line also exceeds that of free space, resulting in a slow-wave structure with a negative insertion phase (see Section 5.2.1).

The MTM Wilkinson balun was implemented in microstrip technology on a Rogers RO3003 substrate with $\varepsilon_r = 3$ and height $h = 0.762$ mm at a design frequency $f_0 = 1.5$ GHz. A five-stage design was chosen for the $+90°$ MTM phase-shifting line as well as the $-90°$ MTM phase-shifting line. The experimental results were compared with the simulated results obtained using Agilent-ADS.

Figure 5.17 shows the measured versus the simulated return loss magnitude response for port 1, showing good agreement between the two, indicating that the device is well matched, especially around $f_0 = 1.5$ GHz. The measured and simulated return losses for ports 2 and 3 exhibit similar responses.

Figure 5.18 shows excellent isolation for the device as well as equal power split between the two output ports.

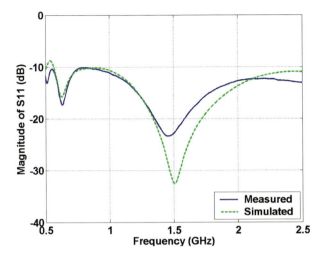

Figure 5.17 Measured and simulated return loss magnitude responses for port 1 of MTM balun. From [34], copyright © 2005 by the Institute of Electrical and Electronics Engineers, Inc.

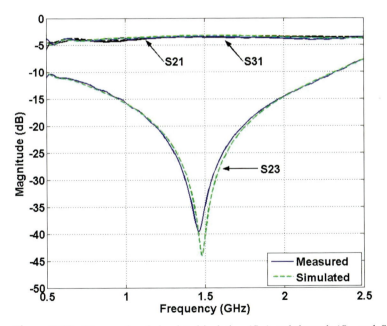

Figure 5.18 Measured and simulated isolation (S_{23}) and through (S_{21} and S_{31}) magnitude responses of MTM balun. From [34], copyright © 2005 by the Institute of Electrical and Electronics Engineers, Inc.

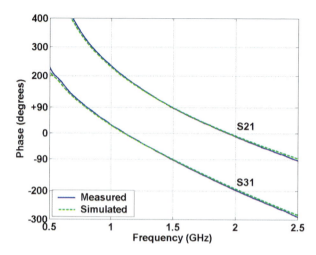

Figure 5.19 Measured and simulated phase responses of S_{21} ($+90°$ MTM line) and S_{31} ($-90°$ MTM line) of MTM balun. From [34], copyright © 2005 by the Institute of Electrical and Electronics Engineers, Inc.

Figure 5.19 shows the measured versus simulated phase responses of the two balun branches. The experimental results agree very closely with the simulated results. It can be observed that the phase of S_{21} is exactly equal to $+90°$ at $f_0 = 1.5$ GHz, while the phase of S_{31} is exactly equal to $-90°$ at $f_0 = 1.5$ GHz and the phase characteristics of the two branches are quite similar.

Figure 5.20 shows the measured and simulated differential output phases of the MTM balun, with excellent agreement between the two. It can be observed that the differential output phase remains flat for a large frequency band, which follows directly from the fact that the phase characteristics of the $+90°$ and $-90°$ lines correspond very closely. The flat differential output phase has a $180° \pm 10°$ bandwidth of 1.16 GHz, from 1.17 to 2.33 GHz. Since the device exhibits excellent return loss and isolation and through characteristics over this frequency range, it can be concluded that the MTM balun can be used as a broadband single-ended to differential converter in the frequency range from 1.17 to 2.33 GHz.

For comparison, a distributed TL Wilkinson balun employing $-270°$ and $-90°$ TLs instead of the $+90°$ and $-90°$ MTM lines was also simulated, fabricated, and measured at $f_0 = 1.5$ GHz, and the differential output phase of the TL balun is also shown in Figure 5.20. It can be observed that the phase response of the TL balun is linear with frequency, with a slope equal to the difference between the phase slopes of the $-270°$ and $-90°$ TLs. Since the gradient of the resulting phase characteristic is quite steep, this renders the output differential phase response of the TL balun narrow band. Thus, the TL balun exhibits a measured differential phase bandwidth of only 11 percent, from 1.42 to 1.58 GHz, compared to 77 percent exhibited by the MTM balun. In addition, the TL balun occupies an area of 33.5 cm^2 compared to 18.5 cm^2 for the MTM balun. Thus, the MTM balun is more compact, occupying only 55 percent of the area that the conventional TL balun occupies. Furthermore, the MTM balun exhibits more than double the bandwidth compared to a lumped-element implementation using low-pass/high-pass lines, which typically exhibits a bandwidth of 32 percent [20]. This can be attributed to the fact that the low-pass line has a linear phase response,

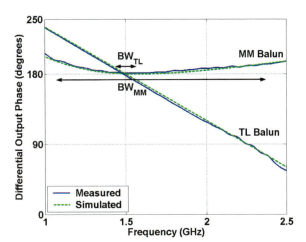

Figure 5.20 Measured and simulated differential phase comparison between MTM balun and TL balun. From [34], copyright © 2005 by the Institute of Electrical and Electronics Engineers, Inc.

while the response of the high-pass line has a varying slope with frequency. Thus, the shapes of the phase responses of the two lines do not match, resulting in a more narrow-band differential output phase.

5.3.4   Low-Profile and Small Ring Antennas

A final example of harnessing the previously described MTM lines is to use a zero-degree phase-shifting line wrapped around in a closed loop in order to implement a small printed antenna. This is shown in Figure 5.21 for a realization at 1.5 GHz. As shown, there are four MTM phase-shifting sections arranged in a square ring. Each constituent section comprises a zero-degree phase-shifting line in a microstrip configuration designed to incur a zero insertion phase at the antenna operating frequency. This allows the inductive posts to ground, which act as the main radiating elements, to be fed in phase. Hence, the antenna operates as a 2D array of closely spaced monopoles that are fed in phase through a compact feed network. This leads to a ring antenna with a small footprint (diameter of $\lambda/25$) and a low profile (height $\lambda/31$) capable of radiating vertical polarization.

The measured return loss obtained indicates a good matching of >25 dB at 1.51 GHz with a -10-dB bandwidth of approximately 2 percent. This bandwidth can be increased to 3 to 4 percent if the dielectric substrate is reduced to about the size of the ring or if the via height is increased. Figure 5.22 shows the measured versus the simulated E- and H-plane patterns obtained with Ansoft's High Frequency Structure Simulator (HFSS), which demonstrates good agreement. It can be observed that the antenna exhibits a radiation pattern with a vertical linear electric field polarization, similar to a short monopole on a finite ground plane. The radiation in the back direction is reduced compared to the forward direction due to the effect of the finite ground plane used; however it is not completely eliminated. Moreover, there is good cross-polarization purity in the E plane, with a maximum measured electric field cross-polarization level of -17.2 dB. In the H plane, the maximum electric field cross-polarization level is only -6.6 dB.

Top View

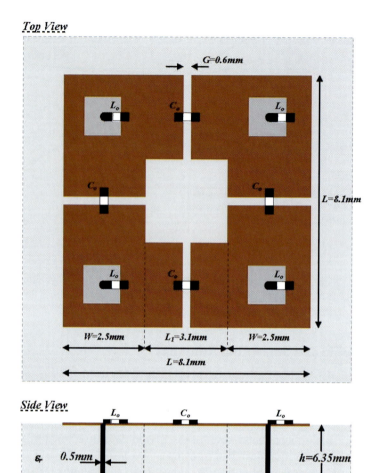

Side View

Figure 5.21 Diagram of MTM ring antenna at 1.5 GHz. The loading capacitance and inductance required to feed the vias in phase are $C_0 = 3.70$ pF and $L_0 = 71.08$ nH.

The loading with chip passive lumped elements is effective at RF and low microwave frequencies. At higher frequencies, these can be replaced by printed lumped elements. For example, a fully printed version of this antenna at 30 GHz, in which the loading lumped-element chip capacitors and inductors were replaced by gaps and vias respectively, was reported in [21]. Finally it should be noted that the measured radiation efficiency of these antennas (but with a truncated ground plane) in the frequency range between 1 and 2 GHz turns out to be between 30 and 50 percent (using the gain comparison method).

Other interesting applications of MTMs for the design of small and efficient antennas can be found in [22].

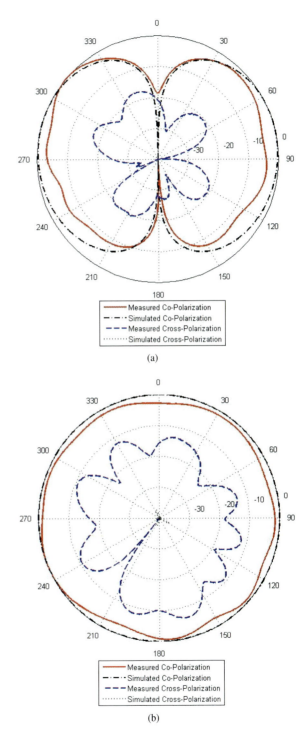

Figure 5.22 Measured and simulated (Ansoft's HFSS) antenna patterns at 1.51 GHz: (*a*) *E* plane and (*b*) *H* plane. From [5], copyright © 2002 by the American Institute of Physics.

5.4 BACKWARD LEAKY-WAVE ANTENNA RADIATING IN ITS FUNDAMENTAL SPATIAL HARMONIC

The TL approach to synthesizing NRI metamaterials has led to the development of a new kind of leaky-wave antenna (LWA). As was described previously, by appropriately choosing the circuit parameters of the dual TL model, a fast-wave structure can be designed that supports a fundamental spatial harmonic which radiates toward the backward direction (i.e., toward the feed) [5, 23].

A CPW implementation of this LWA is shown in Figure 5.23. The gaps in the CPW feed line serve as the series capacitors of the dual TL model, while the narrow lines connecting the center conductor to the coplanar ground planes serve as the shunt inductors (shorted stubs). The capacitive gaps are the radiating elements in this LWA and excite a radiating TM wave. Due to the antiparallel currents flowing on each pair of the narrow inductive lines, they remain non-radiating. Simulated and experimental results for this bidirectional LWA were reported in [5]. Simulation results for a unidirectional LWA design were also presented in [23]. The unidirectional design is simply the LWA described in [5] backed by a long metallic trough, as shown in Figure 5.24. Since the LWA's transverse dimension is electrically small, the backing trough can be narrow (below resonance). The trough used is a quarter wavelength in height and width

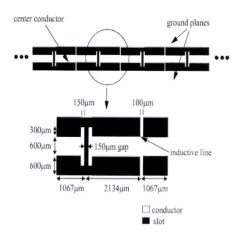

Figure 5.23 Backward LWA based on the TL model at 15 GHz. From [5], copyright © 2002 by the American Institute of Physics.

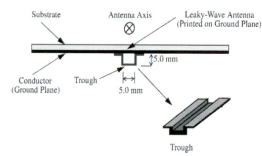

Figure 5.24 Unidirectional BWA antenna design at 15 GHz.

and covers the entire length of the antenna on the conductor side of the substrate. It acts as a waveguide below cutoff and recovers the back radiation, resulting in unidirectional far-field patterns.

A complementary forward unidirectional LWA is also reported in [24]. The periodic structure of [24] also operates on the fundamental spatial harmonic and hence can be thought of as a MTM with a positive phase velocity. Here, we describe experimental results for the unidirectional design proposed in [23]. As noted in [23], a frequency shift of 3 percent, or 400 MHz, was observed in the experiments compared to the method-of-moments simulations of the LWA using Agilent's Advanced Design System (ADS). As a result, the experimental unidirectional radiation patterns are shown at 14.6 GHz while the simulation patterns are shown at 15 GHz. The E- and H-plane patterns are shown in Figures 5.25

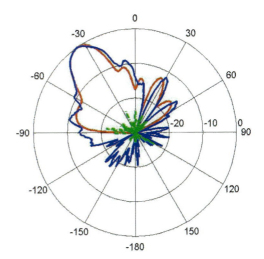

Figure 5.25 E-plane pattern for unidirectional LWA: ($-$, blue) experimental copolarization; ($-\cdot-$, green) experimental cross-polarization; ($---$, red) simulated copolarization using Agilent ADS ($F = 15$ GHz).

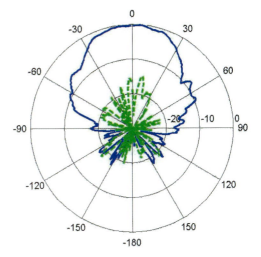

Figure 5.26 H-plane pattern for unidirectional leaky-wave antenna: ($-$, blue) experimental copolarization; ($-\cdot-$, green) experimental cross-polarization.

and 5.26, respectively. A gain improvement of 2.8 dB was observed for the unidirectional design over the bidirectional design, indicating that effectively all of the back radiation is recovered with the trough.

A similar LWA implemented in a microstrip configuration was also reported in [25]. In that implementation varactor diodes were utilized to steer the beam from the backward to the forward direction in a design with a closed stop band [see Eq. (5.4)].

5.5 SUPERRESOLVING NRI TRANSMISSION LINE LENS

Standard diffraction theory imposes a resolution limit when imaging using conventional lenses. This fundamental limit, called the *diffraction limit*, is attributed to the finite wavelength of the electromagnetic waves. The electromagnetic field emanating from a luminous object, lying over the $x-y$ plane, consists of a continuum of plane waves $\exp(-jk_x x - jk_y y)\exp(-jk_z z)$. Each plane wave has a characteristic amplitude and propagates at an angle with respect to the optical z axis given by the direction cosines $(k_x/k_0, k_y/k_0)$, where k_0 is the propagation constant in free space. The plane waves with real-valued direction cosines $(k_x^2 + k_y^2 < k_0^2)$ propagate without attenuation, while the evanescent plane waves with imaginary direction cosines $(k_x^2 + k_y^2 > k_0^2)$ attenuate exponentially along the optical z axis. A conventional lens focuses only the propagating waves, resulting in an imperfect image of the object, even if the lens diameter were infinite. The finer details of the object, carried by the evanescent waves, are lost due to the strong attenuation that these waves experience [$\exp(-z\sqrt{k_x^2 + k_y^2 - k_0^2})$] when traveling from the object to the image through the lens. The Fourier transform uncertainty relation $k_{t\_max}\Delta\rho \sim 2\pi$, relating the maximum transverse wavenumber k_{t_max} to the smallest transverse spatial detail $\Delta\rho$, implies that spatial details smaller than a wavelength are eliminated from the image ($\Delta\rho \sim 2\pi/k_0 = \lambda$). This loss of resolution, which is valid even if the lens diameter were infinite, constitutes the origin of the diffraction limit in its ultimate form. For the typical case of imaging a point source, the diffraction limit manifests itself as an image smeared over an area approximately one wavelength in diameter.

In 2000, John Pendry extended the analysis of Veselago's lens (an NRI slab with a refractive index $n = -1$) to include evanescent waves and observed that such lenses could overcome the classical diffraction limit [11]. Pendry suggested that Veselago's lens would allow "perfect imaging" if it were completely lossless and its refractive index was exactly equal to $n = -1$ relative to the surrounding medium. The left-handed lens achieves imaging with superresolution by focusing propagating waves as would a conventional lens, but in addition it supports growing evanescent waves which restore the decaying evanescent waves emanating from the source. This restoration of evanescent waves at the image plane extends the maximum accessible wavenumbers $k_{t\_max} > k_0$ and allows imaging with superresolution. The physical mechanism behind the growth of evanescent waves is quite interesting: Within the NRI slab multiple reflections result in both growing and attenuating evanescent waves. However, when the index $n = -1$, a

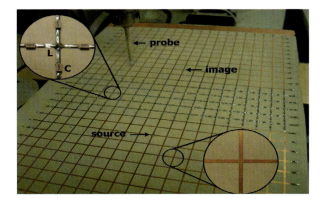

Figure 5.27 Photograph of planar superresolving Veselago lens around 1 GHz [10]. From [10], copyright © 2004 by the American Physical Society.

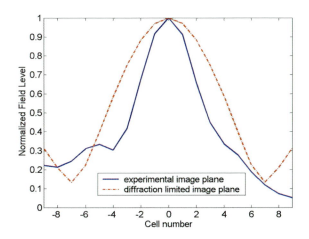

Figure 5.28 Experimental verification of subwavelength focusing. From [10], copyright © 2004 by the American Physical Society.

resonant phenomenon occurs in which the attenuating solution is canceled out, thus leaving only the growing wave present. In a sense, one may think of Veselago's lens as an inverse system that exactly restores propagation in free space.

A picture of a planar version of Veselago's lens that was constructed at the University of Toronto is shown in Figure 5.27. The NRI lens is a slab consisting of a 5×19 grid of printed microstriplines loaded with series capacitors (C_0) and shunt inductors (L_0). This NRI slab is sandwiched between two unloaded printed grids that act as homogeneous media with a positive index of refraction. The first unloaded grid is excited with a monopole (point source) which is imaged by the NRI lens to the second unloaded grid. The vertical electric field over the entire structure is measured using a detecting probe (for details, see [7]).

The measured half-power beam width of the point source image at 1.057 GHz is 0.21 effective wavelengths, which is appreciably narrower than that of the diffraction-limited image corresponding to 0.36 wavelengths (see Fig. 5.28). The enhancement of the evanescent waves for the specific structure under consideration was demonstrated in [26]. Figure 5.29 shows the measured

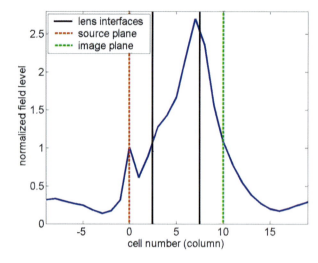

Figure 5.29 Experimental verification of growing evanescent waves in superresolving NRI-TL lens. From [10], copyright © 2004 by the American Physical Society.

vertical electric field above the central row of the lens, which verifies the exponential growth of the fields inside the NRI medium predicted in [26]. Since there is some controversy regarding losses in NRI MTMs, we hereby report that the loss tangent of the NRI medium at 1.05 GHz is estimated to be $\tan(\delta) = 0.062$, which attests to the low-loss nature of the NRI TL lens. However, even such a slight loss is sufficient to deteriorate the growth of evanescent waves to $k_{t\_\max} = 3k_0$ [10]. The superresolving imaging properties of the structure shown in Figure 5.26 have been theoretically investigated by means of a rigorous periodic Green's function analysis in [27]. Furthermore, the corresponding dispersion characteristics for these distributed structures were derived in [28] using periodic 2D TL theory. In the case that the loading is achieved using printed instead of chip loading lumped elements, for example, microstrip gaps and vias or coils to implement series capacitors and inductors, respectively [12], the corresponding dispersion characteristics have been examined using finite-element electromagnetic simulations in [13]. Finally an elegant and insightful explanation for the growth of evanescent waves in NRI media based on successive resonances in equivalent $L-C$ ladder networks can be found in [29].

5.6 DETAILED DISPERSION OF PLANAR NRI-TL MEDIA

To further understand the nature of the modes excited in 2D planar NRI TL media, consider again the loaded TL unit cell of Figure 5.1. To physically implement this unit cell in microstrip, one can use series gaps to realize the series capacitors and vias to the ground to implement the shunt inductors [9, 12, 13]. The resulting structure is shown in Figure 5.30 and consists of patches connected with vias to a ground plane. A top covering plate is also included here; for open structures this plate can be considered placed at infinity. This configuration can be identified

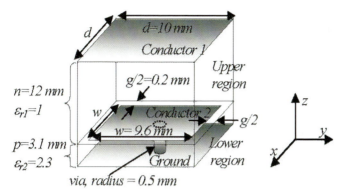

Figure 5.30   Two-dimensional unit cell with relevant dimensions and material parameters. From [31], copyright © 2004 by the Institute of Electrical and Electronics Engineers, Inc.

with the well-known "mushroom" structure proposed by Sievenpiper et al. as a means of implementing an electromagnetic bandgap surface [30].

Consider now the example in which the geometry of Figure 5.30 is loaded with lumped components. Specifically, the via is loaded by an ideal lumped inductor ($L = 11.6$ nH) and the patch edges are loaded by series lumped capacitors ($C = 2$ pF). This additional loading results in a pronounced BW behavior which is easily achievable by adjusting the lumped component values. The corresponding dispersion diagram was computed using the multiconductor transmission line (MTL) theory in [31]. The so-computed results for axial propagation are shown in Figure 5.31, where they are compared with those obtained from a full-wave finite-element method (FEM) commercial solver. Also shown is the dispersion predicted by the simple TL theory of Section 5.2. This diagram is interesting since it reveals an additional stop band that forms at the intersection of the light line and the BW dispersion curve predicted by simple TL theory. With reference to Figure 5.30, the reason that this stop band forms is due to a contradirectional coupling between the BW mode supported by the mushroom surface and the bottom ground plane and the forward transverse electromagnetic (TEM) mode supported by the bottom and top metallic plates. If the top plate is open, this stop band persists since now the TEM mode degenerates to a surface wave mode [13]. The simple TL model does not capture this stop band since it only accounts for the BW mode. On the other hand, the MTL model does predict the stop band since it accounts for both modes.

In fact, the MTL theory enables one to examine the nature of this stop band, which is quite interesting in several ways. In reality, this is a peculiar stop band which describes a mode conversion phenomenon between the backward- and forward-wave (TEM) modes identified above. The two strongly coupled modes are actually characterized by complex-conjugate propagation constants. From Figure 5.31, at the peak of the lowest passband, the value of the complex propagation constant times the period d, (γd), is equal to $(0.125\pi)j$, a pure imaginary number, and at the minimum of the second passband its value is

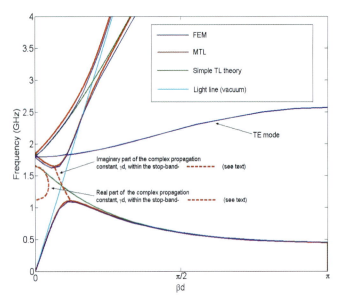

Figure 5.31   On-axis dispersion of loaded version of structure shown in Figure 5.30. From [31], copyright © 2004 by the Institute of Electrical and Electronics Engineers, Inc.

$(0.066\pi)j$. The MTL analysis reveals that within the stop band the imaginary part of γd varies continuously from $(0.125\pi)j$ to $(0.066\pi)j$, but γd also acquires a real part corresponding to exponential growth/decay. For this structure the value of γd at the center of the stop band ($f = 1.38$ GHz) is $(0.089\pi)j \pm 0.1457$. For an infinite structure, one must eliminate the sign corresponding to the exponential growth. This behavior should be taken carefully into account when designing NRI TL media. For example, in the superlens design of the previous section, it was made sure that propagation takes place outside of this contradirectional stop band. On the other hand, this stop band and the underlying mode conversion effect can be exploited to implement novel coupled-line BW couplers in which a microstripline (forward mode) is coupled to a NRI line (backward mode) [32]. The corresponding exponential field variation implied by the complex propagation constants leads to short coupling lengths and large coupling coefficients. Incidentally, this analysis reveals the true nature of the electromagnetic bandgap that is achievable with Sievenpiper's mushroom surface. Indeed, this is a leaky stop band that describes a mode conversion phenomenon and thus care must be exercised in the various intended bandgap applications. In particular, one should be careful such that unwanted radiation leakage effects due to mode conversion do not take place when this surface is operated as an electromagnetic bandgap surface.

Another silent feature of the dispersion diagram in Figure 5.31 is revealed by examining the slope of the dispersion line within the stop band. As shown, this slope becomes negative, which means that the corresponding parallel-plate waveguide mode develops a negative group velocity within the stop band. It is well known that anomalous group velocity effects can be observed in absorptive stop bands (see, e.g., [16]). However, the stop band of Figure 5.31 is associated not with absorption but rather with mismatch (reflection) losses, a situation that is more relevant to the case of propagation within the stop band of the distributed Bragg reflector described in [33].

ACKNOWLEDGMENTS

The author would like to thank Marco Antoniades for his assistance in preparing this manuscript. He also thanks Tony Grbic and Ashwin Iyer. Financial support from the Natural Sciences and Engineering Research Council of Canada (NSERC) through Discovery and Strategic grants as well as through a Steacie Fellowship is gratefully acknowledged. Partial support was also provided by the Department of National Defence/Ottawa and by Nortel Networks.

REFERENCES

1. V. G. Veselago, "The electrodynamics of substances with simultaneously negative values of permittivity and permeability," *Sov. Phys. Usp.*, vol. 10, no. 4, pp. 509–514, Jan. 1968.
2. R. A. Shelby, D. R. Smith, and S. Schultz, "Experimental verification of a negative index of refraction," *Science*, vol. 292, pp. 77–79, Apr. 2001.
3. A. K. Iyer and G. V. Eleftheriades, "Negative refractive index metamaterials supporting 2-D waves," *IEEE MTT-S Int. Microwave Symp. Dig.*, vol. 2, pp. 1067–1070, June 2002.
4. G. V. Eleftheriades, A. K. Iyer, and P. C. Kremer, "Planar negative refractive index media using periodically *L-C* loaded transmission lines," *IEEE Trans. Microwave Theory Tech.*, vol. 50, no. 12, pp. 2702–2712, Dec. 2002.
5. A. Grbic and G. V. Eleftheriades, "Experimental verificiation of backward-wave radiation from a negative refractive index metamaterial," *J. Appl. Phys.*, vol. 92, no. 10, pp. 5930–5935, Nov. 2002.
6. L. Liu, C. Caloz, C. Chang, and T. Itoh, "Forward coupling phenomenon between artificial left-handed transmission lines," *J. Appl. Phys.*, vol. 92, no. 9, pp. 5560–5565, Nov. 2002.
7. A. K. Iyer, P. C. Kremer, and G. V. Eleftheriades, "Experimental and theoretical verification of focusing in a large, periodically loaded transmission line negative refractive index metamaterial," *Opt. Express*, vol. 11, pp. 696–708, Apr. 2003, http://www.opticsexpress.org/abstract.cfm?URI=OPEX-11-7-696.
8. A. Sanada, C. Caloz, and T. Itoh, "Characteristics of the composite right/left-handed transmission lines," *IEEE Microwave Wireless Components Lett.*, vol. 14, pp. 68–70, Feb. 2004.
9. A. Sanada, C. Caloz, and T. Itoh, "Planar distributed structures with negative refractive index," *IEEE Trans. Microwave Theory Tech.*, vol. 52, no. 4, pp. 1252–1263, Apr. 2004.
10. A. Grbic and G. V. Eleftheriades, "Overcoming the diffraction limit with a planar left-handed transmission-line lens," *Phys. Rev. Lett.*, vol. 92, no. 11, 117403, Mar. 2004.
11. J. B. Pendry, "Negative refraction makes a perfect lens," *Phys. Rev. Lett.*, vol. 85, no. 18, pp. 3966–3969, Oct. 2000.
12. G. V. Eleftheriades, "Planar negative refractive index metamaterials based on periodically *L-C* loaded transmission lines," paper presented at the Workshop of Quantum Optics, Kavli Institute of Theoretical Physics, University of Santa Barbara, July 2002, http://online.kitp.ucsb.edu/online/qo02/eleftheriades/(slide 12)

13. A. Grbic and G. V. Eleftheriades, "Dispersion analysis of a microstrip based negative refractive index periodic structure," *IEEE Microwave Wireless Components Lett.,* vol. 13, no. 4, pp. 155–157, Apr. 2003.
14. M. A. Antoniades and G. V. Eleftheriades, "Compact, linear, lead/lag metamaterial phase shifters for broadband applications," *IEEE Antennas Wireless Propag. Lett.*, vol. 2, no. 7, pp. 103–106, July 2003.
15. N. Engheta, "An idea for thin, subwavelength cavity resonators using metamaterials with negative permittivity and permeability," *IEEE Antennas Wireless Propag. Lett.,* vol. 1, pp. 10–13, 2002.
16. O. Siddiqui, M. Mojahedi, and G. V. Eleftheriades, "Periodically loaded transmission line with effective negative refractive index and negative group velocity," *IEEE Trans. Antennas Propag.*, Special Issue on Metamaterials, vol. 51, no. 10, pp. 2619–2625, Oct. 2003.
17. G. V. Eleftheriades, M. A. Antoniades, A. Grbic, A. Iyer, and R. Islam, "Electromagnetic applications of negative-refractive-index transmission-line metamaterials," paper presented at 27th ESA Antenna Technology Workshop on Innovative Periodic Antennas, Santiago, Spain, Mar. 2004, pp. 21–28.
18. J.-S Lim, H.-S. Yang, Y.-T. Lee, S. Kim, K.-S. Seo, and S. Nam, "E-band Wilkinson balun using CPW MMIC technology," *IEE Electron. Lett.*, vol. 40, no. 14, pp. 879–881, July 2004.
19. M. Basraoui and S. N. Prasad, "Wideband, planar, log-periodic balun," *Proc. IEEE Int. Symp. Microwave Theory Tech.*, vol. 2, pp. 785–788, June 1998.
20. H. S. Nagi, "Miniature lumped element 180° Wilkinson divider," *Proc. IEEE Int. Symp. Microwave Theory Tech.*, vol. 1, pp. 55–58, June 2003.
21. G. V. Eleftheriades, A. Grbic, and M. Antoniades, "Negative-refractive-index metamaterials and enabling electromagnetic applications," *IEEE Int. Symp. Antennas Propag. Dig.*, Monterey, CA, June 2004.
22. R. W. Ziolkowski and A. D. Kipple, "Application of double negative materials to increase the power radiated by electrically small antennas," *IEEE Trans. Antennas Propag.*, vol. 51, no. 10, pp. 2626–2640, Oct. 2003.
23. A. Grbic and G. V. Eleftheriades, "A backward-wave antenna based on negative refractive index *L-C* networks," *Proc. IEEE Int. Symp. Antennas Propag.*, vol. IV, San Antonio, TX, June 16–21, 2002, pp. 340–343.
24. A. Grbic and G. V. Eleftheriades, "Leaky CPW-based slot antenna arrays for millimeter-wave applications," *IEEE Trans. Antennas Propag.*, vol. AP-50, no. 11, pp. 1494–1504, Nov. 2002.
25. S. Lim, C. Caloz, and T. Itoh, "Metamaterial-based electronically controlled transmission-line structure as a novel leaky-wave antenna with tunable radiation angle and beamwidth," *IEEE Trans. Microwave Theory Tech.,* vol. MTT 51, pp. 161–173, Jan. 2005.
26. A. Grbic and G. V. Eleftheriades, "Growing evanescent waves in negative-refractive-index transmission-line media," *Appl. Phys. Lett.,* vol. 82, no. 12, pp. 1815–1817, Mar. 2003.
27. A. Grbic and G. V. Eleftheriades, "Negative refraction, growing evanescent waves and subdiffraction imaging in loaded-transmission-line metamaterials," *IEEE Trans. Microwave Theory Tech.*, vol. 51, no. 12, pp. 2297–2305, Dec. 2003 (see also Erratum, *IEEE T-MTT*, vol. 52, no. 5, p. 1580, May 2004).
28. A. Grbic and G. V. Eleftheriades, "Periodic analysis of a 2-D negative refractive index transmission line structure," *IEEE Trans. Antennas Propag.*, Special Issue on Metamaterials, vol. 51, no. 10, pp. 2604–2611, Oct. 2003.
29. A. Alu and N. Engheta, "Pairing an epsilon-negative slab with a mu-negative slab: Resonance, anomalous tunneling and transparency," *IEEE Trans. Antennas Propag.*, Special Issue on Metamaterials, vol. 51, no. 10, pp. 2558–2571, Oct. 2003.
30. D. Sievenpiper, L. Zhang, R. F. J. Broas, N. G. Alexopolous, and E. Yablonovitch, "High-impedance electromagnetic surfaces with a forbidden frequency band," *IEEE Trans. Microwave Theory Tech.*, vol. 47, no. 11, pp. 2059–2074, Nov. 1999.

31. F. Elek and G. V. Eleftheriades, "Dispersion analysis of Sievenpiper's shielded structure using multi-conductor transmission-line theory," *IEEE Microwave Wireless Components Lett.*, vol. 14, no. 9, pp. 434–436, Sept. 2004.

32. R. Islam, F. Elek, and G. V. Eleftheriades, "A coupled-line metamaterial coupler having co-directional phase but contra-directional power flow," *Electron. Lett.*, vol. 40, no. 5, pp. 315–317, Mar. 2004.

33. M. Mojahedi, E. Schamiloglu, K. Agi, and K. J. Malloy, "Frequency domain detection of superluminal group velocity in a distributed Bragg reflector," *IEEE J. Quantum Electron.*, vol. 36, pp. 418–424, Apr. 2000.

34. M.A. Antoniades and G.V. Eleftheriades, "A broadband Wilkinson balun using microstrip metamaterial lines," *IEEE Antennas Wireless Propag. Lett.*, vol. 4, pp. 209–212, Apr. 2005.

CHAPTER 6

RESONANCE CONE ANTENNAS
Keith G. Balmain and Andrea A. E. Lüttgen

6.1 INTRODUCTION

The class of antennas considered here is low profile, consisting of a top-layer grid erected over a ground plane, a geometry realizable using printed-circuit-board techniques. The top layer is a planar network of low-loss, interconnected, reactive elements embedded in a conducting grid (or mesh) of square cells. The reactive elements in each cell are orthogonally aligned, with adjacent elements having reactances of opposite sign, thus giving near-zero net reactance (i.e., resonance) around the periphery of any cell. It is apparent that such a loaded grid is highly anisotropic, a property of central importance for this chapter.

Suppose we define a low-impedance path across the planar, anisotropic grid, a path that follows a trail of approximately canceling reactances. If this grid were not planar but rather three dimensional, continuous, and lossless, then this new medium could be characterized by a permittivity matrix having only diagonal terms that are positive (capacitive) or negative (inductive). For such a case in particular, two of the three diagonal terms would have to be of opposite sign, a condition that would make the medium uniaxial, a wave propagation term familiar in crystallography. In such a medium, a localized signal source generates strong fields that tend to be concentrated over a conical surface with an apex at the source, a surface known as a "resonance cone." For the planar, anisotropic grid medium of primary interest here, a similar phenomenon occurs, with the fields of a small source concentrated along radial lines that lie in the grid plane and extend outward from the source, lines which for convenience we shall continue to call resonance cones.

For application to low-profile antennas which involve an anisotropic grid over and parallel to a ground plane, the horizontal grid currents will be accompanied by the oppositely directed image currents required to produce zero tangential electric field strength on the ground plane. The horizontal currents thus produce zero radiation immediately over the ground plane, a situation which would severely limit the scope of antenna applications.

A solution to this problem is to introduce vertical currents that track the resonance cones, taking advantage of the fact that such vertical currents would have codirected image currents, thus producing additive, vertically polarized far fields close to the ground plane. The introduction of vertical currents tied to

Metamaterials: Physics and Engineering Explorations, Edited by N. Engheta and R. W. Ziolkowski
Copyright © 2006 the Institute of Electrical and Electronics Engineers, Inc.

the resonance cones can be achieved by connecting reactive elements from the anisotropic grid to the ground plane, provided that the reactive elements have sufficiently high impedances that they do not prevent the occurrence of the resonance cones. This approach will be described in the remainder of this chapter, enabling the design of horizontally radiating resonance cone antennas.

6.2   PLANAR METAMATERIAL, CORNER-FED, ANISOTROPIC GRID ANTENNA

Previous moment method simulations and near-field scanning experiments for a corner-fed, anisotropic grid over ground [1, 2] indicated that it performed as expected, with the near field exhibiting resonance cones extending outward from the corner source. These cones are analogous to the resonance cones that are well known for their high fields and outward power flow in studies of antennas in highly anisotropic plasmas [3], as sketched in the example of Figure 6.1.

The relevant plasma parameters are often displayed in the CMA diagram of Figure 6.2 (so named after Clemmow, Mullaly, and Allis), which shows the regions in parameter space in which the relevant partial differential equation describing the wave propagation is hyperbolic in the spatial coordinates, that is, the parameter regions where resonance cones exist. As well, the same figure displays the influence of the plasma medium. Representing the anisotropic plasma in terms of capacitors and inductors, it also indicates in circuit terms how the plasma influences the near fields and input impedance of a small probe immersed in the plasma [4].

Now, returning to the planar, anisotropic, wire-grid-over-ground geometry of immediate concern, we can appreciate better the formation of resonance cones as high-field regions that extend outward over the circuit board from the corner feed point [5]. The cone directions depend on the reactances of the orthogonal inductors and capacitors (or their distributed equivalents) that are embedded in the planar wire grid, which in turn is positioned over a ground plane and is parallel to it. In these early simulations, no circuit elements were connected

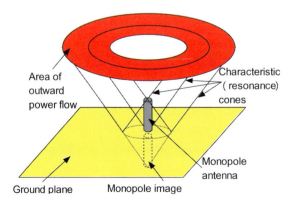

Area of outward power flow

Ground plane   Monopole image

Characteristic (resonance) cones

Monopole antenna

Figure 6.1   Monopole antenna near fields in a highly anisotropic (resonant) plasma, showing characteristic cones (resonance cones) extending from the antenna ends and the feed point at the antenna center, under the assumption that the antenna currents decrease linearly to zero at the antenna ends.

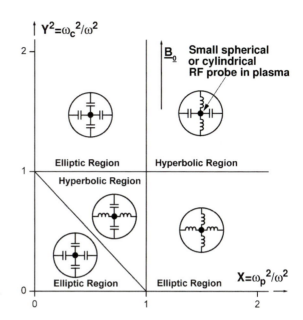

Figure 6.2 CMA plasma parameter diagram showing the elliptic and hyperbolic regions in parameter space, with ω_c indicating the electron cyclotron frequency and ω_p the plasma frequency. The capacitors and inductors are oriented in the directions of positive and negative permittivities, thus enabling characterization of the impedance properties of the plasma medium surrounding a small probe [4].

between the grid and ground, apart from the terminating resistors around the grid periphery.

Early in the course of this research, it was noted that the cone fields of what appears to be a dominant mode exhibit rapid phase reversals as the observation point moves transversely across the cone. In particular, this phase reversal applies to the currents flowing in the cone vicinity such that the net radial, outward currents tend to be equal and opposite on the two sides of the cone, in a manner similar to that of the currents on a two-wire transmission line. Further, one can see in the simulations an apparent second dominant mode having fields and currents resembling those in a parallel-plate waveguide, that is, currents that do not exhibit phase reversal in a cut across a resonance cone. The two dominant modes appear to be coupled and consequently are summed automatically in the course of carrying out the moment method simulation.

In extending the above near-field work on a small structure (one-tenth of a wavelength on a side) to a radiating structure at least an order of magnitude larger, a clue for success arises from the work of Grbic and Eleftheriades [6,7], in which fast-wave radiation from a one-dimensional periodic transmission line was enabled by the addition in each unit cell of an inductor extending from a horizontal wire over ground to the ground plane, together with a series capacitor inserted in the horizontal wire. Accordingly, for the present expanded grid-over-ground configuration, a vertically oriented inductor was connected from each grid–wire intersection to ground, as shown in Figure 6.3.

Even with the much larger square grid, a well-defined resonance cone still exists, as seen in the simulation results of Figure 6.4. Figure 6.5 shows that phase reversal across the resonance cone is maintained and, further, that the phase progression is from the upper left in the figure to the lower right. The

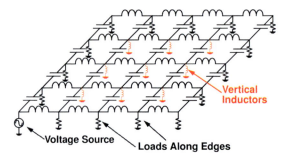

Figure 6.3 Uniform anisotropic planar L-C grid over ground, with corner feed, resistive edge-loading, and inductors to ground. The source is at 750 MHz, and the grid is 48 cm square, 2 cm high over ground, with each cell in the grid measuring 2 cm by 2 cm.

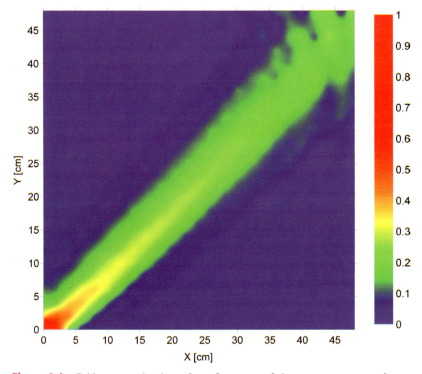

Figure 6.4 Grid-to-ground voltage for a flat, corner-fed antenna over ground at 750 MHz. The horizontal-grid part of the structure is located at 2 cm above the ground plane.

result is vertically polarized, unidirectional radiation in the horizontal plane (see Figs. 6.6 and 6.7) approximately at right angles to the cone (i.e., broadside to the cone), thus corresponding to the direction of phase progression noted above. The unidirectional property of the radiation appears to arise from the superposition of the two modes, one symmetric and one asymmetric, in such a way as to produce cancellation of the radiated fields in one direction together with addition in the opposite direction.

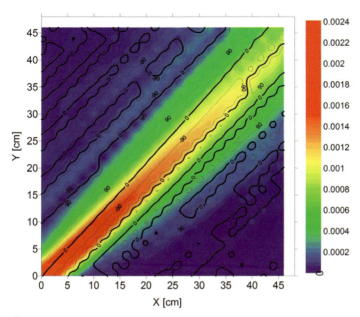

Figure 6.5   Longitudinal (diagonal) current oriented in the direction of the resonance cone, showing magnitude (color) and phase (contour lines). Note the tendency for phase reversal across the resonance cone.

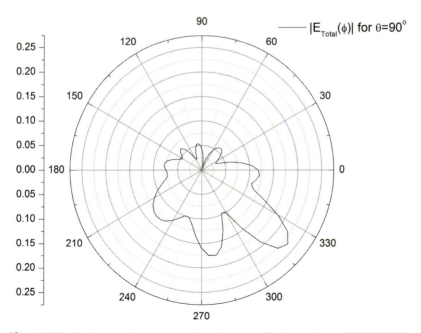

Figure 6.6   Horizontal-plane, vertically polarized radiation pattern at 750 MHz for a flat, corner-fed antenna located 2 cm above the ground plane.

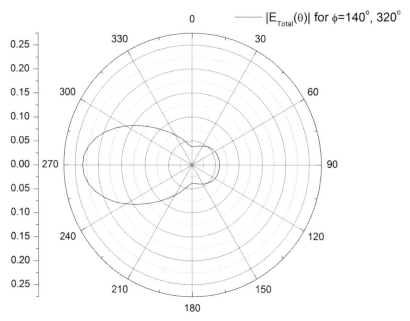

Figure 6.7   Vertical-plane radiation pattern (at angle of maximum horizontal-plane radiation) at 750 MHz for a flat, corner-fed antenna located 2 cm above the ground plane.

Figure 6.8 shows the relative absorbed power distribution around the grid periphery. There are peaks at points 1 and 96 simply because they are adjacent to the feed point. The points more than a free-space wavelength away are numbered from 24 to 72. There is a local maximum at point 48 exactly on the diagonal (i.e., at the upper right corner). Thus this point is over 1.5 free-space wavelengths from the feed point, which is neither near nor far so it can be said to be at an intermediate distance. Strong fringes are visible to one side of the resonance cone maximum, from points 49 to 59. These fringes become stronger with increasing distance from the source as in homogeneous, anisotropic plasmas, as shown in the book by Mareev and Chugunov [8].

A remaining challenge on the way to better antenna performance is the nonuniformity along the cone axis of the vertical current distribution which is weighted in favor of the feed point, a characteristic commonly observed in leaky-wave antennas. Initial results with lowered grid height near the feed point (for reduced radiation there) were promising. A version of this latter idea with a planar but sloped grid was then tried, in expectation of first flattening and ultimately tapering the radiating source distribution, in order to strengthen the main lobe of the radiation pattern and ultimately to reduce the side lobes. A conceptual sketch of the sloped-grid configuration is shown in Figure 6.9.

A plot of grid-to-ground voltage for the sloped grid is shown in Figure 6.10, but this does not properly represent the spatial distribution of the vertical currents that actually generate the vertically polarized radiation in the horizontal plane. A

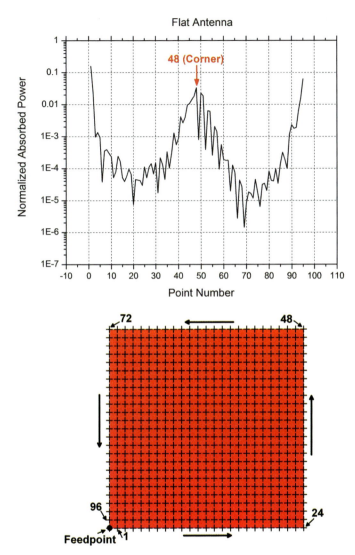

Figure 6.8 Normalized power (i.e., relative power) absorbed at 750 MHz by resistors attached to ground along the grid edges. The maximum occurs at the upper right corner (Point No. 48).

better approximate representation of the radiating source distribution is postulated as the product of each vertical current multiplied by its height, a distribution which we shall call the "excitation," as shown in Figure 6.11, and which is relatively uniform along the cone, thus suggesting an improved radiation pattern arising from the sloped configuration.

The horizontal-plane radiation pattern of Figure 6.12 shows a well-defined main lobe, low back radiation, and symmetric side lobes, indicating that improved

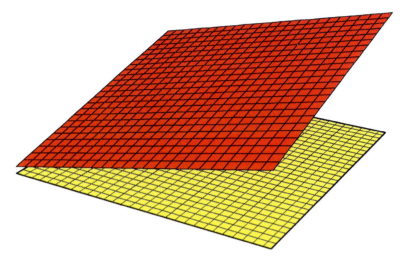

Figure 6.9 Schematic view of sloped antenna. The generator is located between the ground plane and the lower left corner of the grid. Not shown are the grid-to-ground inductors and edge-terminating resistors.

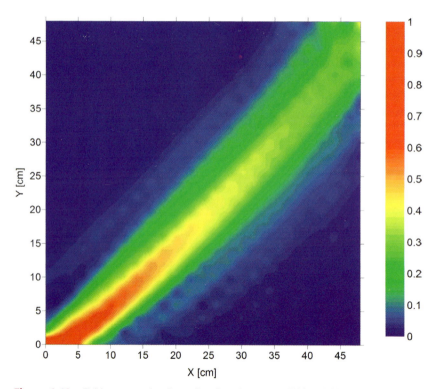

Figure 6.10 Grid-to-ground voltage for sloped antenna. This grid has 24×24 cells of size 2 cm $\times$ 2 cm each. The lower left corner is located 0.25 cm above the ground plane and the slope angle is $3.8°$.

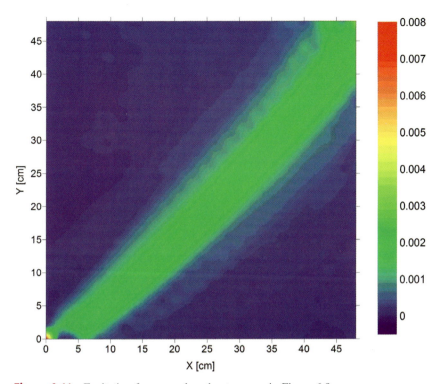

Figure 6.11   Excitation for same sloped antenna as in Figure 6.9.

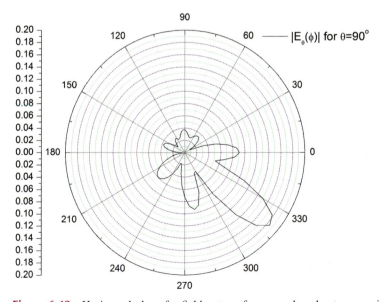

Figure 6.12   Horizontal-plane far-field pattern for same sloped antenna as in Figure 6.9.

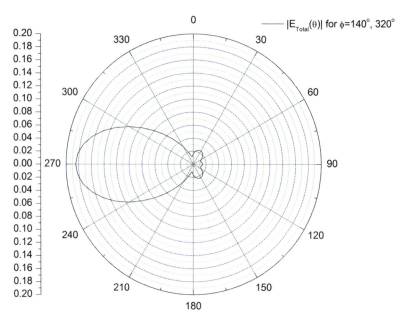

Figure 6.13   Vertical-plane far-field pattern (at angle of maximum radiation ϕ) for sloped antenna shown in Figure 6.9.

functioning has been achieved, albeit with the disadvantage of high side-lobe levels. A lower level secondary lobe is visible in the backfire direction at $223°$, being more sharply defined than in the case of the flat antenna (Fig. 6.6). The vertical-plane pattern of Figure 6.13 exhibits a single maximum in the horizontal direction.

Further insight was sought through the computation of the real part of the Poynting vector at the ground-plane level both beneath the anisotropic grid and for a moderate distance beyond it. These results, along with the vertical electric field on the ground plane, are displayed in Figure 6.14. The E-field magnitude contour at moderate distance strongly resembles the radiation pattern in the far field (Fig. 6.12), indicating the rapid evolution with distance from near field to far field.

A slight additional improvement was achieved by bending the grid in the middle, as indicated in the sketch given in Figure 6.15, making it roof shaped with the peak of the roof running between two opposite corners. This leaves the remaining two opposite corners close to ground level and therefore amenable to source or load attachment, a situation which has the advantage of allowing the main beam to be switched from one side of the array to the other by switching the feed point to the opposite corner. This double-sloped antenna exhibits properties so similar to the single-sloped antenna that little is to be gained by showing them here. Nevertheless, it can be noted that the double-sloped antenna does exhibit slightly lower side lobes adjacent to the main beam, compared with the single-sloped antenna.

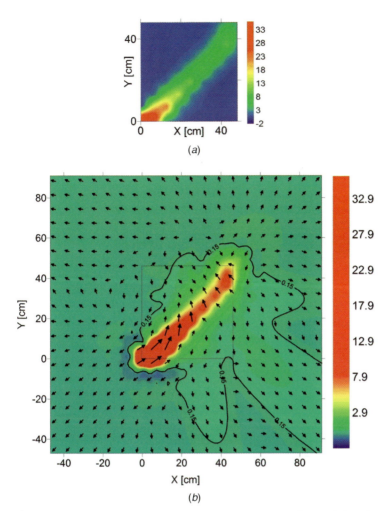

Figure 6.14 (*a*) Near-field and (*b*) intermediate-distance fields of sloped antenna. Both figures show in color the vertical electric field on the ground plane. In the lower figure, areas outside the square outline of the metamaterial are included and the color scale is adjusted to better visualize the lower levels there. The 0.15-V/m contour has approximately the same shape as the horizontal-plane pattern in Figure 6.12. The arrows indicate the Poynting vector on the ground plane.

6.3 RESONANCE CONE REFRACTION EFFECTS IN A LOW-PROFILE ANTENNA

It is postulated that multiple resonance cone refraction has the potential to produce *contained* cone configurations, meaning cone near-field patterns that never reach the edge of the planar anisotropic metamaterial supporting the resonance cones. Computer simulation was carried out for a four-quadrant, anisotropic

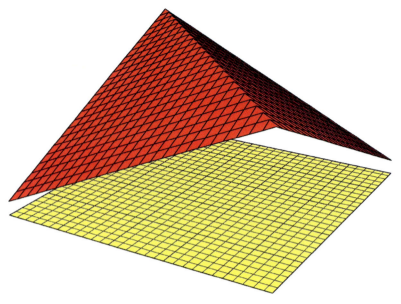

Figure 6.15 Schematic view of double-sloped antenna grid over ground plane.

planar metamaterial with a source centrally located on an outside edge. The result was the expected backward-cone refraction across each inner boundary between any two adjacent metamaterial quadrants with interchanged $L-C$ loads, thus producing generally the closed-square-loop cone configuration. However, for any arbitrary interior source location, there were always cones that reached the edges of the planar metamaterial, thus producing undesirable scattering and foiling the attempt to achieve cone containment.

It was realized that the problem of cones reaching the outer boundary of the anisotropic grid could be solved by introducing a third interior refractive boundary between regions with interchanged L and C [9]. This leads to a checkerboard grid with a 3×2 arrangement of alternating, relatively large, square anisotropic regions which one could call "macrocells," each composed of many of the small, basic cells which one could then call "microcells," as shown in Figure 6.16. The transition regions between macrocells consist of rows and columns of microcells where each half of each microcell has the characteristics of the adjacent medium. With a source as shown at board center, the resonance cone pattern can be established using the principles outlined in the papers by Balmain, Lüttgen, and Kremer [1, 2]. The result is the fully contained, double-square pattern of Figure 6.17. As the frequency departs from the design center value, each square of the cone pattern breaks up into the expanding and contracting square spirals depicted in Figure 6.18, in which perfect cone containment is no longer possible. Nonetheless, containment can still be approximated with the aid of expanding-spiral attenuation through various loss processes, including radiation. As already mentioned in the context of a corner-fed antenna, the generation of

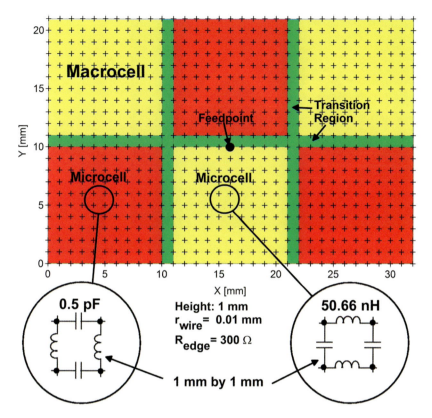

Figure 6.16   Metamaterial configuration designed for resonance cone containment at 1 GHz.

resonance cone radiation can be enabled by dropping inductors to ground from the anisotropic $L-C$ loaded grid [5,6].

In the present case, the same concept was tried as depicted in Figure 6.19 using grid dimensions expanded by an order of magnitude. Resonance cone formation and directional steering were still clearly in evidence, albeit with some additional cone pattern complexity. It was expected that resonance cone radiation would be degraded by cone interaction with the edges of the planar grid (due to power absorption and scattering), a problem that conceivably could be addressed by using refraction to divert the cones away from the edges, as already described for the configuration with no inductors to ground.

For the expanded six-macrocell configuration, the voltage distribution pattern of Figure 6.20 shows evidence of expanding and contracting square spirals. In general, the expanding spirals weaken before reaching the grid edges due to a combination of radiation and ohmic losses, while the contracting spirals curl up until, with sufficiently low losses, they shrink to become a pair of *vortices* with square cross sections, each located near the intersection of two interior boundaries. Each vortex is accompanied by a relatively high current to ground through

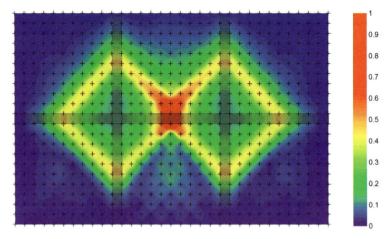

Figure 6.17   Double-square resonance cone pattern simulated at 1 GHz for configuration of Figure 6.16.

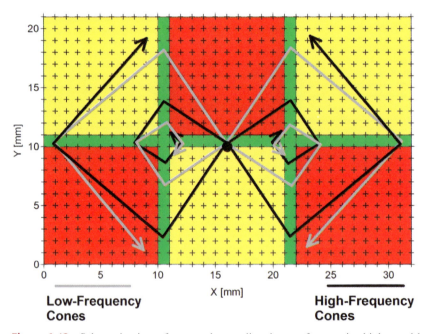

Figure 6.18   Schematic view of expected cone directions at frequencies higher and lower than diagonal cone frequency.

the closest vertical inductors, as can be seen more vividly in Figure 6.21, thus making each vortex an induced source of radiation that is vertically polarized and mainly horizontally directed. Figure 6.22 is included to clarify the relationship between the voltage pattern and the layout of the wire grid and the macrocell boundaries.

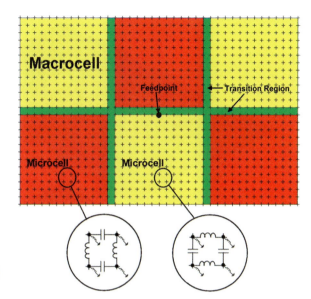

Figure 6.19 Low-profile, six-macrocell antenna configuration composed of microcells 1 cm² for operation at 1 GHz. Adjacent macrocells have microcells with interchanged inductors and capacitors. The crosses indicate the grid nodes that are connected to the ground plane through inductors.

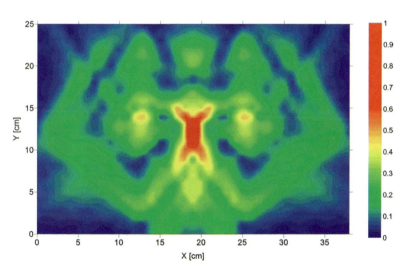

Figure 6.20 Grid-to-ground voltage at 1 GHz for low-profile antenna containing six macrocells, with each inter-macrocell transition involving interchanged L and C loads in its microcells, as in Figure 6.19.

Figure 6.23 combines the voltage pattern with the Poynting vector real parts as calculated in the grid plane. The power flow is seen to track the expanding resonance cones and to follow the contracting cone pattern into a pair of counterrotating vortices.

Clearly the vortices could have a strong influence on the total radiation pattern, governed by the fact that their vertical currents are in phase and separated by a little less than half a free-space wavelength, thus producing near-null radiation

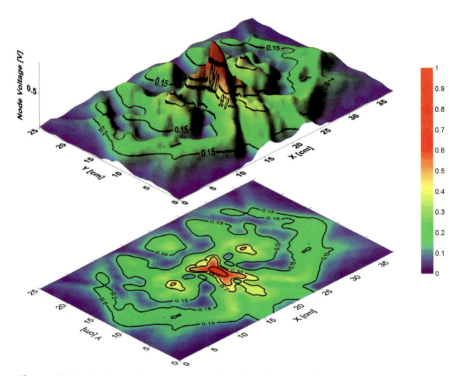

Figure 6.21   Surface and contour plots of node voltage at 1 GHz.

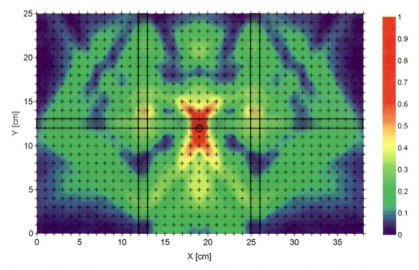

Figure 6.22   Grid-to-ground voltage at 1 GHz as shown in Figure 6.20. Here the wire intersections are indicated by crosses, and the boundary regions (transition bands) separating the six macrocells are clearly marked.

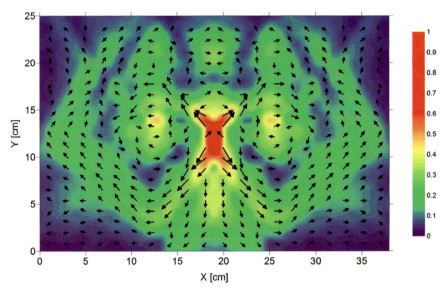

Figure 6.23 Grid-to-ground voltage as shown in Figure 6.20. The arrows indicate the Poynting vectors in the grid plane.

fields in the horizontal plane, in the $0°$ and $180°$ directions, and additive fields in the orthogonal directions. The null-field effects are evident in the horizontal-plane radiation pattern of Figure 6.24. Figure 6.25 shows the vertical-plane pattern for $\phi = 90°$.

The final influence on the radiation pattern comes from the six macro-cell regions on the anisotropic grid, because, for a given propagation direction (especially for directions parallel to the grid edges) the macrocell regions exhibit alternating passband and stop-band behavior. The stop-band property can strongly attenuate the amplitude of any wave attempting to traverse the region in question. In the antenna context it functions as a *stop region* and therefore as an internal *reflector*. Following the same line of thinking, one is led to the view that the pass-band behavior of a region points to its function as a *pass region* or, in the antenna context, as an internal *director*. Finally, in these cases where the anisotropic grid structure over ground is functioning as an antenna, its height is typically 1 cm at a frequency of 1 GHz, or $\frac{1}{30}$th of a wavelength, making the structure clearly a *low-profile antenna*. Here, the outcome of all this is a unidirectional, vertically polarized low-profile antenna with radiation concentrated close to the horizontal plane.

The unidirectional radiation patterns of Figures 6.24 and 6.25 are of special interest. A likely major contributor to them is the macrocell immediately below the source (i.e., in the negative *y* direction from the source). Figure 6.23 shows fairly high power flow in the negative *y* direction spread over most of this macrocell as well as high grid-to-ground voltage implying high currents in the

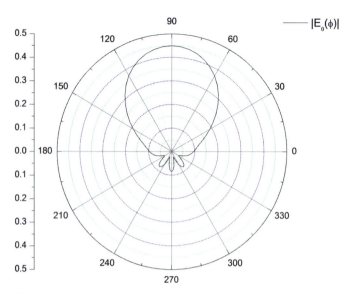

Figure 6.24 Horizontal-plane, vertically polarized, E-field radiation pattern for low-profile antenna in Figure 6.19.

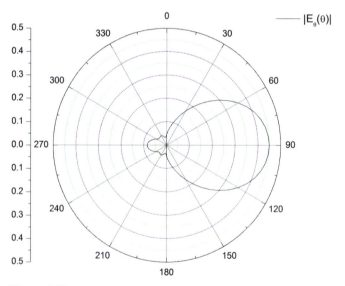

Figure 6.25 Vertical-plane E-field radiation pattern for low-profile antenna of Figure 6.19.

vertical inductors. However, for power flow in the negative y direction, the $L-C$ loading (Fig. 6.19) is indicative of backward-wave propagation in a transmission line mode. This means that the phase velocity is in the *positive* y direction, thus contributing to the unidirectional radiation pattern in Figure 6.24.

6.4 CONCLUSIONS

The basic corner-fed, low-profile, orthogonal $L-C$ grid over ground can be transformed into an effective vertically polarized antenna by the addition of a vertical inductor between each grid–wire intersection and ground such that relatively strong vertical currents are induced along the resonance cone line. This type of antenna exhibits a unidirectional horizontal-plane radiation pattern with its main lobe in a direction approximately at right angles to the resonance cone line.

For a low-profile $L-C$ grid over ground, it is known that resonance cone refraction can occur at an interface defined by a change in grid properties, an effect that is especially strong when the refractive interface is defined by interchange of the $L-C$ elements in the grid. When the refractive interfaces are arranged so as to create six contiguous square regions separated by such interfaces, a frequency exists at which the resonance cones return to the location of a centrally located original source. For a total array sufficiently small, the cones then retrace their paths, forming a pattern of two touching squares with a common point at the source, but the cones never reach the edge of the array. If, as mentioned above, inductors are connected between each grid intersection and ground, the cone directions are altered and radiation is enhanced, especially for a sufficiently large array. When the cone pattern includes an inward spiral, a vortex is formed where the vertical currents reach a peak, thus forming an induced secondary source of radiation. For the particular case considered, there are actually two vortices equidistant from the source, resulting in a pair of in-phase induced point sources of radiation. Moreover, one of the six regions in the array exhibits pronounced backward-wave propagation, thus contributing to the observed unidirectional radiation pattern.

ACKNOWLEDGMENTS

The authors acknowledge the extensive research support provided by Defense Research and Development Canada—Ottawa and by the Natural Sciences and Engineering Research Council of Canada.

REFERENCES

1. K. G. Balmain, A. A. E. Lüttgen, and P. C. Kremer, "Resonance cone formation, reflection, refraction, and focusing in a planar anisotropic metamaterial," *IEEE Antennas Wireless Propag. Lett.*, vol. 1, pp. 146–149, 2002.
2. K. G. Balmain, A. A. E. Lüttgen, and P. C. Kremer, "Power flow for resonance cone phenomena in planar anisotropic metamaterials," *IEEE Trans. Antennas Propag.*, Special Issue on Metamaterials, vol. 51, no. 10, pp. 2612–2618, Oct. 2003.
3. K. G. Balmain, "The impedance of a short dipole antenna in a magnetoplasma," *IEEE Trans. Antennas Propag.*, vol. AP-12, no. 5, pp. 605–617, Sept. 1964.
4. K. G. Balmain and G. A. Oksiutik, "RF probe admittance in the ionosphere: Theory and experiment," in *Plasma Waves in Space and in the Laboratory*. Vol. 1, Edinburgh University Press, 1969, pp. 247–261.

5. K. G. Balmain, A. A. E. Lüttgen, and G. V. Eleftheriades, "Resonance cone radiation from a planar, anisotropic metamaterial," paper presented at the 2003 IEEE AP-S International Symposium and USNC/CNC/URSI North American Radio Science Meeting, Columbus, OH, June 23, 2003, abstract, *URSI Digest*, p. 24.

6. A. Grbic and G. V. Eleftheriades, "Experimental verification of backward-wave radiation from a negative refractive index metamaterial," *J. Appl. Phys.*, vol. 92, no. 10, pp. 5930–5935, Nov. 2002.

7. A. Grbic and G. V. Eleftheriades, "Leaky CPW-based slot antenna arrays for millimeter-wave applications," *IEEE Trans. Antennas Propag.*, vol. 50, no. 11, pp. 1494–1504, Nov. 2002.

8. E. A. Mareev and Y. V. Chugunov, *Antennas in Plasmas* (in Russian), Nizhny Novgorod: Institute of Applied Physics, Academy of Science of the USSR, 1991.

9. K. G. Balmain, A. A. E. Lüttgen, and P. C. Kremer, "Using resonance cone refraction for compact RF metamaterial devices," paper presented at the International Conference on Electromagnetics in Advanced Applications 2003 (ICEAA '03), Special Session on "Metamaterials: EBG Structures," Torino, Italy, Sep. 8–12, 2003.

CHAPTER 7

MICROWAVE COUPLER AND RESONATOR APPLICATIONS OF NRI PLANAR STRUCTURES

Christophe Caloz and Tatsuo Itoh

7.1 INTRODUCTION

From the first experimental demonstration of a *left-handed (LH)* material [1], constituted by negative-permittivity (ε) thin wires (TWs) [2] and negative-permeability (μ) split-ring resonators (SRRs) [3], *effectively homogeneous*[1] *artificial structured materials with unusual properties*, called *metamaterials (MTMs)*, have emerged as a new paradigm of physics and engineering. The LH structures, which are characterized by backward-wave propagation or negative refractive index (NRI) and many related properties, have been the most popular MTMs. In fact, backward-wave propagation has been known for decades in periodic structures [4,5], but the novelty of LH materials is the fact that they are effectively homogeneous fundamental-mode structures (hence the term "MTMs"), fully characterizable in terms of their constitutive parameters ε and μ, whereas previously known backward-wave structures were scattering media based on the propagation of (negative) space harmonics.

The TW SRR structures of the type presented in [1] are inherently narrow band or severely lossy due to their resonant nature. In contrast, *nonresonant* transmission line (TL) MTMs, exhibiting simultaneously broad bandwidth and low losses, have been recently introduced [6–8]. The LH TLs have then been extended and generalized to the concept of composite right/left-handed (CRLH) structures where mixed contributions of both the LH and right-handed (RH) structures occur in practice, have been accurately described and applied to a vast of suite guided-, radiated-, and refracted-wave applications [9, 10].

---

[1] That is, seen as homogeneous by electromagnetic waves, which implies a structural unit size p much smaller than wavelength λ, $p \ll \lambda$.

Metamaterials: Physics and Engineering Explorations, Edited by N. Engheta and R. W. Ziolkowski
Copyright © 2006 the Institute of Electrical and Electronics Engineers, Inc.

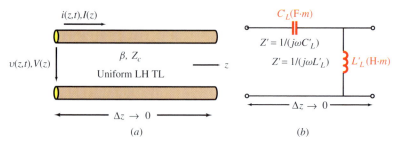

Figure 7.1   Incremental circuit model of uniform LH TL: (*a*) TL section; (*b*) incremental circuit model.

7.2   COMPOSITE RIGHT/LEFT-HANDED TRANSMISSION LINE METAMATERIALS

7.2.1   Left-Handed Transmission Lines

A LH TL is the dual of a RH TL, exhibiting series capacitance (C) and shunt inductance (L). A *uniform* LH TL[2] with its incremental circuit model is shown in Figure 7.1 [11].

The transmission characteristics of this TL are easily obtained from the telegrapher equations [12]. Considering a lossless medium for simplicity,[3] we have

$$\gamma = j\beta = \sqrt{Z'Y'} = -j\frac{1}{\omega\sqrt{L'_L C'_L}} \rightarrow \beta = -\frac{1}{\omega\sqrt{L'_L C'_L}} < 0 \tag{7.1}$$

$$Z_c = \sqrt{\frac{Z'}{Y'}} = +\sqrt{\frac{L'_L}{C'_L}} > 0 \tag{7.2}$$

$$v_p = \frac{\omega}{\beta} = -\omega^2\sqrt{L'_L C'_L} < 0 \qquad v_g = \left(\frac{\partial\beta}{\partial\omega}\right)^{-1} = +\omega^2\sqrt{L'_L C'_L} > 0 \tag{7.3}$$

where it is unambiguously found that the propagation constant β is negative, indicating backward-wave propagation, the characteristic impedance Z_c is positive, and the phase velocity v_p and the group velocity v_g are respectively negative and positive, that is, antiparallel. These characteristics are the attributes of a LH transmission medium.

7.2.2   Composite Right/Left-Handed Structures

Figure 7.2 shows microstrip planar examples of series-C/shunt-L TL structures. In addition to the LH series-C/shunt-L contributions, these structures also exhibit

[2] In strict terms, a uniform TL would require a *strictly homogeneous* LH material, which is not known to exist as a natural substance. However, *effectively homogeneous* LH TLs are available in structured artificial architectures, as will be shown further.

[3] The loss mechanism in a LH TL is identical to that in a RH medium and is treated in [13].

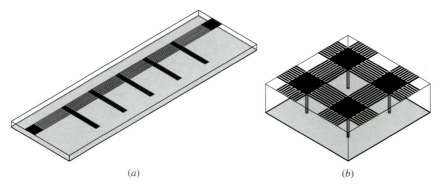

(a)   (b)

Figure 7.2 Planar CRLH TL structures in microstrip technology, constituted of series interdigital capacitors and shunt stub inductors (including via connections to ground). The gray areas represent the ground planes and the black areas represent the metal traces. The unit cell size p is much smaller than the guided wavelength (at least, $p < \lambda_g/4$) to ensure effective homogeneity of the structure and subsequent effective-uniformity behavior of the TL. (a) One-dimensional structure. (b) Two-dimensional structure.

RH series-L/shunt-C natural effects due to the currents in the fingers of the interdigital capacitors and to the trace-to-ground capacitances. In fact, any physically realizable LH structure includes combined LH and RH contributions. Therefore, the concept of CRLH materials was introduced to describe practical planar LH MTMs [9].

Figure 7.3 shows the equivalent incremental circuit model and the dispersion diagram of a corresponding microwave network (Section 7.2.3) CRLH TL.

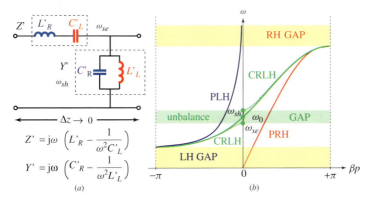

(a)   (b)

Figure 7.3 Incremental circuit model of CRLH TL. The subscripts R and L stand for RH and LH, respectively. (a) Unit-cell prototype. (b) Corresponding microwave network dispersion diagram computed by (7.12) [corresponding to idealized uniform TL relations (7.4) and (7.7) (bandwidth extending from $\omega = 0$ to $\omega = \infty$), except for existence of LH and RH gaps]. According to (7.5), $\omega_{\mathrm{se}} < \omega_{\mathrm{sh}}$ or $\omega_{\mathrm{se}} > \omega_{\mathrm{sh}}$, depending on the LC parameters. The curves for a purely LH (PLH) structure ($L_R = C_R = 0$) and for a purely RH (PRH) structure ($L_L = C_L = \infty$) are also shown for comparison.

The model reveals that a CRLH MTM is LH at lower frequencies (only L_L, C_L as $\omega \to 0$) and RH at higher frequencies (only L_R, C_R as $\omega \to \infty$), as illustrated in the dispersion diagram, which is straightforwardly obtained from the telegrapher equations as

$$\gamma = \alpha + j\beta = js(\omega)\sqrt{\chi'} \qquad (7.4a)$$

with

$$\chi' = \omega^2 L_R' C_R' + \frac{1}{\omega^2 L_L' C_L'} - \frac{1}{L_L' C_L'}\left(\frac{1}{\omega_{se}^2} + \frac{1}{\omega_{sh}^2}\right) \qquad (7.4b)$$

where ω_{se} and ω_{sh} are the series and shunt resonances, respectively, given by

$$\omega_{se} = \frac{1}{\sqrt{L_R' C_L'}} \quad \text{(rad/s)} \qquad \omega_{sh} = \frac{1}{\sqrt{L_L' C_R'}} \quad \text{(rad/s)} \qquad (7.5)$$

and where $s(\omega)$ is a sign function equal to $+1$ if $\omega < \min(\omega_{se}, \omega_{sh})$ (LH range) and to -1 if $\omega > \max(\omega_{se}, \omega_{sh})$ (RH range). In general, a gap exists between the LH and RH ranges due to the distinct eigenfrequencies ω_{se}, ω_{sh}, where $v_g = 0$ (v_g is the slope of the dispersion curves). It can be shown that maximum attenuation α in the gap occurs at the frequency $\omega_0 = 1/\sqrt[4]{L_R' C_R' L_L' C_L'} = \sqrt{\omega_{se}\omega_{sh}}$ [10]. In addition, the characteristic impedance is given by

$$Z_c = Z_L\sqrt{\frac{(\omega/\omega_{se})^2 - 1}{(\omega/\omega_{sh})^2 - 1}} \qquad Z_L = \sqrt{\frac{L_L'}{C_L'}} \qquad Z_R = \sqrt{\frac{L_R'}{C_R'}} \qquad (7.6)$$

In the particularly interesting situation where $\omega_{se} = \omega_{sh}$, called the *balanced* case, the series/shunt resonances cancel each other, which closes up the center gap (Figure 7.3a). We have then $\omega_0 = \omega_{se} = \omega_{sh}$ and

$$\beta^{(bal)} = \omega\sqrt{L_R' C_R'} - \frac{1}{\omega\sqrt{L_L' C_L'}} \qquad Z_c^{(bal)} = Z_L = Z_R \qquad (7.7)$$

where the LH and RH contributions clearly decouple in the expression of $\beta(\omega)$ and Z_c is seen to be *frequency independent*, meaning that broadband matching is possible. In addition, we have, at the transition frequency ω_0, $\beta^{(bal)}(\omega_0) = 0$, $v_p^{(bal)}(\omega_0) = \infty$, $\lambda^{(bal)}(\omega_0) = \infty$, $v_g^{(bal)}(\omega_0) = 1/\left(2\sqrt{L_R' C_R'}\right) = \sqrt{L_L'}/\left(2C_R'\sqrt{C_L'}\right) \neq 0$, showing that *infinite-wavelength nonzero group velocity is achieved at the arbitrarily designed transition frequency ω_0. The phase origin is transferred from direct current (DC) to ω_0 from a RH to a CRLH TL.*[4] The equivalent MTM constitutive parameters are obtained by mapping the telegrapher equations to Maxwell equations in a transverse electromagnetic (TEM) waveguide [15, 16],

$$\mu = \mu(\omega) = L_R' - \frac{1}{\omega^2 C_L'} \qquad \varepsilon = \varepsilon(\omega) = C_R' - \frac{1}{\omega^2 L_L'} \qquad (7.8)$$

---

[4] This bears resemblance with low-pass to bandpass filter transformation, but conventional bandpass filters are not backward-wave and effective structures.

from which the refractive index $n = \sqrt{\mu_r \varepsilon_r}$ is seen to be negative and positive for frequencies below and above ω_0, respectively. The CRLH constitutive parameters are observed to be *frequency dispersive*.[5]

7.2.3 Microwave Network Conception and Characteristics

An effectively uniform CRLH TL, related to an effectively homogeneous CRLH MTM and corresponding to the model of Figure 7.3a, can be implemented in structured configurations, such as those shown in Figure 7.2. In this case, real LC components are used and the TL is realized by the (periodic or not) repetition of a unit cell with small average electrical length, $\Delta\phi < \pi/2$, or $p < \lambda_g/4$, where p is the average cell size. The transformation from the idealized uniform CRLH TL of Section 7.2.1 to a practical microwave network CRLH TL is based on the relations[6]

$$Z' = \frac{Z}{p} = j\left[\omega\left(\frac{L_R}{p}\right) - \frac{1}{\omega(C_L p)}\right] \quad \rightarrow \quad L'_R = \frac{L_R}{p} \qquad C'_L = C_L p \quad (7.9a)$$

$$Y' = \frac{Y}{p} = j\left[\omega\left(\frac{C_R}{p}\right) - \frac{1}{\omega(L_L p)}\right] \quad \rightarrow \quad C'_R = \frac{C_R}{p} \qquad L'_L = L_L p \quad (7.9b)$$

The resulting microwave network TL may be conveniently analyzed as a circuit by using the transmission (or *ABCD*) matrix formalism [12] for a line including an arbitrary number of cells N, and all the transmission characteristics are then easily obtained from standard conversion into scattering parameters [12]. In particular, the dispersion and attenuation diagrams are obtained from the transmission parameter S_{21} by[7]

$$\beta = \begin{cases} -\varphi^{\text{unwrapped}}(S_{21,N})/d + \varsigma & (7.10a) \\ \alpha = -\ln|S_{21,N}|/d & (7.10b) \end{cases}$$

A structured CRLH TL exhibits, as apparent in Figure 7.3a and shown in Figure 7.3b, a *passband* behavior, with a transmission bandwidth delimited by the cutoff frequencies [10][8]

$$\omega_{cL} = \omega_0 \sqrt{\frac{\xi\omega_0^2 - \sqrt{(\xi\omega_0^2)^2 - 4}}{2}} \overset{(\text{bal})}{=} \omega_R\left|1 - \sqrt{1 + \frac{\omega_L}{\omega_R}}\right| \overset{\omega_L \ll \omega_R}{\approx} \frac{\omega_L}{2} \quad (7.11a)$$

---

[5] Frequency dispersion is a necessary condition, imposed by the (dispersive) entropy condition $W = [\partial(\omega\varepsilon)/\partial\omega]E^2 + [\partial(\omega\mu)/\partial\omega]H^2 > 0$ [14], for $\varepsilon < 0$, $\mu < 0$ to be possible. We verify here that $W = [C'_R + 1/(\omega^2 L'_L)]E^2 + [L'_R + 1/(\omega^2 C'_L)]H^2 > 0$.

[6] Per-unit-length immittances (impedance Z' in Ω/m and admittance Y' in S/m) are replaced by real immittances (impedance Z in Ω and admittance Y in S). Prime variables represent per- or time-unit-length quantities, while nonprimed variables represent real quantities (inductances in H and capacitances in F).

[7] Here ς is a phase offset associated with ω_0 so that $\beta(\omega_0) = 0$.

[8] These formulas are based on the assumption that the line is infinite. However, if $N > 3 - 5$, the cutoffs are sharp enough so that they are clearly defined and these formulas are then extremely accurate.

$$\omega_{cR} = \omega_0 \sqrt{\frac{\xi\omega_0^2 + \sqrt{(\xi\omega_0^2)^2 - 4}}{2}} \overset{(bal)}{=} \omega_R\left(1 + \sqrt{1 + \frac{\omega_L}{\omega_R}}\right) \overset{\omega_L \ll \omega_R}{\approx} 2\omega_R + \frac{\omega_L}{2}$$

(7.11b)

where $\xi = L_R C_L + L_L C_R + (2/\omega_L)^2$. Characteristics of unbalanced and a balanced CRLH TLs obtained in this manner are exemplified in Figure 7.4.

In addition, the following analytical dispersion relation is obtained by applying the Bloch–Floquet theorem [12, 17]:

$$\gamma = \alpha + j\beta = \frac{1}{p}\cosh^{-1}\left(1 - \frac{\chi}{2}\right) \overset{|\gamma p| \ll 1}{\approx} \frac{1}{p}\cosh^{-1}\left(1 + \frac{\gamma p}{2}\right)$$

(7.12)

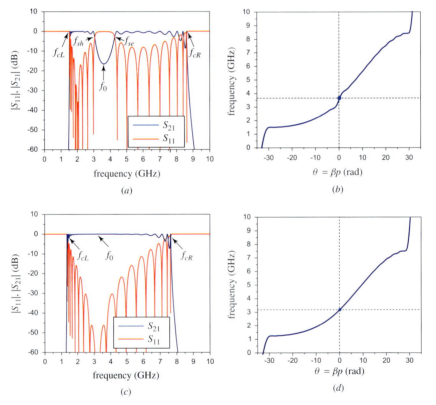

Figure 7.4 Scattering parameters of an ideal $N = 10$-cell microwave LC network CRLH TL. (*a*) Unbalanced case, with parameters $L_R = 2$ nH, $C_R = 1$ pF, $L_L = 2.5$ nH, $C_L = 0.75$ pF, $Z_c = 50\ \Omega$ ($Z_R = \sqrt{L_R/C_R} = 44.72\ \Omega$, $Z_L = \sqrt{L_L/C_L} = 57.54\ \Omega$). From (7.11), $f_{cL} = 1.51$ GHz, $f_0 = 3.63$ GHz, and $f_{cR} = 8.69$ GHz. (*b*) Dispersion relation for the line of (*a*) computed by (7.10a). (*c*) Balanced case, with parameters $L_R = 2.5$ nH, $C_R = 1$ pF, $L_L = 2.5$ nH, $C_L = 1$ pF, $Z_c = 50\ \Omega$ ($Z_L = \sqrt{L_L/C_L} = Z_R = \sqrt{L_R/C_R} = 50\ \Omega$). From (7.11), $f_{cL} = 1.32$ GHz, $f_0 = 3.18$ GHz, and $f_{cR} = 7.69$ GHz. (*d*) Dispersion relation for the line of (*c*) computed by (7.10a).

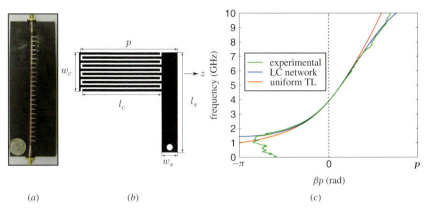

(a)   (b)   (c)

Figure 7.5   Microstrip CRLH TL. (*a*) Prototype, including $N = 24$ cells. (*b*) Unit-cell layout and parameters $p = 6.1$ mm, $\ell_c = 5.0$ mm, $w_c = 2.4$ mm, $\ell_s = 8.0$ mm, $w_s = 1.0$ mm, width of digits 0.15 mm, and all spacings 0.1 mm. (*c*) Dispersion diagram obtained: (1) experimentally from S_{21} measurement using (7.10a); (2) by *LC* network simulation with extracted parameters $L_R = 2.45$ nH, $C_R = 0.50$ pF, $L_L = 3.38$ nH, $C_L = 0.68$ pF ($Z_L/Z_R = 1.01$, $f_0 = 3.9$ GHz); (3) by the uniform TL approximation (7.4) with $L'C'$ parameters computed by (7.9a).

where χ is expression (7.4) without the primes. Application of Taylor's approximation in the last term shows the equivalence between the network realization [Eq. (7.12)] and the uniform idealization [Eq. (7.9)] via the relations (7.9). This relation is plotted in Figure 7.3*b*. Extension to a 2D (or even 3D) CRLH TL is straightforward [16] by writing $\overline{\gamma} = \gamma_x \hat{x} + \gamma_y \hat{y} = (\alpha_x + j\beta_x)\hat{x} + (\alpha_x + j\beta_x)\hat{y}$, $\beta = \sqrt{\beta_x^2 + \beta_y^2}$ and using a solid-state physics crystallographic formalism for the Brillouin zones [18].

7.2.4   Microstrip Technology Implementation

In principle, CRLH TLs and CRLH MTMs can be implemented in any technology. We consider here (distributed) *microstrip* implementations of the type shown in Figure 7.2 [19]. The four parameters (L_R, C_R, L_L, C_L) of a balanced[9] CRLH TL of transition frequency ω_0 matched to ports of impedance Z_0 and with a given fractional bandwidth are obtained from Eq. (7.6) (nonprimed), (7.11), and (7.12) (with $\omega_0 = \omega_{se} = \omega_{sh}$). These components are then synthesized by using the parameter extraction procedure exposed in [20] from full-wave simulations of one isolated cell of the distributed structure. Figure 7.5 shows a microstrip balanced CRLH TL prototype with its dispersion relation. Near fields at a few fractions of wavelength distance from the top of the line are shown in Figure 7.6 and demonstrate the unusual behavior of the guided wavelength, which is increasing in the LH range to ω_0 where it becomes infinite and then decreasing in the LH range, as expected from (7.7) with $\beta = s(\omega)2\pi/\lambda$.

---

[9] A balanced design is most often desired as it allows broadband matching.

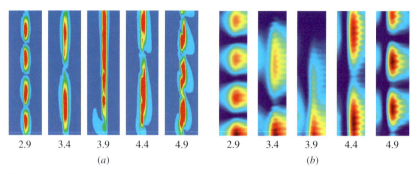

Figure 7.6   Distribution of magnitude of the electric field just above line of Figure 7.5a: (*a*) full-wave simulation (method of moments); (*b*) near-field measurement with vertical monopole.

7.3   METAMATERIAL COUPLERS

Two novel edge-coupled coupled-line CRLH MTM couplers are presented in this section, a symmetric "impedance coupler" (IC) [20] and an asymmetric "phase coupler" (PC) [21].[10] The geometry and anticipated port designation for these two couplers are shown in Figure 7.7. These two couplers are based on fundamentally different principles but exhibit the advantage of providing *arbitrary coupling levels* (up to quasi-complete coupling), whereas conventional edge-coupled couplers are typically limited to less than 10-dB maximum coupling, while conserving the broad-bandwidth benefit of their conventional counterparts. Detailed information on conventional microwave couplers is available in [22].

7.3.1   Symmetric Impedance Coupler

In a coupler constituted of two identical (symmetric) TEM TLs, the field solutions may be represented by the superposition of an *even (e) mode* and an *odd (o) mode*, which are both also TEM. As a consequence (of their TEM nature), these two modes have the same propagation constant $\beta_e = \beta_o = nk_0$ (n is the refractive index of the dielectric medium of the TL). In contrast, these two modes have different characteristic impedances, $Z_{ce} \neq Z_{co}$ because their equivalent TL capacitances are different. The matching condition to ports of impedance Z_0 is achieved with the condition

$$Z_c = \sqrt{Z_{ce} Z_{co}} \tag{7.13}$$

from which the scattering parameters, referred to Figure 7.7, are $S_{11} = 0$, $S_{41} = 0$,

$$S_{21} = \frac{\sqrt{1 - k^2}}{\sqrt{1 - k^2} \cos\theta + j \sin\theta} \qquad S_{31} = \frac{jk \sin\theta}{\sqrt{1 - k^2} \cos\theta + j \sin\theta} = C_Z \tag{7.14}$$

---

[10] These terms (IC and PC) are introduced to avoid confusion possibly arising from the unusual coupling phenomena occurring in the couplers described here.

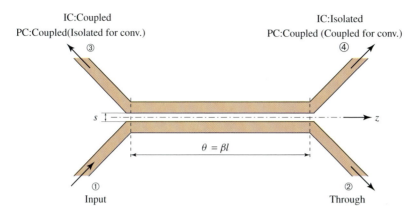

IC:Coupled
PC:Coupled(Isolated for conv.)

IC:Isolated
PC:Coupled (Coupled for conv.)

Figure 7.7   Geometry and port designation of general coupled-line coupler. The coupled and isolated ports are exchanged between the cases of an IC (conventional backward-wave coupler) and the case of a PC (conventional forward-wave coupler) in the conventional (conv.) case. In contrast, the MTM PC presented here has the same port configurations as the conventional and the MTM ICs.

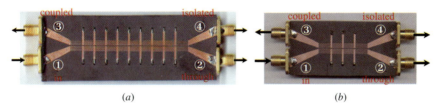

Figure 7.8   Impedance coupling edge-coupled directional coupler, constituted of two interdigital/stub CRLH TLs with same unit cell as TL of Figure 7.5a: (*a*): complete coupling nine-cell prototype; (*b*) 3-dB three-cell prototype. In both cases, the line spacing $s = 0.3$ mm.

where $\theta = \beta\ell$ is the electrical length of the coupler and $k = (Z_{ce} - Z_{co})/(Z_{ce} + Z_{co})$ is the coupling factor, so called because it corresponds to maximum coupling C_Z, obtained for $\theta = \pi/2$, or $\ell = \lambda/4$.[11] If the two lines are not perfectly TEM but *quasi*-TEM, we have $\beta_e \approx \beta_o$, so that $\theta_e \approx \theta_o \, (= \theta)$ and, therefore, the above relations are still approximately valid.[12]

Let us consider the coupled-line structure composed of two (identical) quasi-TEM microstrip CRLH TLs similar to that of Figure 7.5a, shown in Figure 7.8 and corresponding to the even−odd models of Figure 7.9. These

---

[11] This coupler may be called an IC, since coupling depends on the difference between the even−odd *impedances*.

[12] Better approximations would be
$Z_c = \left[(Z_{ce} \sin\theta_e + Z_{co} \sin\theta_o)/(Z_{ce} \sin\theta_o + Z_{co} \sin\theta_e) \right] \sqrt{Z_{ce} Z_{co}}$ and
$\theta = (\theta_e + \theta_o)/2 = (2\pi\ell/\lambda_0)(\varepsilon_{ee} + \sqrt{\varepsilon_{eo}})/2$ [22]. The nonperfectly TEM nature of the TLs alters isolation ($S_{41} \neq 0$).

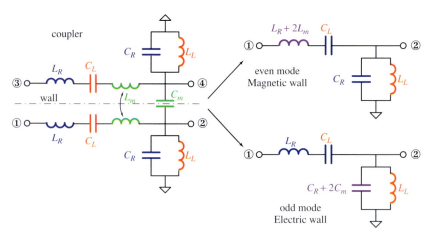

Figure 7.9   Equivalent circuit model for unit cell of IC CRLH coupler, e.g., the one of Figure 7.8, and corresponding even–odd mode TL models. The topology of the models of both the even and the odd TLs are identical to that of a simple CRLH TL (Fig. 7.3a), where L_R has been replaced by $L_{Re} = L_R + 2L_m$ and C_L has been replaced by $C_{Lo} = C_L + 2C_m$, L_m and C_m being the mutual inductance and coupling capacitance, respectively.

even–odd models are seen to be identical to that of an isolated CRLH TL (Fig. 7.3a without primes) under the substitutions

$$L_R \rightarrow L_R + 2L_m = L_{Re} \quad \text{(even)} \qquad C_R \rightarrow C_R + 2C_m = C_{Ro} \quad \text{(odd)} \quad (7.15)$$

It follows, using (7.6) (without primes), that the even–odd characteristic impedances read

$$Z_{ce} = Z_L \sqrt{\frac{1 - (\omega/\omega_{\mathrm{se},e})^2}{1 - (\omega/\omega_{\mathrm{sh}})^2}} \qquad Z_{co} = Z_L \sqrt{\frac{1 - (\omega/\omega_{\mathrm{se}})^2}{1 - (\omega/\omega_{\mathrm{sh},o})^2}} \qquad (7.16)$$

where $\omega_{\mathrm{se},e} = 1/\sqrt{L_{Re}C_L}$, $\omega_{\mathrm{sh},e} = 1/\sqrt{L_L C_R}$, $\omega_{0e} = \sqrt{\omega_{\mathrm{se},e}\omega_{\mathrm{sh}}}$ and $\omega_{0o} = \sqrt{\omega_{\mathrm{se}}\omega_{\mathrm{sh},o}}$. Because each of the two lines in isolation is balanced with the parameters (L_R, C_R, L_L, C_L), the even–odd equivalent lines, having different parameters, are necessarily unbalanced, which results in the emergence of even–odd gaps. The IC operates in these gaps, which are designed to overlap each other. Since matching is obtained within the even–odd TL gaps, we need to generalize the expression of the IC coupling coefficient in (7.14) by changing $\theta = \beta\ell$ into $\theta = \gamma\ell = (\alpha + j\beta)\ell$, where $\alpha_e \approx \alpha_o \approx \alpha$ and $\beta_e \approx \beta_o \approx \beta$. The IC coupling coefficient C_Z then becomes

$$C_Z = S_{31} = \frac{(Z_{ce} - Z_{co})\tanh\left[(\alpha + j\beta)\ell\right]}{2Z_c + (Z_{ce} + Z_{co})\tanh\left[(\alpha + j\beta)\ell\right]} \quad \begin{matrix} |\beta| \approx 0, \\ \alpha\ell > 1 \\ \approx \end{matrix}$$

$$\times \frac{Z_{ce}/Z_c - Z_{co}/Z_c}{2 + (Z_{ce}/Z_c + Z_{co}/Z_c)} \qquad (7.17)$$

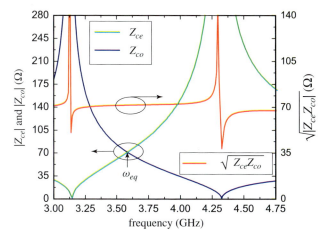

Figure 7.10 Circuit simulated even–odd characteristic impedances (magnitude) for coupler of Figure 7.8. Left-hand axis: magnitudes of impedances. Right-hand axis: square root of product of impedances appearing in (7.13). The parameters of the even–odd TLs are $L_m = 1.0$ nH, $C_m = 0.8$ pF, $f_{se,e} = 3.14$ GHz, $f_{0e} = 3.67$ GHz, $f_{sh,e} = 4.30$ GHz, $f_{se,o} = 3.13$ GHz, $f_{0o} = 3.68$ GHz, and $f_{sh,o} = 4.33$ GHz. The Z_{ce}/Z_{co} crossing frequency is $\omega_{eq} - 3.59$ GHz and $Z_0 = 50$ Ω. Here $Z_{0e} = Z_{0o}$ at a frequency ω_{eq}, which can be determined from (7.16) [20].

where the approximation in the last term holds in the even–odd gaps ($\beta \approx 0$) if the length of the coupler is sufficient so that $\alpha \ell > 1$. By using (7.13) to eliminate either Z_{ce} or Z_{co} and also taking into account the fact that $Z_{ci} = j\mathrm{Im}(Z_{ci})$ ($i = e, o$) (Figure 7.10), this expression is further transformed, by defining $\xi = Z_c/\mathrm{Im}(Z_{co}) = \mathrm{Im}(Z_{ce})/Z_c$, to[13]

$$C_Z \approx \frac{\xi + \xi^{-1}}{2j + (\xi - \xi^{-1})} \quad \text{with} \quad |C_Z| \approx \frac{\xi + \xi^{-1}}{\sqrt{4 + (\xi - \xi^{-1})^2}} = 1 \quad (7.18)$$

This final relation demonstrates that *complete backward coupling* is achieved in this IC if its length is such that $\alpha \ell > 1$ over a bandwidth which depends essentially on the even–odd bandwidth via the parameters in (7.15).[14] Figure 7.10 illustrates the highly unusual behavior of the even–odd characteristic impedances.

The performances of the quasi-0-dB[15] IC coupler of Figure 7.8a are presented in Figure 7.11: Close-to-zero coupling is achieved in the range from 3.2 to 4.6 GHz (36 percent) with a directivity of approximately 25 dB. The coupler

---

[13] The last equality is easily verified by developing the denominator as $\sqrt{4 + (\xi - \xi^{-1})^2} = \sqrt{\xi^2 + \xi^{-2} + 2} = \sqrt{(\xi + \xi^{-1})^2} = \xi + \xi^{-1}$, which is equal to the numerator.

[14] Note that the length corresponding to maximal coupling is not necessarily $\lambda/4$ (or odd multiples of this quantity) as for the conventional coupler case.

[15] Although it is not directly useful in practice (a simple strip connection would be more reasonable!), this quasi-complete power coupler shows that any level of coupling can be easily obtained, by reducing the length of the coupler and increasing its spacing, since coupling up to almost 0 dB is achievable.

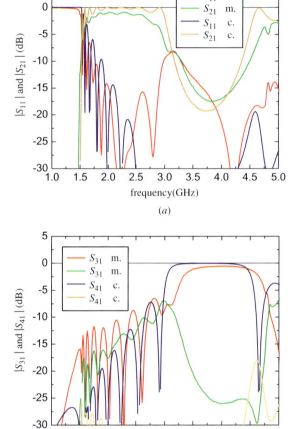

Figure 7.11 Circuit model (c.) simulated (using extracted parameters $L_m = 1.0$ nH and $C_m = 0.8$ pF) and measured (m.) scattering parameters for quasi-0-dB coupler of Figure 7.8a: (a) S_{11} and S_{21}; (b) S_{31} and S_{41}.

length is around $1.05\lambda_e$ (λ_e is the effective permittivity of the conventional microstrip on the same substrate). The IC shown in Figure 7.8b is a 3-dB coupler with an amplitude balance of 2 dB over a bandwidth of 50 percent from 3.5 to 5.8 GHz and quadrature phase balance of $90° \pm 5°$ from 3.5 to 4 GHz [20].

7.3.2 Asymmetric Phase Coupler

When the two lines constituting the coupler are different, the coupled-line structure is asymmetric, and the even–odd analysis has to be replaced by the more involved c/π mode analysis [22]. An intermediate problem is the symmetric (identical line) coupler constituted of nonperfect TEM ($\beta_e \neq \beta_o$) TLs. If the spacing in this coupler is sufficiently large so that $Z_{0e} \approx Z_{0o}$ (due to negligible edge capacitance between the lines) and therefore $C_Z = S_{31} \approx 0$ from (7.14),

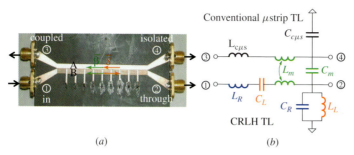

Figure 7.12   Phase coupling edge-coupled directional coupler. (*a*) The 0-dB coupling prototype constituted of one conventional microstripline (line *A*) and one interdigital/stub CRLH TL (line *B*) operated exclusively in its LH range. (*b*) Equivalent circuit model for the unit cell. The design parameters of the (nine-cell) CRLH TL (bottom line) are the same as in Figure 7.5. The microstripline (cμs) is modeled by the parameters $L_{\text{cμs}}$ and $C_{\text{cμs}}$ representing the equivalent series inductance and shunt capacitance, respectively, for one (lumped-implemented) unit cell. The parameters C_m and L_m represent the coupling capacitance and mutual inductance, respectively. The spacing between the lines is $s = 0.3$ mm and the length of the coupler is $\ell = 62$ mm.

coupling is based on even–odd velocities[16] and occurs at port 4 while port 3 becomes isolated[17] (Fig. 7.7). We have in general for the coupling coefficient $C_\theta = S_{41}$ [22]

$$
S_{41} = -2j\frac{\sqrt{p}}{1+p}\exp\left(\frac{-j(\beta_c+\beta_\pi)\ell}{2}\right)\sin\left[\frac{(\beta_c-\beta_\pi)\ell}{2}\right] \overset{A\to B}{\approx}
$$
$$
- j\exp\left(\frac{-j(\beta_e+\beta_o)\ell}{2}\right)\sin\left[\frac{(\beta_e-\beta_o)\ell}{2}\right] \tag{7.19}
$$

where the last expression is an approximation for the symmetric coupler when the two lines (*A* and *B*) tend to become identical (then $\beta_{c,\pi} \to \beta_{e,o}$ and $p \to 1$). This equation reveals that maximum coupling occurs for the coupler length[18]

$$
\ell = \frac{\pi}{|\beta_c-\beta_\pi|} \overset{A\to B}{\approx} \frac{\pi}{|\beta_e-\beta_o|} = \frac{\lambda_0}{2\left|\sqrt{\varepsilon_{ee}}-\sqrt{\varepsilon_{eo}}\right|} \tag{7.20}
$$

Figure 7.12 shows the CRLH of interest, which consists of a CRLH TL identical to that of Figure 7.5*a* coupled to a conventional microstripline. In this coupler, the coupled/isolated ports are inverted due to the propagation constant $\overline{\beta}$ and Poynting $\overline{S}$ vector orientations shown in Figure 7.12*a*. In addition, we assume that the CRLH TL is *operated exclusively in its LH range*. Since polarities in an *isolated* RH TL and in an *isolated* LH TL are opposite ($\beta_{\text{LH}} - \|\beta_{\text{RH}}\|$), the isolated

---

[16] This coupler may be called a PC, since coupling depends on the difference between the even–odd *phases*.

[17] It is therefore also called a *forward coupler*.

[18] At microwaves, in practice, this length is prohibitively large (hundreds of λ's!) due to the typical quasi-TEM nature of the lines and their large spacing, and this type of coupler is therefore not used in its conventional form. In contrast, forward-wave coupling is very common in photonics.

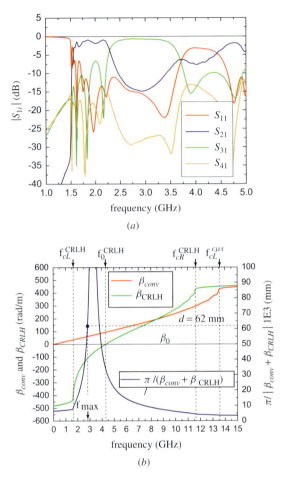

(a)

(b)

Figure 7.13   Results for PC of Figure 7.12. (*a*) Measured scattering parameters. (*b*) Phase constants in isolated conventional and CRLH TLs (left-hand axis) and right-hand term of (7.21) (right-hand axis). The extracted parameters are $L_R = 2.21$ nH, $C_R = 0.45$ pF, $L_L = 3.04$ nH, $C_L = 0.61$ pF, $L_{c\mu s} = 1.64$ nH, $C_{c\mu s} = 0.33$ pF, $L_m = 0.27$ nH, and $C_m = 0.33$ pF. The corresponding frequencies computed from (7.11) (with $L_L = C_L = 0$ and $L_R = L_{c\mu s}$, $C_R = C_{c\mu s}$ for the microstripline) are $f_{c\mu s} = 13.6$ GHz, $f_{cL}^{CRLH} = 1.8$ GHz, $f_0^{CRLH} = 4.3$ GHz, $f_{cR}^{CRLH} = 11.9$ GHz. The frequency computed from the coherence condition (7.21) is $f_{max} = 2.8$ GHz.

microstrip (RH) and CRLH (LH) TLs may be considered an approximation of the c/π equivalent TLs: $\beta_{\mu sp} \to \beta_c \to \beta_e$ and $\beta_{CRLH} \to \beta_\pi \to \beta_o$. While the small difference $|\beta_e - \beta_o|$ leads to poor coupling in the conventional case, we have here $\beta_o \to -|\beta_{CRLH}|$, so that the difference in the denominator of (7.20) is turned into a *sum*,

$$\ell_{max} = \frac{\pi}{\beta_{\mu sp} + |\beta_{CRLH}|} \tag{7.21}$$

which shows that despite $\beta_{\mu sp} \approx |\beta_{CRLH}|$, tight coupling can be achieved over a short length!

The performances of the PC shown in Figure 7.12*a* are presented in Figure 7.13*a*. Quasi-0-dB coupling is achieved over the range from 2.2 to 3.8 GHz (53 percent) with the excellent directivity of 30 dB. Figure 7.13*b* shows that the value of (7.21) has the expected maximum in the center of the coupler bandwidth for its actual length *d*.

7.4 METAMATERIAL RESONATORS

7.4.1 Positive, Negative, and Zero-Order Resonance in CRLH Resonators

Being effectively uniform or effectively homogeneous structures, CRLH MTMs can also be used as resonators when open or short ended.[19] As in conventional resonators, CRLH structures resonate when their length is a multiple of half a wavelength. But because of the transfer of the phase origin from frequency zero to the transition frequency ω_0, a CRLH structure supports negative (LH-band) resonances and a unique zero-order resonance at ω_0 in addition to the conventional positive resonances (RH band) [10,23]

$$\ell = |m|\frac{\lambda}{2} \quad \text{or} \quad \theta_m = \beta_m \ell = \left(\frac{2\pi}{\lambda}\right) \cdot \left(\frac{m\lambda}{2}\right) = m\pi \tag{7.22a}$$

with

$$m = 0, \pm1, \pm2, \ldots \pm\infty \tag{7.22b}$$

as illustrated in Figure 7.14. An interesting feature of the dual modes $\pm m$ is their similar field and impedance characteristics due to their identical magnitude of $\beta_{\pm}$.

If the CRLH structure were strictly uniform, an infinite number of modes would exist, equidistant for $\omega \to \infty$ (linear RH dispersion) and strongly compressed for $\omega \to 0$ (hyperbolic LH dispersion), as apparent in Figure 7.14a. In practice, a CRLH TL is constructed with a finite number N of finite size

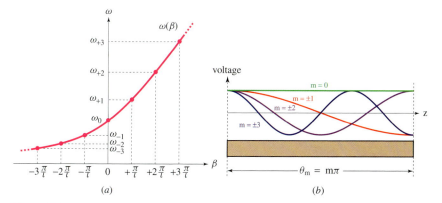

Figure 7.14 CRLH TL resonator: (*a*) dispersion relation and resonance frequencies ω_m of corresponding resonator based on uniform TL (Section 7.2.2); (*b*) typical field distributions of resonance modes (voltage/current distribution for an open/short-circuited metal/slot TL).

[19] It will be essential to realize that what resonates in these structures is not the unit cell, which is invisible to electromagnetic waves due to effective uniformity, but the overall structure constituted of several cells.

cells, which leads to passband characteristics (Section 7.2.3) and consequently to the existence of a finite number $2N - 1$ of resonant modes, as illustrated in Figure 7.15.[20]

Let us focus now on the particularly interesting mode $m = 0$ of the CRLH TL resonator [23].[21] For this mode, the impedances for open- and short-ended coupling excitations, illustrated in Figure 7.16, are derived as

$$Z_{in}^{open} = -jZ_c \cot(\beta\ell) \overset{\beta\to0}{\approx} -j\frac{Z_c}{\beta\ell} = \frac{\sqrt{Z'/Y'}}{\sqrt{Z'Y'}\ell} = \frac{1}{Y'(Np)} = \frac{1}{NY} \quad (7.23a)$$

$$Z_{in}^{short} = jZ_c \tan(\beta\ell) \overset{\beta\to0}{\approx} -jZ_c\beta\ell = \sqrt{\frac{Z'}{Y'}}\sqrt{Z'Y'}\ell = Z'(Np) = NZ \quad (7.23b)$$

which correspond to N times the admittance and impedance of a unit cell, respectively, and show that the zero resonant angular frequency is ω_{sh} and ω_{se} for open- and short-ended conditions, respectively.[22] The quality factors are then

$$Q_0^{open} = \frac{1/(NG)}{\omega_{sh}(L_L/N)} = \frac{1}{G}\sqrt{\frac{C_R}{L_L}} \qquad Q_0^{short} = \frac{NR}{\omega_{se}(NL_R)}R\sqrt{\frac{C_L}{L_R}} \quad (7.24)$$

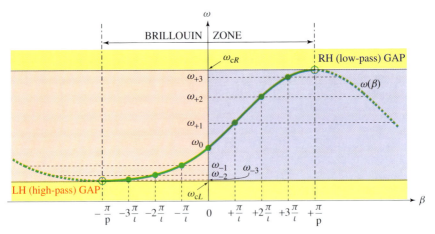

Figure 7.15 Resonances of network (Section 7.2.3) balanced periodic LC network CRLH TL resonator constituted of N unit cells (here $N = 4$). The field distributions are similar to those shown in Figure 7.14b. The length of the resonator ℓ and the period p are related by $\ell = Np$, which results in Brillouin zone edges of $\pm N\pi/\ell = \pm\pi/p$.

[20] In this figure, a periodic implementation is assumed. The points $\beta = \pm\pi/p$ are not "overall structure" resonances but correspond to unit-cell resonances in the Bragg regime.

[21] For the resonator application, the balance condition is not necessary, i.e., we may have $\omega_{se} \neq \omega_{sh}$

[22] Consequently, if the structure is balanced ($\omega_{se} = \omega_{sh}$), the zero-order mode can be excited with either with short or open terminations.

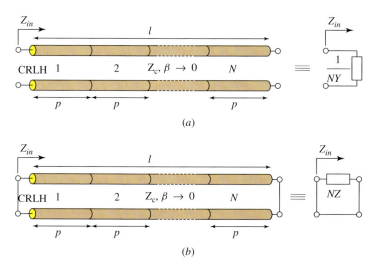

Figure 7.16 CRLH TL resonator and equivalent circuits in terms of unit-cell immittances Z and Y. Each unit cell is identical to the LC cell shown in Figure 7.3a (without primes). (a) Open-ended case. (b) Short-ended case.

This zero-order resonator exhibits the remarkable property that *constant magnitude and phase resonance can be achieved at an arbitrary frequency, not depending on the physical length of the structure but only on the LC loadings.*[23] Another interesting property of this resonator is that, as seen in (7.24), the open-ended resonator is only affected by shunt losses (extremely small in a planar structure with standard substrates) while the short-ended resonator is only affected by series losses.

7.4.2 Zero-Order Antenna

As an illustration of the application of a zero-order CRLH resonator, we consider the zero-order antenna demonstrated in Figure 7.17 [24].[24] The zero-order resonating antenna shown in the inset of Figure 7.17a is open ended (i.e., $\omega_{res} = \omega_{sh}$) and features about a 75 percent footprint reduction in comparison with a corresponding conventional patch antenna. Figure 7.17b shows the radiation patterns for the 4-cell antenna of Figure 7.17a; the patterns for a much larger, 30-cell antenna *with identical unit cell and resonating at the same frequency* show that

---

[23] In low-frequency lumped resonators, resonance also only depends on the LC values but the phase origin is still at $\omega = 0$ whereas it is at the arbitrary frequency ω_0 here.

[24] It should also be noted that one of the most remarkable applications of the CRLH concept is a novel and unique backfire-to-endfire (including broadside) *leaky-wave* antenna, described, for instance, in [9, 10].

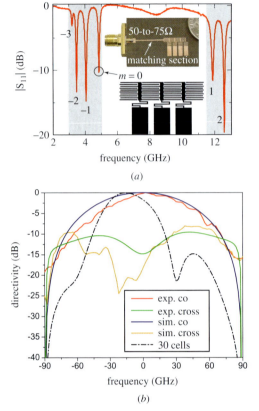

(a)

(b)

Figure 7.17   Zero-order resonant-type CRLH antenna composed of 4 cells and excited by induction at $f = 7.61$ GHz. (*a*) Return loss with antenna picture and layout shown as insets [here the stubs are terminated by virtual ground (large) patch capacitors instead of direct via connections to the ground]. (*b*) Measured radiation patterns (exp., experimental; sim., simulated; co, copolarization; cross, cross-polarization). Also shown is the simulated patterns for a 30-cell antenna composed of the same cells and resonating at a very close frequency ($f = 7.75$ GHz).

resonance does not depend on the physical length but only on the CRLH parameters. The perfectly uniform current distribution in the zero-order mode may mitigate the loss resistance R_l with respect to the radiation resistance R_r and therefore provide higher efficiency $\eta = R_r/(R_r + R_l)$.

7.4.3   Dual-Band Ring Antenna

After an application of the CRLH mode $m = 0$, we will consider another application, for the mode $m = \pm 2$, which is the dual-band resonating ring antenna presented in Figure 7.18 [25]. In this resonator, the mode $m = 1$ is not allowed due to the boundary condition $\xi(\varphi_0 + 2\pi) = \xi(\varphi_0)$ (ξ is the voltage or current along the loop) associated with the loop configuration (Fig. 7.18*a*). The mode $m = 0$ is also prohibited because the center point where all stubs are interconnected (without via ground connection) is not a virtual ground for this specific mode. The interesting feature here is the aforementioned fact that the dual $\pm m$ modes (here ± 2) have similar field and impedance characteristics, which allows dual-band operation with a single and simple feeding mechanism. Figure 7.18*b* shows that the radiation patterns of the two modes are very similar except for the lower gain due to smaller aperture in the lower mode.

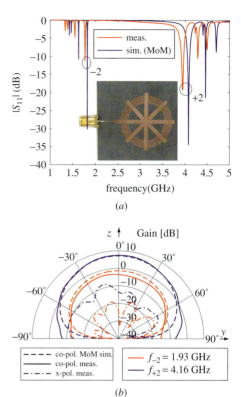

Figure 7.18   Dual-band ring resonator antenna (meas., measured; sim., simulated; co, copolarization; cross, cross-polarization). (*a*) Return loss with antenna prototype in inset. (*b*) Radiation patterns at the two operation frequencies $f_{-2} = 1.93$ GHz and $f_{+2} = 4.16$ GHz.

7.5   CONCLUSIONS

Novel coupler and resonator applications of the powerful CRLH MTM concept have been presented. A diversity of other CRLH practical devices and structures have also been developed [10]. It is expected that many more MTM effects and components will be discovered in the near future.

REFERENCES

1. D. R. Smith, W. J. Padilla, D. C. Vier, S. C. Nemat-Nasser, and S. Schultz, "Composite medium with simultaneously negative permeability and permittivity," *Phys. Rev. Lett.,* vol. 84, no. 18, pp. 4184–4187, May 2000.
2. J. B. Pendry, A. J. Holden, W. J. Stewart, and I. Youngs, "Extremely low frequency plasmons in metallic mesostructure," *Phys. Rev. Lett.,* vol. 76, no. 25, pp. 4773–4776, June 1996.
3. J. B. Pendry, A. J. Holden, D. J. Robbins, and W. J. Stewart, "Magnetism from conductors and enhanced nonlinear phenomena," *IEEE Trans. Microwave Theory Tech.,* vol. 47, no. 11, pp. 2075–1084, Nov. 1999.
4. L. Brillouin, *Wave Propagation in Periodic Structures,* McGraw-Hill, New York, 1946.
5. J. R. Pierce, *Traveling-Wave Tubes,* Van Nostrand, New York, 1950.

6. A. K. Iyer and G. V. Eleftheriades, "Negative refractive index metamaterials supporting 2-D waves," in *IEEE-MTT Int'l Symp. Digest,* Vol. 2, Seattle, WA, June 2002, pp. 412–415.

7. C. Caloz and T. Itoh, "Application of the transmission line theory of left-handed (LH) materials to the realization of a microstrip LH transmission line," in *Proc. IEEE-AP-S USNC/URSI National Radio Science Meeting,* Vol. 2, San Antonio, June 2002, pp. 412–415.

8. A. A. Oliner, "A periodic-structure negative-refractive-index medium without resonant elements," in *URSI Digest, IEEE-AP-S USNC/URSI National Radio Science Meeting,* San Antonio, TX, June 2002, p. 41.

9. C. Caloz and T. Itoh, "Novel microwave devices and structures based on the transmission line approach of meta-materials," in *IEEE-MTT Int'l Symp. Digest,* Vol. 1, Philadelphia, PA, June 2003, pp. 195–198.

10. C. Caloz and T. Itoh, Electromagnetic Metamaterials: Transmission Line Theory and Microwave Applications, IEEE Press and Wiley, New York, 2005.

11. S. Ramo, J. R. Whinnery, and T. Van Duzer, Fields and Waves in Communication Electronics, 3rd ed., Wiley, New York, 1994.

12. M. D. Pozar. Microwave Engineering, 3rd. ed., Wiley, New York, 2004.

13. C. Caloz and T. Itoh, "Lossy transmission line metamaterials," *Microwave Opt. Technol. Lett.,* vol. 43, no. 2, pp. 112–114, Oct. 2004.

14. E. M. Lifshitz, L. D. Landau, and L. P. Pitaevskii. Electrodynamics of Continuous Media, Vol. 8, 2nd ed., Butterworth-Heinemann, Oxford, 1984.

15. G. V. Eleftheriades, A. K. Iyer, and P. C. Kremer, "Planar negative refractive index media using periodically *L-C* loaded transmission lines," *IEEE Trans. Microwave Theory Tech.,* vol. 50, no. 12, pp. 2702–2712, Dec. 2002.

16. C. Caloz and T. Itoh, "Positive/negative refractive index anisotropic 2D metamaterials," *IEEE Trans. Microwave Wireless Components Lett.,* vol. 13, no. 12, pp. 547–549, Dec. 2003.

17. R. E. Collin, Foundations for Microwave Engineering, 2nd ed., McGraw-Hill, New York, 1992.

18. N. W. Ashcroft, N. D. Mermin, and D. Mermin, Solid State Physics, Brooks Cole, New York, 1976.

19. C. Caloz, and T. Itoh, "Transmission line approach of left-handed (LH) structures and microstrip realization of a low-loss broadband LH filter," *IEEE Trans. Antennas Propag.,* vol. 52, no. 5, pp. 1159–1166, May 2004.

20. C. Caloz, A. Sanada, and T. Itoh, "A novel composite right/left-handed coupled-line directional coupler with arbitrary coupling level and broad bandwidth," *IEEE Trans. Microwave Theory Tech.,* vol. 52, no. 3, pp. 980–992, Mar. 2004.

21. C. Caloz and T. Itoh, "A novel mixed conventional microstrip and composite right/left-handed backward-wave directional coupler with broadband and tight coupling characteristics," IEEE Microwave Wireless Components Lett., vol. 14, no. 1, pp. 31–33, Jan. 2004.

22. R. Mongia, I. Bahl, and P. Bhartia, RF and Microwave Coupled-Line Circuits, Artech House, Norwood, MA, 1999.

23. A. Sanada, C. Caloz, and T. Itoh, "Zeroth order resonance in composite right/left-handed transmission line resonators," paper presented at Asia-Pacific Microwave Conference, Vol. 3, Seoul, Korea, Nov. 2003, pp. 1588–1592.

24. A. Sanada, K. Murakami, I. Awai, H. Kubo, C. Caloz, and T. Itoh, "A planar zeroth order resonator antenna using a left-handed transmission line," paper presented at 34th European Microwave Conference, Amsterdam, Netherlands, Oct. 2004, pp. 1341–1344.

25. S. Otto, A. Rennings, C. Caloz, P. Waldow, I. Wolff, and T. Itoh, "Composite right/left-handed λ-resonator ring antenna for dual-frequency operation," in Proc. IEEE AP-S USNC/URSI National Radio Science Meeting, Washington, DC, June 2005.

PART II

ELECTROMAGNETIC BANDGAP (EBG) METAMATERIALS

THREE-DIMENSIONAL VOLUMETRIC EBG MEDIA

HISTORICAL PERSPECTIVE AND REVIEW OF FUNDAMENTAL PRINCIPLES IN MODELING THREE-DIMENSIONAL PERIODIC STRUCTURES WITH EMPHASIS ON VOLUMETRIC EBGs

Maria Kafesaki and Costas M. Soukoulis

8.1 INTRODUCTION

8.1.1 Electromagnetic (Photonic) Bandgap Materials or Photonic Crystals

Electromagnetic bandgap (EBG) materials [known as photonic crystals (PCs) or photonic bandgap (PBG) materials] are a novel class of artificially fabricated structures which have the ability to control and manipulate the propagation of electromagnetic (EM) waves [1–3]. Properly designed photonic crystals can prohibit the propagation of light, or allow it along only certain directions, or localize light in specified areas. They can be constructed in one, two, and three dimensions (1D, 2D, and 3D) with either dielectric or/and metallic materials.

The ability of PCs to control the propagation of light has its origin in their photonic band structure. The concept of photonic band structure [4, 5] arises in analogy to the concept of electronic band structure. Just as electron waves traveling in the periodic potential of a crystal are arranged into energy bands separated by bandgaps, we expect the analogous phenomenon to occur when EM waves propagate in a medium in which the dielectric constant varies periodically in space. Photonic bandgap materials or PCs are the structures which show such a phenomenon, that is, produce a forbidden frequency gap in which all propagating states are prohibited. The investigation of these materials is a topic of intensive studies by many groups, both theoretically and experimentally [1–3].

Metamaterials: Physics and Engineering Explorations, Edited by N. Engheta and R. W. Ziolkowski
Copyright © 2006 the Institute of Electrical and Electronics Engineers, Inc.

The PBG property of PCs makes them the EM analog of the electronic semiconductor crystals, although in the EM case the periodicity alone does not guarantee the existence of a full PBG. Nonetheless, a great advantage of the PCs is that although in semiconductors the periodicity is predetermined, the periodicity in the PCs can be changed at will, thus changing the frequency range of the PBG. Such structures have been built in the microwave and recently in the far-infrared regime, and their potential applications continue to be examined. However, the greatest scientific challenge in the field of PCs is to fabricate composite structures possessing spectral gaps at frequencies up to the optical region.

The first prescription for a periodic dielectric structure [6] that possesses a full PBG rather than a pseudogap was given by Ho, Chan, and Soukoulis at Iowa State University (ISU). This proposed structure was a periodic arrangement of dielectric spheres in a lattice-like diamond. It was found that PBGs exist over a wide region of filling ratios, for both dielectric spheres in air and air spheres in a dielectric, and for refractive index contracts between spheres and host as low as 2. However, this diamond dielectric structure is not easy to fabricate, especially in the micrometer- and submicrometer-length scales for infrared or optical devices. In the same time frame as ISU's findings about the diamond structure [6], Yablonovitch was devising [7] an ingenious way of constructing a structure with the symmetry of the diamond lattice. This was achieved by properly drilling cylindrical holes through a dielectric block. Such a structure with only three sets of holes (three-cylinder structure) became the first experimental structure [7] that demonstrated the existence of a (full or complete) PBG, in agreement with the predictions [8] of the theoretical calculations. This is a successful example where the theory was used to design dielectric structures with desired properties. It is very interesting to note that after 15 years since the introduction [6] of the diamond lattice by the ISU group, it still possesses [9] the largest PBG.

Another example of a successful synergy between theory and experiment is encountered in the layer-by-layer structure (see Fig. 8.1), the so-called wood pile structure. The layer-by-layer structure was designed by the ISU group [10] and has a full 3D PBG over a wide range of structural parameters. The structure consists of layers of rods, with a stacking sequence that repeats every fourth layer. It was first fabricated [11] by stacking alumina cylinders, and it was demonstrated to have a full 3D PBG at 12 to 14 GHz.

Another interesting class of PCs is the A7 class of structures [12]. These structures have rhombohedral symmetry and can be generated by connecting lattice points of the A7 structure by cylinders. The A7 class of structures can be described by two structural parameters that can be varied to optimize the gap. For special values of the parameters the structure reduces to simple cubic, diamond, and the Yablonovitch three-cylinder structure. Gaps as large as 50 percent are found [12] in the A7 class of structures for well-optimized values of the structural parameters; fabrication of these structures would be a very interesting task.

The fabrication and the testing of PC structures is a task that has attracted intensive efforts, dating back to the original efforts by Yablonovitch [15]. Fabrication can be either easy or extremely difficult, depending upon the desired

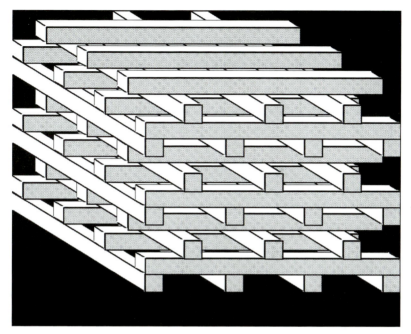

Figure 8.1   Layer-by-layer structure, producing full 3D PBG. The structure is constructed by an orderly stacking of dielectric rods with a simple 1D pattern of rods in each layer. Although rods of rectangular cross section are shown here, the rods may also be of circular or elliptical cross sections.

wavelength of the bandgap and the level of dimensionality. Since the wavelength of the bandgap scales directly with the lattice constant of the PC, lower frequency structures that require larger dimensions are easier to fabricate. At the other extreme, optical wavelength PBGs require PC lattice constants less than 1 μm. Building PCs in the optical regime is a major challenge in PBG research and requires methods that push the current state-of-the-art micro- and nanofabrication techniques. Clearly, the most challenging PBG structures are fully 3D structures with bandgaps in the infrared or optical regions of the spectrum. This area of PBG research has been one of the most active, and perhaps most frustrating, in recent years.

The first attempts toward PBG structures operating in the infrared or optical regime have targeted the miniaturization of the existing microwave PBG structures. Since 1991, both Yablonovitch and Scherer have been working toward reducing the size of Yablonovitch's [7] three-cylinder structure to micrometer-length scales [16]. However, it is very difficult to drill uniform holes of appreciable depth with micrometer diameters. Thus, Scherer's efforts were only partially successful in producing a PC with a gap at optical frequencies. Another approach for the miniaturization of Yablonovitch's three-cylinder structure was undertaken by a group at the Institute of Microtechnology in Mainz, Germany, in collaboration with FORTH, in Greece, and the ISU, using deep X-ray lithography

(LIGA) [17]. In this method polymethylmethacrylate (PMMA) resist layers with thickness of 500 μm were irradiated to form a "three-cylinder" structure. Since the dielectric constant of the PMMA is not large enough for the formation of a PBG, the holes in the PMMA structure were filled with a ceramic material. After the evaporation of the solvent, the samples were heat treated, and a lattice of ceramic rods corresponding to the holes in the PMMA structure remained. A few layers of this structure were fabricated; it was measured to have a bandgap centered at 2.5 THz. A scanning electron microscopy (SEM) view of this structure, with a lattice constant of 114 μm is shown in Figure 8.2. Recent experiments are currently trying to fill the PMMA holes with a metal.

Attempts at the miniaturization of the layer-by-layer wood pile structure shown in Figure 8.1 include a miniature version that was fabricated [18] by laser rapid prototyping using a laser-induced direct-write deposition from the gas phase. The structure consisted of oxide rods that were submicrometer in size; the measured PBG was centered at 2 THz. Recent work at Sandia National Laboratory by Lin [19] and at Kyoto University by Noda [20] has demonstrated growth up to five layers of the layer-by-layer wood pile structure at both the 10- and 1.5-μm wavelengths. The measured transmittance of these structures showed bandgaps centered at 30 and 200 THz, respectively. These are really spectacular achievements. They were able to overcome very difficult technological challenges in planarization, orientation, and 3D growth at the required micrometer-length scales.

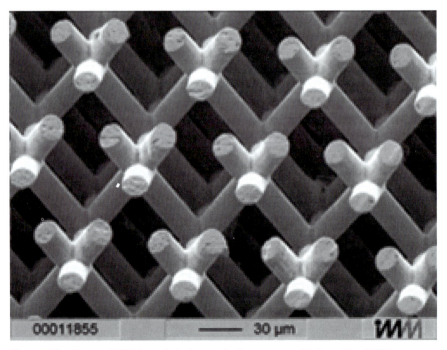

Figure 8.2 "Inverse" Yablonovitch three-cylinder structure fabricated by LIGA.

Another approach to obtain PCs in submicrometer regimes is by using colloidal suspensions. Colloidal suspensions have the ability to spontaneously form bulk 3D crystals with lattice parameters on the order of 1 to 10 nm. Also, 3D dielectric lattices have been developed from a solution of artificially grown monodisperse spherical SiO_2 particles. However, both these procedures give structures with a quite small dielectric contrast ratio (less than 2), which is not enough to achieve a full bandgap. Much effort is going into finding new methods for increasing the dielectric contrast ratio in these structures. Several groups [21–28] are trying to produce ordered macroporous materials of titania, silica, and zirconia by using the emulsion droplets as templates, around which material is deposited through a sol–gel process. Subsequent drying and heat treatment have yielded solid materials with spherical pores left behind the emulsion droplets. Another very promising technique in fabricating PCs at optical wavelengths is 3D holographic lithography [29]. Very recently, high-quality large-scale wood pile structures operating at 1.5 μm have been fabricated by direct laser writing [30].

Since the fabrication of 3D PCs at optical wavelengths is still a difficult process, an alternative method has been proposed: A three-layer dielectric structure is created in the vertical direction, with the central layer having a higher dielectric constant than the upper and lower dielectric layers, and a 2D PC is patterned in that layered structure. In such a structure light is confined in the vertical direction by traditional waveguiding due to dielectric index mismatch and in the lateral direction by the presence of a 2D PC. There are two routes that have been followed, one where the upper and lower dielectric layers are air and the other where the upper and lower dielectric layers have dielectric constants smaller than the central layer but much higher than 1. The first structure is called a self-supported membrane [31], while the second is referred to as a regular waveguide [32]. It is not yet resolved which structure has lower losses [31–35]. It is clear, however, that for optoelectronic applications the membrane-based PCs might not be easy to use. It is therefore of considerable importance to find out what type of structure has the lowest losses and the best efficiency of bends.

One of the most challenging applications of the miniature PCs is in the telecommunication regime, for the construction of fully photonic integrated circuits (PICs). Essential building blocks for the realization of PC-based PICs are PC waveguides, waveguide bends, and combiners, which are constructed by properly forming defects in the PCs. Light then is confined in the defects' path and is guided along this 1D channel, the PC waveguide, because the 3D PC prevents it from escaping into the bulk crystal. Simulations have predicted very exciting results that would have significant impact on applications, but the inclusion of defects in an already difficult-to-build 3D PC further complicates the fabrication requirements.

8.1.2 Left-Handed Materials or Negative-Index Materials

Recently, there have been many studies about metamaterials that have a negative refractive index n. These materials, called left-handed materials (LHMs),

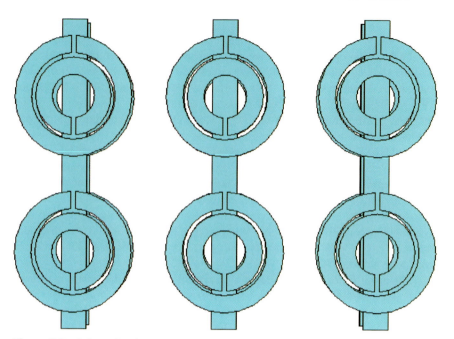

Figure 8.3 Schematic of a combination of SRRs and continuous wires. Such a combination is the most common way up to now to obtain LHMs.

theoretically discussed first by Veselago [36], have simultaneously negative electrical permittivity ε and magnetic permeability μ. A practical realization of such metamaterials, employing split-ring resonators (SRRs) and continuous wires (see Fig. 8.3), was first proposed by Pendry [37, 38] who also suggested that a slab of metamaterial with $n = -1$ could act as a perfect lens [39].

The first realization of some of Pendry's ideas was achieved by Smith et al. in 2000 [40], and since then various new samples (composed of SRRs and wires) have been prepared [41,42], all of which have been shown to exhibit a passband in which it was assumed that ε and μ are both negative. This assumption was based on measuring independently the transmission T of the wires alone and then the T of the SRRs alone. If the peak in the combined metamaterial composed of SRRs plus wires was in the stop bands for the SRRs alone (which is thought to correspond to negative μ) and for the wires alone (which is thought to correspond to negative ε), the peak was considered to be left handed (LH). Further support for this interpretation was provided by the demonstration that some of these materials exhibit negative refraction of EM waves [43].

Subsequent experiments [44] have reaffirmed the property of negative refraction, giving strong support to the interpretation that these metamaterials can be correctly described by negative permeability due to the SRRs and negative permittivity due to the wires. However, as was shown in [45], this is not always the case since the SRRs, in addition to their magnetic response, which was first described by Pendry [38], exhibit also a resonant electric response in frequencies not far from the magnetic response frequency. The electric response of the SRRs, which is demonstrated by closing their air gaps (destroying their resonant magnetic response), is identical to that of cut wires and it is added to the electric response (ε) of the wires. Consequently, the effective plasma frequency ω'_p of the combined system of wires and SRRs (or closed SRRs) is always lower than the plasma frequency of only the wires, ω_p. With this consideration and the analytical expressions for ε and μ which stem from it [45], one is able to explain and reproduce all of the low-frequency transmission T and reflection R characteristics of the SRRs plus wires based LHMs.

Moreover, considering the electric response of the SRRs and combining it with the fact that closing of the SRR gaps leaves this response unchanged, an easy criterion [45] to identify if an experimental transmission peak is LH or right handed (RH) is readily obtained: If closing the gaps of the SRRs in a given LHM structure removes from the T spectrum the peak close to the position of the SRR dip, this is strong evidence that the T peak is indeed LH. If the gap above the peak is removed, the peak is most likely RH. This criterion is very valuable in experimental studies, where one cannot easily obtain the effective ε and μ. The criterion is used experimentally and is found that some T peaks that were thought to be LH turned out to be RH [46].

There has also been a significant amount of numerical work [47–52] in which the complex transmission and reflection amplitudes for a finite length of metamaterial were calculated. Using these data a retrieval procedure was applied to obtain the effective permittivity ε and permeability μ under the assumption that the metamaterial can be treated as homogeneous. This procedure confirmed [53,54] that a medium composed of SRRs and wires could indeed be characterized by effective ε and μ with negative real parts over a finite frequency band and thus a refractive index also having a negative real part.

Recently, efforts have been made to fabricate LH structures at the terahertz frequency range. A magnetic response has been observed from SRRs at 2 THz [55], 6 THz [56] and 100 THz [57]. This response was experimentally observed through the electric excitation of the magnetic resonance (EEMR) [58], that is, the excitation of the magnetic resonance through the external electric field. This EEMR effect occurs for given orientations of the gaps of the SRR with respect to the external electric field, independently of the propagation direction, and makes possible the experimental characterization of small artificial magnetic structures [58] as it eliminates the necessity of in-plane incidence of an external EM field.

8.2 THEORETICAL AND NUMERICAL METHODS

To study theoretically and numerically the propagation of EM waves in PCs and LH materials, a variety of theoretical and numerical methods have been employed. These methods are used to calculate either the *band structure* of such materials (considering them as infinite) or the *transmission properties* of finite PC or LH slabs.

The most widely used methods, which can be applied to both PCs and LH materials, are the plane-wave (PW) method, the transfer matrix method (TMM), and the finite-difference time-domain (FDTD) method. In the following we will describe these methods and present their capabilities and main disadvantages.

The starting point in all these methods is Maxwell's equations in isotropic materials:

$$\nabla \cdot \mathbf{D} = 0 \qquad \nabla \cdot \mathbf{H} = 0 \tag{8.1}$$

$$\nabla \times \mathbf{E} = -\frac{\partial \mathbf{B}}{\partial t} \qquad \nabla \times \mathbf{H} = \frac{\partial \mathbf{D}}{\partial t} \tag{8.2}$$

where

$$\mathbf{D}(\mathbf{r}) = \varepsilon_0 \varepsilon(\mathbf{r}) \mathbf{E}(\mathbf{r}) \qquad \mathbf{B}(\mathbf{r}) = \mu_0 \mu(\mathbf{r}) \mathbf{H}(\mathbf{r}) \tag{8.3}$$

8.2.1   Plane-Wave Method

The PW method [59,60] is mainly used to calculate the dispersion relation, and hence the band structure of perfect PCs, considering them as infinite systems, or of PCs with isolated defects, in combination with a supercell scheme [2]. It is usually applied to lossless, dielectric, nonmagnetic media. The dispersion relation is calculated by transforming the problem into an eigenvalue problem, which gives the eigenfrequencies $\omega(\mathbf{k})$ for each wave vector $\mathbf{k}$.

Since the media under study are characterized by a spatially varying dielectric function $\varepsilon(\mathbf{r})$, Maxwell's equations (8.2), considering a harmonic time dependence of the form $e^{+j\omega t}$ and $\mu = 1$, are recast to their time-harmonic form

$$\nabla \times \mathbf{E} = -j\omega\mu_0 \mathbf{H} \qquad \nabla \times \mathbf{H} = j\omega\varepsilon(\mathbf{r})\varepsilon_0 \mathbf{E} \tag{8.4}$$

The two equations in (8.4) can be combined to generate equations containing only the magnetic or only the electric field:

$$\nabla \times (\varepsilon^{-1}(\mathbf{r})\nabla \times \mathbf{H}) = \frac{\omega^2}{c_0^2}\mathbf{H} \tag{8.5}$$

and

$$\nabla \times (\nabla \times \mathbf{E}) = \frac{\omega^2}{c_0^2}\varepsilon(\mathbf{r})\mathbf{E} \tag{8.6}$$

with $c_0^2 = 1/\mu_0\varepsilon_0$. The eigenfrequencies ω are obtained by the solution of either Eq. (8.5) or Eq. (8.6). Here we will proceed using Eq. (8.5).

At this point, we have to note that the vector nature of the wave equations (8.5) and (8.6) is of crucial importance. Early attempts [2] adopting the scalar wave approximation led to qualitatively wrong results, as unphysical longitudinal modes appeared in the solutions.

In the simplest and most common case, where $\varepsilon(\mathbf{r})$ is a real and frequency-independent periodic function of $\mathbf{r}$, the solution of the problem scales with the spatial period of $\varepsilon(\mathbf{r})$: For example, reducing the size of the structure by a factor of 2 will not change the spectrum of EM modes other than scaling all frequencies up by a factor of 2.

Because of the periodicity of the problem, we can translate the periodic function $\varepsilon^{-1}(\mathbf{r})$ of (8.5) into the reciprocal space, writing it as a sum of plane waves with their wave vectors being given by the reciprocal lattice vectors $\mathbf{G}$, that is,

$$\varepsilon^{-1}(\mathbf{r}) = \sum_{\mathbf{G}} \varepsilon^{-1}(\mathbf{G}) \, \exp(-j\mathbf{G} \cdot \mathbf{r}) \tag{8.7}$$

Moreover, we can make use of Bloch's theorem to expand the magnetic field of (8.5) in terms of Bloch waves:

$$\mathbf{H}(\mathbf{r}) = \sum_{\mathbf{K}} \mathbf{H}_{\mathbf{K}} \, \exp(-j\,\mathbf{K} \cdot \mathbf{r}) \tag{8.8}$$

where $\mathbf{K} = \mathbf{k} + \mathbf{G}$, $\mathbf{k}$ is a vector in the first Brillouin zone (BZ), $\mathbf{H}_{\mathbf{K}}$ are the Fourier components of the periodic amplitude of the $\mathbf{k}$ Bloch wave, and the summation is taken in fact over the vectors $\mathbf{G}$.

The substitution of Eqs. (8.7) and (8.8) into Eq. (8.5) leads to the eigenvalue problem

$$\sum_{\mathbf{K}'} \varepsilon_{\mathbf{K},\mathbf{K}'}^{-1} \mathbf{K} \times (\mathbf{K}' \times \mathbf{H}_{\mathbf{K}'}) = -\frac{\omega^2}{c_0^2} \mathbf{H}_{\mathbf{K}} \tag{8.9}$$

where $\varepsilon_{\mathbf{K},\mathbf{K}'}^{-1} = \varepsilon^{-1}(\mathbf{K} - \mathbf{K}') = \varepsilon^{-1}(\mathbf{G} - \mathbf{G}')$ [see (8.7)].

At this point we have to note that dielectric functions with sharp spatial discontinuities require an infinite number of plane waves in their Fourier expansion; this cannot be achieved in realistic calculations where the sums have to be truncated. To avoid this problem, we smear out the interfaces of the dielectric objects in the unit cell. For example, for modeling a cylinder of radius a with a dielectric function ε, we employ the smeared dielectric function

$$\varepsilon(\mathbf{r}) = 1 + \frac{\varepsilon - 1}{1 + \exp[(r - a)/w]} \tag{8.10}$$

where the width w of the interface is chosen as a small fraction of the radius a ($\approx 0.01a - 0.05a$). In practice, we incorporate the smearing and define the dielectric function $\varepsilon(\mathbf{r})$ over a grid in real space; then we compute its transform in our finite plane-wave basis set to obtain $\varepsilon(\mathbf{G} - \mathbf{G}')$; then the term $\varepsilon^{-1}(\mathbf{G} - \mathbf{G}')$ of (8.9) is obtained by the inversion of the $\varepsilon(\mathbf{G} - \mathbf{G}')$ matrix. This procedure yields much better convergence than the alternative method of determining $\varepsilon^{-1}(\mathbf{r})$ in real space and then performing a Fourier transform to obtain $\varepsilon^{-1}(\mathbf{G} - \mathbf{G}')$.

The transversality of the $\mathbf{H}$ field implies that $\mathbf{K} \cdot \mathbf{H}_{\mathbf{K}} = 0$; thus, $\mathbf{H}_{\mathbf{K}}$ can be written as

$$\mathbf{H}_{\mathbf{K}} = h_{\mathbf{K},1}\mathbf{e}_1 + h_{\mathbf{K},2}\mathbf{e}_2 \tag{8.11}$$

where the unit vectors $\mathbf{e}_1$ and $\mathbf{e}_2$ form with $\mathbf{K}$ an orthogonal triad $(\mathbf{e}_1, \mathbf{e}_2, \mathbf{K})$. The solution of (8.9) for the magnetic field (8.11) then reduces to the eigenvalue system

$$\sum_{\mathbf{K}'} \mathbf{M}_{\mathbf{K},\mathbf{K}'} h_{\mathbf{K}'} = \frac{\omega^2}{c_0^2} h_{\mathbf{K}} \tag{8.12}$$

which gives the allowed frequencies $\omega(\mathbf{k})$. In (8.12)

$$\mathbf{M}_{\mathbf{K},\mathbf{K}'} = |\mathbf{K}||\mathbf{K}'| \, \varepsilon_{\mathbf{K},\mathbf{K}'}^{-1} \begin{pmatrix} \mathbf{e}_2 \cdot \mathbf{e}_2' & -\mathbf{e}_2 \cdot \mathbf{e}_1' \\ -\mathbf{e}_1 \cdot \mathbf{e}_2' & \mathbf{e}_1 \cdot \mathbf{e}_1' \end{pmatrix}$$

$$h_{\mathbf{K}} = \begin{pmatrix} h_{\mathbf{K},1} \\ h_{\mathbf{K},2} \end{pmatrix} \qquad h_{\mathbf{K}'} = \begin{pmatrix} h_{\mathbf{K}',1} \\ h_{\mathbf{K}',2} \end{pmatrix} \qquad (8.13)$$

and the unit vectors $\mathbf{e}_1'$ and $\mathbf{e}_2'$ form an orthogonal triad with $\mathbf{K}'$.

As was mentioned earlier, in the above eigenfrequency calculation we used the wave equation for the magnetic field, Eq. (8.5), and not Eq. (8.6) for the electric field. In principle, we also could follow the same procedure for Eq. (8.6). However, the resulting eigenvalue problem would then be either an eigenvalue problem with a non-Hermitian matrix $\mathbf{M}$ or a generalized (instead of a simple) eigenvalue problem, which requires, in both cases, a more demanding computational procedure for its solution than the one associated with Eq. (8.12). Consequently, it is advantageous to use Eq. (8.5) rather than Eq. (8.6) to obtain the band structure of a PC.

In practice, the photonic band structure given by the frequencies $\omega(\mathbf{k})$ is computed over several sets of high-symmetry points in the Brillouin zone or on a grid in the Brillouin zone if the density of states is needed. A plane-wave convergence check is an essential step in that computation.

The first structure [2] considered by researchers with the plane-wave approach was a face-centered-cubic (fcc) structure composed of low-index dielectric spheres in a high-index dielectric (ε) background. There is no full bandgap (i.e., gap for all directions in the BZ and thus for all directions of propagation of the EM waves) between the second and third bands, while a sizable complete gap exists between the eighth and ninth bands (8–9 gap). The 8–9 position of the gap is a generic feature of the band structure of fcc PBG materials that is worth mentioning. The size (gap width over midgap frequency) of the full bandgap is about 8 percent for a refractive index contrast of 3.1.

A structure that has been investigated thoroughly, as was mentioned in the introduction, is the diamond structure [3, 6, 8, 9]. The diamond structure presents a full 3D PBG between the second and third bands (2–3 gap) for a wide range of filling ratios. This gap exists for (i) high-dielectric spheres on the sites of the diamond lattice, (ii) low-dielectric spheres on the diamond sites, and (iii) the diamond structure connected by dielectric rods. The best performing gap (29 percent) is reached for the diamond structure with 89 percent air spheres, that is, a multiply connected sparse structure. A similar large gap (30 percent) is also found for the diamond structure connected with dielectric rods with about a 30 percent dielectric filling fraction. These gap magnitudes have been obtained for a refractive index contrast of 3.6, appropriate for a GaAs background and air spheres.

The band structure and the corresponding density of states (DOS) for a diamond lattice is shown in Figure 8.4, for a system of dielectric spheres of $n = 3.6$ and a filling ratio 0.34. This filling ratio corresponds to the diamond close packing, where the 2–3 full bandgap ceases to exist. The system shown in Figure 8.4 was first studied by Ho, Chan, and Soukoulis [6] by the PW method. It was soon realized [2, 13] that for a 0.34 filling ratio with such high-index spheres

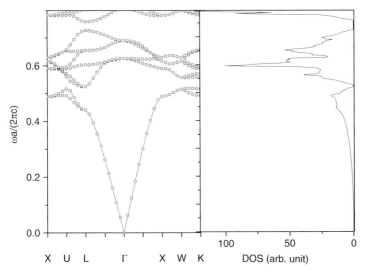

Figure 8.4 Band structure (left plot) and DOS calculation (right plot) for diamond lattice of dielectric spheres in air. The spheres' index of refraction is $n = 3.6$ and their filling ratio is 34 percent. Both the band structure and the DOS are calculated through the PW method, employing a very large number of PWs.

the PW method is very difficult to converge. The same conclusion was reached by Moroz [61]. When one is using the PW method, one has to exercise extreme care when handling dielectric spheres having a high index of refraction, that is, one needs a lot of terms in the Fourier transform to obtain accurate results. Even today's PW methods, especially the MIT photonic bands (MPBs) package [62], still need an extrapolation to infinitely many PWs to yield convergent results. Band structure calculations of PCs with high-index dielectric spheres might give better convergent results if the multiple scattering (or photonic-KKR) method [61] was used.

8.2.2 Transfer Matrix Method

While the method described in the previous section focuses on a particular wave vector [i.e., gives $\omega = \omega(\mathbf{k})$], there are complementary methods that focus on a single frequency [i.e., give $\mathbf{k} = \mathbf{k}(\omega)$], like the TMM. The TMM was first used to calculate the band structure of a PC by Pendry and MacKinnon [63].

The TMM is able to calculate the band structure of PC-based structures, including structures of complex or frequency-dependent dielectric functions (like metallic ones). This feature is not readily available through the PW method. The main power of the TMM, though, is its ability to calculate the stationary scattering properties, that is, the complex transmission (t) and reflection (r) amplitudes, of finite slabs of PCs and of LH materials. Such calculations are extremely useful in the interpretation of experimental measurements of the transmission and reflection data.

The calculation of the transmission and reflection coefficients for PW scattering from a slab of PC or LH metamaterial is performed by assuming that the slab, which is finite along the direction of the incoming incident wave (z direction here), is placed between two semi-infinite slabs of vacuum and employing the time-harmonic Maxwell equations (8.4). (By imposing periodic boundary conditions, the slab is considered infinite along the directions perpendicular to that of the propagation direction of the incident wave.)

The approach used with the TMM consists of the calculation of the EM field components at a specific z plane (e.g., after the slab) from the field components at a previous z plane (e.g., before the slab). For the implementation of this procedure Eqs. (8.4) are discretized, employing a rectangular grid on which the fields and the material parameters are defined. The result is a system of local difference equations:

$$
\begin{aligned}
E_x(i, l, k+1) = {} & E_x(i, l, k) + jc\omega\mu_0\mu(i, l, k)H_y(i, l, k) \\
& + \frac{jc}{a\omega\varepsilon_0\varepsilon(i, l, k)}\{a^{-1}[H_y(i-1, l, k) - H_y(i, l, k)] \\
& - b^{-1}[H_x(i, l-1, k) - H_x(i, l, k)]\} \\
& - \frac{jc}{a\omega\varepsilon_0\varepsilon(i+1, l, k)}\{a^{-1}[H_y(i, l, k) - H_y(i+1, l, k)] \\
& - b^{-1}[H_x(i+1, l-1, k) - H_x(i+1, l, k)]\} \quad (8.14)
\end{aligned}
$$

$$
\begin{aligned}
E_y(i, l, k+1) = {} & E_y(i, l, k) - jc\omega\mu_0\mu(i, l, k)H_x(i, l, k) \\
& + \frac{jc}{b\omega\varepsilon_0\varepsilon(i, l, k)}\{a^{-1}[H_y(i-1, l, k) - H_y(i, l, k)] \\
& - b^{-1}[H_x(i, l-1, k) - H_x(i, l, k)]\} \\
& - \frac{jc}{b\omega\varepsilon_0\varepsilon(i, l+1, k)}\{a^{-1}[H_y(i-1, l+1, k) - H_y(i, l+1, k)] \\
& - b^{-1}[H_x(i, l, k) - H_x(i, l+1, k)]\} \quad (8.15)
\end{aligned}
$$

$$
\begin{aligned}
H_x(i, l, k+1) = {} & H_x(i, l, k) - jc\omega\varepsilon_0\varepsilon(i, l, k+1)E_y(i, l, k+1) \\
& + \frac{jc}{a\omega\mu_0\mu(i-1, l, k+1)}\{a^{-1}[E_y(i, l, k+1) \\
& - E_y(i-1, l, k+1)] - b^{-1}[E_x(i-1, l+1, k+1) \\
& - E_x(i-1, l, k+1)]\} - \frac{jc}{a\omega\mu_0\mu(i, l, k+1)} \\
& \times \{a^{-1}[E_y(i+1, l, k+1) - E_y(i, l, k+1)] \\
& - b^{-1}[E_x(i, l+1, k+1) - E_x(i, l, k+1)]\} \quad (8.16)
\end{aligned}
$$

$$
\begin{aligned}
H_y(i, l, k+1) = {} & H_y(i, l, k) + jc\omega\varepsilon_0\varepsilon(i, l, k+1)E_x(i, l, k+1) \\
& + \frac{jc}{b\omega\mu_0\mu(i, l-1, k+1)}\{a^{-1}[E_y(i+1, l-1, k+1) \\
& - E_y(i, l-1, k+1)] - b^{-1}[E_x(i, l, k+1) - E_x(i, l-1, k+1)]\} \\
& - \frac{jc}{b\omega\mu_0\mu(i, l, k+1)}\{a^{-1}[E_y(i+1, l, k+1) - E_y(i, l, k+1)] \\
& - b^{-1}[E_x(i, l+1, k+1) - E_x(i, l, k+1)]\} \quad (8.17)
\end{aligned}
$$

In the above equations $\varepsilon(i, l, k)$ and $\mu(i, l, k)$ are the relative electrical permittivity and magnetic permeability at the grid cell (i, l, k) and a, b, c are the dimensions of each grid cell along the x, y, z directions, respectively. We have to mention that the components E_z, H_z are eliminated from further consideration, due to the transversality of the fields, and that the field components $E_x(i, l, k)$, $E_y(i, l, k)$, $H_x(i, l, k)$, $H_y(i, l, k)$ are defined at different points of their associated grid cell (i, l, k) (they are mutually displaced by a half grid cell). Special attention has to be taken with the material discretization because the symmetries of the structure also have to be maintained in the discretized system.

Equations (8.14) to (8.17) connect the field components at the $k + 1$ plane with those at the k plane. After rearrangement of terms they can take the form

$$\begin{pmatrix} \mathbf{E}(k+1) \\ \mathbf{H}(k+1) \end{pmatrix} = \mathbf{T} \begin{pmatrix} \mathbf{E}(k) \\ \mathbf{H}(k) \end{pmatrix} \tag{8.18}$$

The matrix $\mathbf{T}$ is the transfer matrix, which allows one to compute the whole solution from a previously known z slice. In the vacuum the matrix $\mathbf{T}$ can be diagonalized exactly; its left and right eigenvectors define the PW basis for the scattering problem. By propagating the vacuum basis vectors through the sample and by subsequent decomposition of the results with respect to the vacuum basis again, one obtains the $\mathbf{T}$ matrix of the slab. With $\mathbf{T}$ known, the scattering amplitudes r and t can be obtained by using the relation between $\mathbf{T}$ and the scattering matrix $\mathbf{S}$ in this basis:

$$\mathbf{T} = \begin{pmatrix} t_+ - r_+ t_-^{-1} r_- & r_+ t_-^{-1} \\ -t_-^{-1} r_- & t_-^{-1} \end{pmatrix} \qquad \mathbf{S} = \begin{pmatrix} t_+ & r_+ \\ r_- & t_- \end{pmatrix} \tag{8.19}$$

($\mathbf{S}$ defines the transmission and reflection amplitudes for waves incident from the left or right of a slab, $t_{\mp}$ and $r_{\pm}$.) For economy of computer time and memory, the transfer matrices of the sample slices can be applied consecutively and algorithmically, that is, not as matrix multiplications. Intermediate renormalization steps account for the exponential growth of some modes inside the sample and keep the simulation stable [48]. Implementations of the TMM can be made to be quite efficient because they rely mainly on linear algebra operations such as matrix factorization and successive inversion.

For the calculation of the band structure $\mathbf{k}(\omega)$ of a system, one has to compute the eigenvalues of $\mathbf{T}$ while also applying periodic boundary conditions along the direction of propagation. Details about this procedure can be found in [63].

The TMM method has been extensively applied to band structure calculations of PCs containing absorptive and frequency dispersive (e.g., metallic) materials. It has been applied also to the simulation of the scattering properties of finite PCs, PCs with defects, PCs with complex and frequency-dependent dielectric functions [64], and LHMs composed of SRRs of various shapes and metallic wires [45, 50]. In all these cases the agreement between the theoretical calculations and the experimental results, where available, has been very good.

In Figure 8.5 we show an example of the application of the TMM method to the calculation of the transmission coefficient through a slab of a metamaterial composed of rectangular SRRs printed on a dielectric board and of closed SRRs

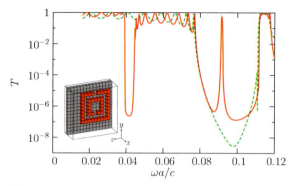

Figure 8.5   TMM calculation for transmission coefficient versus frequency for system composed of rectangular SRRs (solid red line) and a system of closed SRRs (SRRs with no gaps) (green dashed line). The system length along the propagation direction is 10 unit cells. The inset shows the geometry of the unit cell (1 SRR attached on a dielectric board). The relative permittivity for the metal is taken to be $\varepsilon_m = (-3 + j5.88)10^5$ and for the dielectric board $\varepsilon_b = 12.3$. All relative permeabilities are 1. The unit-cell size is $6a \times 14a \times 14a$, where a is the discretization length and c the light velocity in air. Figure from [45], copyright © 2004 by the American Physical Society.

[45]. As has been already mentioned, when the gaps of the SRRs are closed, their magnetic response is switched off while their electric response remains unchanged. This can be seen clearly in Figure 8.5; the spectrum of the closed SRRs is almost identical to that of the SRRs, with the exception of the dip at $\omega a/c \approx 0.04$ (where $\mu < 0$) and the peak at $\omega a/c \approx 0.095$ (where $\mu < 0$, $\varepsilon < 0$, i.e., a LH peak; the $\mu < 0$ here is due to the presence of the inner ring, which also exhibits a magnetic response [65]).

8.2.3   Finite-Difference Time-Domain Method

Like the TMM, the FDTD method can be also used to calculate both band structure and scattering properties of PCs and LHM, and it also involves discretization of the Maxwell equations. The difference here is that, while the TMM is employed for steady-state solutions, the FDTD method is used for general time-dependent solutions. The steady-state solutions then are obtained through fast Fourier transforming the time-domain results. This permits the study of both the transient and the steady-state response of a system. An additional advantage is the possibility of obtaining a broadband steady-state response with just a single calculation, as the excitation signal can be a pulse rather than a monochromatic wave.

Since FDTD is a time-domain method, the starting point for its implementation is the time-dependent Maxwell equations, specifically Eqs. (8.2). The curl equations (8.2) are discretized using a rectangular grid (which stores the field components and the material properties ε and μ) and central differences for the space and time derivatives. The procedure results in a set of finite-difference equations, which updates the field components in time. The equations for the update of E_x and H_x read as follows:

$$E_x^{n+1}(i, l, k) = E_x^n(i, l, k)$$
$$+ \frac{\Delta t}{\varepsilon_0 \varepsilon(i, l, k)} \left[\frac{H_z^{n+1/2}(i, l + 1/2, k) - H_z^{n+1/2}(i, l - 1/2, k)}{b} \right.$$
$$\left. - \frac{H_y^{n+1/2}(i, l, k + 1/2) - H_y^{n+1/2}(i, l, k - 1/2)}{c} \right] \quad (8.20)$$

$$H_x^{n+1/2}(i, l, k) = H_x^{n-1/2}(i, l, k)$$
$$+ \frac{\Delta t}{\mu_0 \mu(i, l, k)} \left[\frac{E_y^n(i, l, k + 1/2) - E_y^n(i, l, k - 1/2)}{c} \right.$$
$$\left. - \frac{E_z^n(i, l + 1/2, k) - E_z^n(i, l - 1/2, k)}{b} \right] \quad (8.21)$$

The corresponding equations for E_y, E_z, H_y, H_z are similar to the above. The FDTD equations for various types of materials, together with computational procedure, source incorporation procedure, stability criteria, and so on, are presented in a very clear and complete way in [66]. Here we just review some of the main points of the FDTD calculation procedure as it is applied to PCs and LHMs, to familiarize the reader with the method and to facilitate the comparison with the other methods. In Eqs. (8.20) and (8.21) $E_x^n(i, l, k)$ and $H_x^n(i, l, k)$ are the x components of the electric and magnetic field in the $(i, l, k) \equiv (ia, lb, kc)$ grid cell at the n time step (where $t = t_n = n\Delta t$), and so on; a, b, c are, respectively, the dimensions of the grid cell along the x, y, z directions and Δt is the time step.

Here, as in the TMM, the different EM field components are located at different points of their associated grid cell, following the well known Yee scheme [66]: the **E**-field components, which are calculated at times $n\Delta t$, are located at the face-centered points of the grid cell, while the **H**-field components, which are calculated at times $(n + 1/2)\Delta t$, are located at the edges of the grid cell (every **E** component is surrounded by four circulating **H** components and vice versa). This scheme results in second-order accuracy and a complete fulfillment of all four of Maxwell's equations, although only two equations [Eqs. (8.2)] are directly employed.

Using the FDTD equations [e.g., (8.20) and (8.21)], one can obtain $\mathbf{E}(t)$ and $\mathbf{H}(t)$ at any point within a finite slab and, through fast Fourier transforming, $\mathbf{E}(\omega)$ and $\mathbf{H}(\omega)$. The transmission (reflection) coefficient, T (R), is then calculated by dividing the Fourier transform of the transmitted (reflected) Poynting vector $\mathbf{S} = \text{Re}[\mathbf{E}(\omega) \times \mathbf{H}^*(\omega)]/2$ by the incident Poynting vector. Note that what is calculated is the power coefficient $T = |t|^2$ ($R = |r|^2$), a real quantity, and not the complex transmission (reflection) amplitude.

The transmission calculation procedure usually consists of sending a pulse (e.g., a Gaussian) and then obtaining the transmitted frequency-domain fields[1] $\mathbf{E}(\omega)$, $\mathbf{H}(\omega)$ and thus T. The slab along the directions perpendicular to the

---

[1] Usually the transmitted fields at different detection points after the sample are detected, and an average of the resulting pointing vectors is taken.

direction of propagation of the incident wave can be considered as either infinite or finite. The first case is achieved by using periodic boundary conditions at the associated boundaries and the second by using absorbing boundary conditions (i.e., the incident wave at the boundaries is absorbed by them). Absorbing boundary conditions are also used to close the computational cell in the propagation direction. The most efficient absorbing boundary conditions that have been applied to date are the perfectly matched layer (PML) conditions [66], while Liao's conditions [66, 67] are also efficient and widely used.[2]

Equations (8.20) and (8.21) and the corresponding ones for the other field components describe dielectric media with no losses. The FDTD study of *dielectric media with losses* ($\varepsilon = \varepsilon_r - j\varepsilon_i$) is achieved usually by introducing a conductivity $\sigma = \omega\varepsilon_0\varepsilon_i$ through an external current ($\mathbf{J} = \sigma\mathbf{E}$) added to the first of the equations given in Eqs. (8.2). This leads to a modification of the terms appearing in the standard finite-difference equations but leaves the computational procedure unaltered (see [66]).

To model *dispersive materials*, such as metals, as is required, for example, in the study of metallic PCs or of LH materials, one has to introduce a specific dispersion model [e.g., Drude model, $\varepsilon(\omega) = 1 - \omega_p^2/(\omega^2 - j\omega\gamma)$] and translate the equation $\mathbf{D} = \varepsilon_0\varepsilon(\omega)\mathbf{E}$ [see Eqs. (8.3)] into the time domain [68]. The result is an additional FDTD equation on the top of the standard FDTD equations. (Note that the relation $\mathbf{D} = \varepsilon_0\varepsilon\mathbf{E}$ does not hold in the time domain when dispersive materials are involved; thus $\mathbf{D}$ and $\mathbf{E}$ have to be calculated independently within the FDTD procedure.) A similar procedure is employed also for magnetic materials, $\mu = \mu(\omega)$ [68].

The FDTD method [66, 69] is an excellent tool for the study of transmission through finite slabs, as it can model almost arbitrary material combinations and microstructure configurations. It has been utilized in many systems, containing dielectric or metallic components [70–74] as well as in materials with nonlinear dielectric properties [75–77]. Methods to transform the output near fields to radiating far fields have also been employed [66]; this is particularly necessary for antenna problems, where far-field radiation patterns are desired.

The FDTD method, as was mentioned earlier, can also be utilized for band structure calculations [73, 78, 79]. In this case the computational domain is usually a single unit cell of the periodic structure, with periodic boundary conditions in all its boundaries. An excitation containing a wide frequency range is used to excite the allowed modes for each wave vector. These modes appear as spikes in the Fourier transform of the time-domain fields.

An example of the application and the potential of the FDTD method is shown in Figure 8.6. Figure 8.6(*a*) shows the magnetic field (at a specific time point) of a Gaussian (in-space) beam which undergoes reflection and refraction at the interface between air and a hexagonal PC constructed with dielectric rods in air at a frequency belonging to the convex photonic band, that is, the band in

---

[2] Liao boundary conditions are based on extrapolation of the fields in space and time by use of a Newton backward-difference polynomial. They are introduced in [67]. Liao boundary conditions are also described in detail in [66].

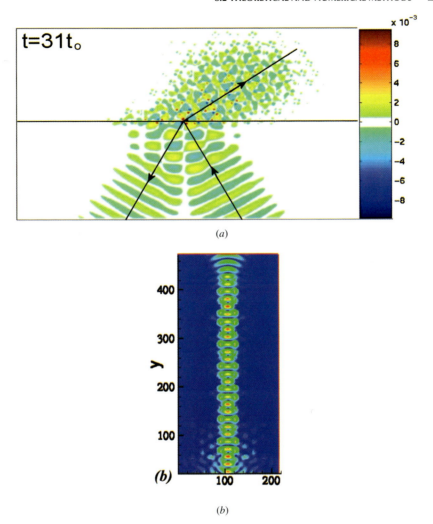

(*a*)

(*b*)

Figure 8.6 (*a*) FDTD picture showing magnetic field of transverse electric (TE) Gaussian (in-space) beam which undergoes reflection and refraction at interface between air and PC (hexagonal lattice of dielectric rods with $\varepsilon = 12.96$ and radius over lattice constant 0.35) for $t = 31t_0$. The frequency of the beam belongs to a "negative" (convex) band of the PC ($a/\lambda = 0.58$, a being the lattice constant, λ the free-space wavelength), close to the Γ inverse-lattice point. Here $2t_0$ is the time difference between the outer and the inner rays to reach the interface; $t_0 \approx 1.5T$, where T is the period $2\pi/\omega$ of the wave. Figure from [80] copyright © 2003 by the American Physical Society. (*b*) Electric field of TE wave which is guided through PC waveguide. The PC waveguide is formed by removing one row of holes along the ΓK direction from a hexagonal 2D PC (made of cylindrical holes, with radius over lattice constant 0.2463, patterned in a host of $\varepsilon = 10.56$). The a/λ dimensionless frequency of the guided wave is 0.24. The units in the axes are grid cells of the FDTD discretization scheme.

which the group velocity in the PC is opposite to the **k** vector [80]. In such a frequency band the PC should behave as a LH system; indeed the refracted beam in Figure 8.6(*a*), which undergoes negative refraction, unambiguously proves the validity of this consideration. Figure 8.6(*b*) demonstrates the guiding of an EM wave through a straight PC waveguide formed by removing one row of holes from a hexagonal 2D photonic crystal.

8.3   COMPARISON OF DIFFERENT NUMERICAL TECHNIQUES

As we mentioned in the previous section, the PW method is usually used to treat infinite periodic systems, giving their dispersion properties. Although it can be applied only in systems with nondispersive components (frequency-independent ε and μ), the PW method is the fastest and the easiest to apply. It can give within a single calculation all the spectrum ω for a given wave vector. Its main disadvantages are its inability to treat systems with dispersive components and finite media and its relative difficulty to treat systems with defects. In the last case a supercell scheme has to be employed, which, in many cases, leads to calculations that are very computer time and memory consuming.

The TMM, on the other hand, is able to calculate the band structure of systems with dispersive components but it is less easy to apply than the PW method. It is usually used for the calculation and analysis of the stationary scattering properties of finite-in-length samples. Among the most important advantages of the TMM is its ability to treat samples with almost arbitrary internal structure and arbitrary material combinations (e.g., metallic, lossy), giving the complex transmission and reflection amplitudes, that is, magnitude, phase, and polarization information. The simultaneous amplitude and the phase knowledge can be used in the inversion of the transmission and reflection data to obtain the effective material parameters (ε and μ) for the systems under study (provided that the effective medium approach is valid).

Among the drawbacks of the TMM is the necessity of the discretization of the unit cell, which introduces some numerical artifacts and some constraints into the shape and size of the components inside the unit cell. For example, to simulate "tiny" components, as is usually required in the study of LHMs, one needs very fine discretization, practically possible only within a nonuniform discretization scheme. Otherwise large calculation times and large memory requirements are unavoidable.

The FDTD method, like the TMM, can also model finite slabs with almost arbitrary internal structure and material combinations. Its main advantages compared to the TMM is that it can give the transmission properties over a wide spectral range with just a single calculation. It also can give time-domain pictures of the fields and the currents over the entire computational domain. Moreover, it can treat defects with no additional computational complications.

Concerning the disadvantages of the FDTD method, part of them stem from the inherent discretizations required, and they were discussed above in connection with the TMM. In the case of dispersive materials, though, one encounters additional problems, coming from the time scale that the dispersion model introduces, as the time step of the method (Δt) cannot be much larger than the characteristic time scale of the dispersion model. This constraint imposes restrictions in the frequency regimes that can be studied and, through them, in the size of the structures involved.

Concerning the application of the FDTD method in band structure calculations, the level of difficulty is similar to that of the TMM. The advantages and disadvantages compared to the TMM are essentially those mentioned in the two previous paragraphs in connection with the calculations of the scattering parameters.

Apart from the three methods that we have described and analyzed in this chapter, additional methods have been applied to the study of PCs and LHMs, although less extensively. Some of those are variations of the PW, TMM, and FDTD methods. Among the existing methods, one worth mentioning is the multiple-scattering (MS) or photonic-KKR method [61, 81], which is a vectorial extension of the well-known electronic band structure calculation method KKR, and its modification known as the layered-MS method [82]. They can both give band structure and transmission properties of PCs and LHMs, treating accurately the dispersive components, defects, as well as high-index contrast systems. Their main disadvantages are the heavy formalism, the difficulties in the computational procedure, and the large calculation times.

8.4 CONCLUSIONS

We presented a brief historical review of the theoretical and experimental efforts in designing and fabricating PCs and LHMs, starting from the first successful designs and arriving at the latest developments. The latter included PBGs in the infrared or optical regime and materials with negative magnetic permeability at around 100 THz. We also presented the theoretical and experimental challenges and the problems of the field as well as its current status and several current research directions.

We also reviewed the three most successful and widely used numerical techniques employed in the studies of PCs and LHMs. These are the PW method, the TMM, and the FDTD method. We presented the key ideas and equations of each method and discussed their capabilities and disadvantages. Finally we presented a few representative results from each method. We are excited about the future applications of PCs and LHMs and the prospects for using these computational techniques to help design, fabricate, and test these PCs and LHMs.

ACKNOWLEDGMENTS

It is a pleasure to thank our colleagues E. N. Economou, P. Markos, T. Koschny, N. Katsarakis, M. Sigalas, E. Ozbay, S. Foteinopoulou, E. Lidorikis, and D. R. Smith for their collaboration and insights. Financial support by the EU projects DALHM, Metamorphose and Phoremost, and by NATO CLG 981471 and DARPA (Contract No. MDA972-01-2-0016) are acknowledged. This work was partially supported by Ames Laboratory (Contract. No. W-7405-Eng-82). Financial support by the Greek Ministry of Education, through the PYTHAGO-RAS project, is also acknowledged.

REFERENCES

1. C. M. Soukoulis, (Ed.), *Photonic Crystals and Light Localization in the 21st Century*, NATO ASI, Series C, Vol. 563, 2001.
2. C. M. Soukoulis, Ed., *Photonic Band Gaps and Localization*, New York: Plenum, 1993; *Photonic Band Gap Materials*, Kluwer, Dordrecht, The Netherlands, 1996.
3. J. D. Joannopoulos, R. D. Mead, and J. N. Winn, *Photonic Crystals*, Princeton University Press, Princeton, NJ, 1995.
4. E. Yablonovitch, "Inhibited spontaneous emission in solid-state physics and electronics," *Phys. Rev. Lett.*, Vol. 58, pp. 2059–2062, May 1987.
5. S. John, "Strong localization of photons in certain disordered dielectric superlattices," *Phys. Rev. Lett.*, Vol. 58, pp. 2486–2489, June 1987; S. John, *Physics Today*, Vol. 32, pp. 33–38, May 1991.
6. K. M. Ho, C. T. Chan, and C. M. Soukoulis, "Existence of a photonic gap in periodic dielectric structures," *Phys. Rev. Lett.*, Vol. 65, pp. 3152–3155, Dec. 1990.
7. E. Yablonovitch, T. J. Gmitter, and K. M. Leung, "Photonic band structure:The face-centered-cubic case employing nonspherical atoms," *Phys. Rev. Lett.*, Vol. 67, pp. 2295–2298, Oct. 1991.
8. C. T. Chan, K. M. Ho, and C. M. Soukoulis, "Photonic band gaps in experimentally realizable periodic dielectric structures," *Europhys. Lett.*, Vol. 16, no. 6, pp. 563–565, 1991.
9. M. Maldovan and E. L. Thomas, "Diamond-structured photonic crystals," *Nature Materials*, Vol. 3, pp. 593–600, Sept. 2004.
10. K. M. Ho, C. T. Chan, C. M. Soukoulis, R. Biswas, and M. Sigalas, "Photonic band gaps in three dimensions: New layer-by-layer periodic structures," *Solid State Comm.*, Vol. 89, pp. 413–416, Feb. 1994.
11. E. Ozbay, A. Abeyta, G. Tuttle, M. C. Tringides, R. Biswas, M. Sigalas, C. M. Soukoulis, C. T. Chan, and K. M. Ho, "Measurement of a three-dimensional photonic band gap in a crystal structure made of dielectric rods," *Phys. Rev. B*, Vol. 50, pp. 1945–1948, July 1994.
12. C. T. Chan, S. Datta, K. M. Ho, and C. M. Soukoulis, "A7 structure: A family of photonic crystals," *Phys. Rev. B*, Vol. 50, pp. 1988–1991, July 1994.
13. H. S. Sozuer, J. W. Haus, and R. Inguva, "Photonic bands: Convergence problems with the plane-wave method," *Phys. Rev. B*, Vol. 45, pp. 13962–13972, June 1992.
14. T. Suzuki and P. Yu, "Dispersion relation at point L in the photonic band structure of the face-centered cubic lattice with active or conductive dielectric media," *J. Opt. Soc. of Am. B*, Vol. 12, pp. 570–582, Apr. 1995.
15. E. Yablonovitch, and T. J. Gmitter, "Photonic band structure: The face-centered-cubic case," *Phys. Rev. Lett.*, Vol. 63, pp. 1950–1953, Oct. 1989.
16. C. Cheng and A. Scherer, "Fabrication of photonic band-gap crystals," *J. Vac. Sci. Tech. B*, Vol. 13, pp. 2696–2700, Nov.-Dec. 1995; C. Cheng, V. Arbet-Engels, A. Scherer and E. Yablonovitch, "Nanofabricated three dimensional photonic crystals operating at optical wavelengths," *Physica Scripta*, Vol. 68, pp. 17–19, 1996.
17. G. Feiertag, W. W. Ehrfeld, H. Freimuth, G. Kiriakidis, H. Lehr, T. Pedersen, M. Schmidt, C. M. Soukoulis, and R. Weiel, "Fabrication three-dimensional photonic bandgap material by deep x-ray lithography," in *Photonic Band Gap Materials* ed. by C. M. Soukoulis, Kluwer, Dordrecht, The Netherlands, 1996,

pp. 63–69; G. Feiertag, W. W. Ehrfeld, H. Freimuth, H. Kolle, H. Lehr, M. Schmidt, M. M. Sigalas, C. M. Soukoulis, G. Kiriakidis and T. Pedersen, "Fabrication of photonic crystals by deep x-ray lithography," *Appl. Phys. Lett.*, Vol. 71, pp. 1441–1443, Sept. 1997.

18. M. C. Wanke, O. Lehmann, K. Muller, Q. Wen, and M. Stuke, "Laser rapid prototyping of photonic band-gap microstructures," *Science*, Vol. 275, pp. 1284–1286, Feb. 1997.

19. S. Y. Lin, J. G. Fleming, D. L. Hetherington, B. K. Smith, R. Biswas, K. M. Ho, M. M. Sigalas, W. Zubrzycki, S. R. Kurtz, Jim Buret, "A three-dimensional photonic crystal operating at infrared wavelengths," *Nature*, Vol. 394, pp. 251–254, July 1998; J. G. Fleming and S. Y. Lin, "Three-dimensional photonic crystal with a stop band from 1.35 to 1.95 micrometers," *Opt. Lett.*, Vol. 24, pp. 49–51, Jan. 1999.

20. N. Yamamoto, S. Noda and A. Chutinan, "Development of one period of a three-dimensional photonic crystal in the 5-0 micrometers wavelength region by wafer fusion and laser beam diffraction pattern observation techniques," *Jpn. J. Appl. Phys.*, Vol. 37, pp. L1052–L1054, September 1998; S. Noda, N. Yamamoto, H. Kobayashi, M. Okano, and K. Tomoda, "Optical properties of three-dimensional photonic crystals based on III-V semiconductors at infrared to near-infrared wavelengths," *Appl. Phys. Lett.*, Vol. 75, pp. 905-907, Aug. 1999; S. Noda, K. Tomoda, N. Yamamoto, and A. Chutinan, "Full three-dimensional photonic crystals at near-infrared wavelengths," *Science*, Vol. 289, pp. 604–606, July 2000.

21. J. E. G. J. Wijnhoven and W. L. Vos, "Preparation of photonic crystals made of air spheres in titania," *Science*, 281, pp. 802–804, Aug. 1998.

22. A. Imhof and D. J. Pine, "Ordered macroporous materials by emulsion templating," *Nature*, Vol. 389, pp. 948–951, Oct. 1997.

23. B. T. Holland, C. F. Blanford, and A. Stein, "Synthesis of macroporous minerals with highly ordered three-dimensional arrays of spheroidal voids," *Science*, Vol. 281, pp. 538–540, July 1998.

24. A. A. Zakhidov, R. H. Baughman, Z. Iqbal, C. Cui, I. Khayrullin, S. O. Dantas, J. Marti, and V. G. Ralchenko, "Carbon structures with three-dimensional periodicity at optical wavelengths," *Science*, Vol. 282, pp. 897–901, Oct. 1998.

25. G. Subramania, K. Constant, R. Biswas, M. M. Sigalas, and K.-M. Ho, "Optical photonic crystals fabricated from colloidal systems," *Appl. Phys. Lett.*, Vol. 74, pp. 3933–3935, June 1999; G. Subramania, K. Constant, R. Biswas, M. M. Sigalas, K.-M. Ho, "Inverse face-centered cubic thin film photonic crystals," *Adv. Mater.* Vol. 13, 443–446, Mar. 2001.

26. Y. Vlasov, X. Z. Bo, J. C. Sturm and D. J. Norris, "On-chip natural assembly of silicon photonic bandgap crystals," *Nature*, Vol. 414, pp. 289–293, Nov. 2001.

27. A. Blanco, E. Chomski, S. Grabtchak, M. Ibisate, S. John, S. W. Leonard, C. Lopez, F. Meseguer, H. Miguez, J. P. Mondia, G. A. Ozin, O. Toader and H. M. van Driel, "Large-scale synthesis of a silicon photonic crystal with a complete three-dimensional bandgap near 1.5 micrometres," *Nature*, Vol. 405, pp. 437–440, May 2000.

28. O. D. Velev and E. Kaler, "Structured porous materials via colloidal crystal templating: From inorganic oxides to metals," *Adv. Mater.*, Vol. 12, pp. 531–534, Apr. 2000, and references therein.

29. M. Campbell, D. N. Sharp, M. T. Harrison, R. G. Denning, A. J. Turberfield, "Fabrication of photonic crystals for the visible spectrum by holographic lithography," *Nature*, Vol. 404, pp. 53–56, Mar. 2000.

30. M. Deubel, G. von Freymann, M. Wegener, S. Pereira, K. Busch, C. M. Soukoulis, "Direct laser writing of three dimensional photonic-crystal templates for telecommunications," *Nature Materials*, Vol. 3, pp. 444–447, July 2004.

31. O. Painter, J. Vuckovic, and A. Scherer, "Defect modes of a two-dimensional photonic crystal in an optically thin dielectric slab," *J. Opt. Soc. Am. B*, Vol. 16, pp. 275–285, Feb. 1999; B. D'Urso, O. Painter, J. O'Brien, T. Tombrello, A. Yariv, A. Scherer, "Modal reflectivity in finite-depth two-dimensional photonic-crystal microcavities," *J. Opt. Soc. Am. B*, Vol. 15, pp. 1155–1159, Mar. 1998.

32. H. Benisty, C. Weisbuch, D. Labilloy, M. Rattier, C. J. M. Smith, T. F. Krauss, Richard M. De La Rue, R. Houdre, U. Oesterle, C. Jouanin, D. Cassagne, "Optical and confinement properties of two-dimensional photonic crystals," *J. Light Technol.*, Vol. 17, pp. 2063–2077, Nov. 1999.

33. M. Kafesaki, M. Agio, and C. M. Soukoulis, "Waveguides in finite-height two-dimensional photonic crystals," *J. Opt. Soc. Am. B*, Vol. 19, pp. 2232–2240, Sept. 2002; "Losses and transmission in two-dimensional slab photonic crystals," *J. Appl. Phys.*, Vol. 96, pp. 4033–4038, Oct. 2004.

34. W. Bogaerts, P. Bienstman, D. Taillaert, R. Baets, D. De Zutter, "Out-of-plane scattering in photonic crystal slabs," *Photonics Technol. Lett.*, Vol. 13, pp. 565–567, June 2001.

35. E. Chow, S. Y. Lin, S. G. Johnson, P. R. Villeneuve, J. D. Joannopoulos, J. R. Wendt, G. A. Vawter, W. Zubrzycki, H. Hou, A. Alleman, "Three-dimensional control of light in a two-dimensional photonic crystal slab," *Nature*, Vol. 407, pp. 983–986, Oct. 2000; E. Chow, S. Y. Lin, J. R. Wendt, S. G. Johnson, J. D. Joannopoulos, "Quantitative analysis of bending efficiency in photonic-crystal waveguide bends at 1.55 micrometer wavelengths," *Opt. Lett.*, Vol. 26, pp. 286–288, Mar. 2001.

36. V. G. Veselago, "The electrodynamics of substances with simultaneously negative values of ε and μ," *Sov. Phys. Usp.*, Vol. 10, pp. 509–514, Jan.–Feb. 1968 [*Usp. Fiz. Nauk*, Vol. 92, pp. 517, 1967].

37. J. B. Pendry, A. J. Holden, W. J. Stewart, and I. Youngs, "Extremely low frequency plasmons in metallic mesostructures," *Phys. Rev. Lett.*, Vol. 76, pp. 4773–4776, June 1996; "Low frequency plasmons in thin-wire structures," *J. Phys.: Condens. Matt.*, Vol. 10, pp. 4785–4809, June 1998.

38. J. B. Pendry, A. J. Holden, D. J. Robbins, and W. J. Stewart, "Magnetism from conductors and enhanced nonlinear phenomena," *IEEE Trans. Microwave Theory Tech.*, Vol. 47, pp. 2075–2084, Nov. 1999.

39. J. B. Pendry, "Negative refraction makes a perfect lens," *Phys. Rev. Lett.*, Vol. 85, pp. 3966–3969, Oct. 2000.

40. D. R. Smith, W. J. Padilla, D. C. Vier, S. C. Nemat-Nasser, and S. Schultz, "Composite medium with simultaneously negative permeability and permittivity," *Phys. Rev. Lett.*, Vol. 84, pp. 4184–4187, May 2002.

41. A. Shelby, D. R. Smith, S. C. Nemat-Nasser, and S. Schultz, "Microwave transmission through a two-dimensional, isotropic, left-handed metamaterial," *Appl. Phys. Lett.*, Vol. 78, pp. 489-491, Jan. 2001; M. Bayindir, K. Aydin, E. Ozbay, P. Marko›s, and C. M. Soukoulis, "Transmission properties of composite metamaterials in free space," *Appl. Phys. Lett.*, Vol. 81, pp. 120–122, July 2002.

42. K. Li, S. J. McLean, R. B. Greegor, C. G. Parazzoli, and M. Tanielian, "Free-space focused-beam characterization of left-handed materials," *Appl. Phys. Lett.*, Vol. 82, pp. 2535–2537, Apr. 2003.

43. R. A. Shelby, D. R. Smith, and S. Schultz, "Experimental verification of a negative index of refraction," *Science*, Vol. 292, pp. 77–79, Apr. 2001.

44. C. G. Parazzoli, R. Greegor, K. Li, B. E. C. Koltenbach, and M. Tanielian, "Experimental verification and simulation of negative index of refraction using Snell's Law," *Phys. Rev. Lett.*, Vol. 90, 107401, Mar. 2003; A. A. Houck, J. B. Brock, and I. L. Chuang, "Experimental observations of a left-handed material that obeys Snell's Law," *Phys. Rev. Lett.*, Vol. 90, 137401, Apr. 2003.

45. T. Koschny, M. Kafesaki, E. N. Economou, and C. M. Soukoulis, "Effective medium theory of left-handed materials," *Phys. Rev. Lett.*, Vol. 93, 107402, Sept. 2004.

46. N. Katsarakis, T. Koschny, M. Kafesaki, E. N. Economou, E. Ozbay, and C. M. Soukoulis, "Left- and right-handed transmission peaks near the magnetic resonance frequency in composite metamaterials," *Phys. Rev. B*, Vol. 70, 201101, Nov. 2004.

47. T. Weiland, R. Schumann, R. B. Greegor, C. G. Parazzoli, A. M. Vetter, D. R. Smith, D. V. Vier, and S. Schultz, "Ab initio numerical simulation of left-handed metamaterials: Comparison of calculations and experiments," *J. Appl. Phys.*, Vol. 90, pp. 5419–5424,

2001; P. Markoš and C. M. Soukoulis, "Numerical studies of left-handed materials and arrays of split ring resonators," *Phys. Rev. E*, Vol. 65, 036622, Mar. 2002.

48. P. Markoš, and C. M. Soukoulis, "Transmission losses in left-handed materials," *Phys. Rev. E*, Vol. 66, 045601, Oct. 2002; "Transmission studies of left-handed materials," *Phys. Rev. B*, Vol. 65, 033401, Jan. 2002.

49. J. Pacheco, T. M. Grzegorczyk, B.-I. Wu, Y. Zhang, and J. A. Kong, "Power propagation in homogeneous isotropic frequency-dispersive left-handed media," *Phys. Rev. Lett.*, Vol. 89, 257401, Dec. 2002.

50. P. Markoš and C. M. Soukoulis, "Transmission properties and effective electromagnetic parameters of double negative metamaterials," *Optics Express*, Vol. 11, pp. 649–661, Apr. 2003; P. Markoš and C. M. Soukoulis, "Absorption losses in periodic arrays of thin metallic wires," *Opt. Lett.*, Vol. 28, pp. 846–848, May 2003.

51. S. O'Brien and J. B. Pendry, "Photonic band-gap effects and magnetic activity in dielectric composites," *J. Phys.: Condens. Matter*, Vol. 14, pp. 4035-4044, Apr. 2002; *ibid*, "Magnetic activity at infrared frequencies in structured metallic photonic crystals," Vol. 14, pp. 6383–6394, July 2002.

52. S. O'Brien, D. McPeake, S. A. Ramakrishna, and J. B. Pendry, "Near-infrared photonic band gaps and nonlinear effects in negative magnetic metamaterials," *Phys. Rev. B*, Vol. 69, 241101, June 2004.

53. D. R. Smith, S. Schultz, P. Marko›s, and C. M. Soukoulis, "Determination of effective permittivity and permeability of metamaterials from reflection and transmission coefficients," *Phys. Rev. B*, Vol. 65, 195104, Apr. 2002.

54. T. Koschny, P. Markoš, D. R. Smith, and C. M. Soukoulis, "Resonant and antiresonant frequency dependence of the effective parameters of metamaterials," *Phys. Rev. E*, Vol. 68, 065602, Dec. 2003.

55. T. J. Yen, W. J. Padilla, N. Fang, D. C. Vier, D. R. Smith, J. B. Pendry, D. N. Basov, and X. Zhang, "Terahertz magnetic response from artificial materials," *Science*, Vol. 303, pp. 1494–1496, Mar. 2004.

56. N. Katsarakis, G. Konstantinidis, A. Kostopoulos, R. S. Penciu, T. F. Gundogdu, M. Kafesaki, E. N. Economou, Th. Koschny, C. M. Soukoulis, "Magnetic response of split-ring resonators in the far-infrared frequency regime," *Opt Lett.,* Vol. 30,. pp. 1348–1350, June 2005.

57. S. Linden, C. Enrich, M. Wegener, J. Zhou, Th. Koschny, and C. M. Soukoulis, "Magnetic response of metamaterials at 100 terahertz," *Science*, Vol. 306, pp. 1351–1353, Nov. 2004.

58. N. Katsarakis, T. Koschny, M. Kafesaki, E. N. Economou, and C. M. Soukoulis, "Electric coupling to the magnetic resonance of split ring resonators," *Appl. Phys. Lett.*, Vol. 84, pp. 2943–2945, Apr. 2004.

59. K. Sakoda, *Optical Properties of Photonic Crystals*, Springer, Berlin, 2001.

60. P. R. Villeneuve and M. Piche, "Photonic bandgaps in periodic dielectric structures," *Progr. Quantum Electron.*, Vol. 18, No. 2, pp. 153–200, 1994.

61. A. Moroz, "Metallo-dielectric diamond and zinc-blende photonic crystals," *Phys. Rev. B*, Vol. 66, 115109, Sept. 2002.

62. The MIT Photonic Bands (MPB) package can be found at http://ab-initio.mit.edu/mpb/

63. J. B. Pendry and A. MacKinnon, "Calculation of photon dispersion relations," *Phys. Rev. Lett.*, Vol. 69, pp. 2772-2775, Nov. 1992; J. B. Pendry, " Photonic band structures," *J. Mod. Opt.*, Vol. 41, pp. 209-229, Feb. 1994.

64. M. M. Sigalas, C. T. Chan, K. M. Ho, and C. M. Soukoulis, "Metallic photonic bandgap materials," *Phys. Rev. B*, Vol. 52, pp. 11744–11751, Oct. 1995; D. R. Smith, S. Shultz, N. Kroll, M. M. Sigalas, K. M. Ho, and C. M. Soukoulis, "Experimental and theoretical results for a two-dimensional metal photonic band-gap cavity," *Appl. Phys. Lett.*, Vol. 65, pp. 645–647, Aug. 1994.

65. M. Kafesaki, Th. Koschny, R. S. Penciu, T. F. Gundogdu, E. N. Economou, and C. M. Soukoulis, "Left-handed metamaterials: detailed numerical studies of the transmission properties," *J. of Opt. A: Pure Appl. Opt.*, Vol. 7, pp. S12–S22, Jan. 2005.

66. A. Taflove and S. C. Hagness, *Computational Electrodynamics: The Finite Difference Time Domain Method*, Artech House, Boston, 2000.

67. Liao boundary conditions are based on extrapolation of the fields in space and time by use of a Newton backward-difference polynomial. They are introduced in Z. P. Liao, H. L. Wong, B. P. Yang, and Y. F. Yuan, *Sci. Sin., Ser. A*, Vol. 27, pp. 1063–1076, 1984. Liao boundary conditions are also described in detail in [66].

68. R. W. Ziolkowski, "Pulsed and CW Gaussian beam interactions with double negative metamaterial slabs," *Optics Express*, Vol. 11, pp. 662–681, Apr. 2003.

69. K. Kunz and R. Luebbers, (Eds.), *The Finite Difference Time Domain Method for Electromagnetics*, CRC Press, Boca Raton, FL, 1993.

70. A. Lavrinenko, P. I. Borel, L. H. Frandsen, M. Thorhauge, A. Harpoth, M. Kristensen, and T. Niemi, "Comprehensive FDTD modelling of photonic crystal waveguide components," *Optics Express*, Vol. 12, pp. 234–248, Jan. 2004.

71. R. W. Ziolkowski and M. Tanaka, "FDTD analysis of PBG waveguides, power splitters, and switches," *Optical and Quantum Electronics*, Vol. 31, pp. 843–855, Oct. 1999; "FDTD modeling of dispersive material photonic band-gap structures," *J. Opt. Soc. A*, Vol. 16, No. 4, pp. 930-940, Apr. 1999; "FDTD modeling of photonic nanometer-sized power splitters and switches," in Integrated Photonics Research (IPR), *OSA Technical Digest*, pp. 175–177, Optical Society of America, Washington, DC, 1998.

72. A. Chutinan and S. Noda, "Highly confined waveguides and waveguide bends in three-dimensional photonic crystal," *Appl. Phys. Lett*, Vol. 75, pp. 3739–3741, Dec. 1999.

73. A. Chutinan and S. Noda, "Waveguides and waveguide bends in two-dimensional photonic crystal slabs," *Phys. Rev. B*, Vol. 62, pp. 4488-4492, Aug. 2000.

74. A. Mekis, J. C. Chen, I. Kurland, S. Fan, P. R. Villeneuve, and J. D. Joannopoulos, "High transmission through sharp bends in photonic crystal waveguides," *Phys. Rev. Lett.*, Vol. 77, pp. 3787–3790, Oct. 1996.

75. R. W. Ziolkowski, "The incorporation of microscopic material models into the FDTD approach for ultrafast optical pulse simulations," *IEEE Trans. Antennas Propagat.*, Vol. 45, pp. 375–391, Mar. 1997; R. W. Ziolkowski, J. A. Arnold, and D. M. Gogny, "Ultrafast pulse interactions with two-level atoms," *Phys. Rev. A*, Vol. 52, pp. 3082–3094, Oct. 1995; R. W. Ziolkowski and J. B. Judkins, "NL-FDTD modeling of linear and nonlinear corrugated waveguides," *J. Opt. Soc. Am. B*, Vol. 11, pp. 1565–1575, Sept. 1994; R. W. Ziolkowski and J. B. Judkins, "Applications of discrete methods to pulse propagation in nonlinear media: Self-focusing and linear-nonlinear interfaces," *Radio Science*, Vol. 28, pp. 901–911, Oct. 1993.

76. Marin Soljacic, Chiyan Luo, J. D. Joannopoulos, and S. Fan, "Nonlinear photonic crystal microdevices for optical integration," *Opt. Lett.*, Vol. 28, pp. 637–639, Apr. 2003.

77. E. P. Kosmidou and T. D. Tsiboukis, "An FDTD analysis of photonic crystal waveguides comprising third-order nonlinear materials," *Opt. Quant. Electr.*, Vol. 35, pp. 931–946, Aug. 2003.

78. C. T. Chan, Q. L. Yu, and K. M. Ho, "Order-N spectral method for electromagnetic waves," *Phys. Rev. B*, Vol. 51, pp. 16635–16642, June 1995.

79. E. Lidorikis, M. M. Sigalas, E. N. Economou, and C. M. Soukoulis, "Gap deformation and classical wave localization in disordered two-dimensional photonic-band-gap materials," *Phys. Rev B*, Vol. 61, pp. 13458–13464, May 2000.

80. S. Foteinopoulou, E. N. Economou, and C. M. Soukoulis, "Refraction in media with a negative refractive index," *Phys. Rev. Lett.* 90, 107402, Mar. 2003.

81. X. D. Wang, X.-G. Zhang, Q. Yu, and B. N. Harmon, "Multiple-scattering theory for electromagnetic waves," *Phys. Rev. B*, Vol. 47, pp. 4161–4167, Feb. 1993.

82. N. Stefanou, V. Karathanos, and A. Modinos, "Scattering of electromagnetic waves by periodic structures," *J. Phys.: Condens. Matter*, Vol. 4, pp. 7389–7400, Sept. 1992.

CHAPTER 9

FABRICATION, EXPERIMENTATION, AND APPLICATIONS OF EBG STRUCTURES

Peter de Maagt and Peter Huggard

9.1 INTRODUCTION

Much of the fundamental understanding of electromagnetic bandgap (EBG) materials can be derived from Brillouin's work in the 1940s. His pivotal book [1] demonstrated that a periodic lattice imposes restrictions on the *k* vectors of waves than may propagate within it. Although Brillouin focused on mechanical waves, some of his concepts can be directly transferred to the EBG domain. From a simplistic viewpoint, media with periodically changing dielectric properties impose periodic boundary conditions on propagating electromagnetic modes. Electromagnetic waves which do not satisfy these boundary conditions cannot propagate. Brillouin's legacy was that the concept of an energy bandgap became an integral part of solid-state physics. A parallel development took place within the field of microwave engineering, where the interaction of electromagnetic waves with periodic structures has been studied and applied for many years. One- and two-dimensional periodic structures in both closed metallic and open waveguides have been used as filters and traveling-wave tubes. Furthermore, planar periodic structures have found widespread application as frequency-selective surfaces (FSSs) and phased array antennas.

The recent revival of scientific interest in the electromagnetic properties of periodically structured materials was initiated by the pioneering work of Yablonovitch [2] and John [3] in 1987. The fully 3D periodic structure "Yablonovite" [4] was manufactured by mechanically drilling holes into a block of dielectric material. This processing prevented the propagation of microwave radiation in any 3D direction provided the frequency lay within a certain range, that is, within the bandgap. In contrast, the undrilled dielectric was transparent in the same frequency range. Since Yablonovitch first explored taking "photonic" bandgap technology from the optical spectrum into the microwave region, the pace of research has increased rapidly. One can even argue that the widespread

Metamaterials: Physics and Engineering Explorations, Edited by N. Engheta and R. W. Ziolkowski

familiarity with the concepts underlying EBG design has accelerated progress in the field by allowing a synergy to develop between people from very different backgrounds. Researchers from such disparate areas as solid-state physics, computational electromagnetics, material science, and microwave engineering have come to collaborate in EBG engineering. The products of their research are the artificially engineered materials which are generically known as photonic bandgap (PBG) materials or photonic crystals.[1]

The rapid advances in both theory and experiment, together with a substantial technological potential, have driven the development of EBG technology. Most of the initial results in this field have been achieved in academic and basic research environments. In contrast, industrial involvement has been lagging because successful near-term commercialization of EBG approaches has not been obvious. Currently, the industrialization and standardization of this technology are being considered and new startup companies have been founded solely to exploit the commercial potential of EBG materials. It now appears that EBG concepts can, in many cases, act as improved replacements for conventional solutions to electromagnetic problems.

This chapter concentrates upon techniques for the realization, characterization, and application of 3D EBG structures for the RF, microwave, and submillimeter regions. Emphasis is given to EBG manufacturing methods, for although computational techniques are delivering increasingly accurate predictions, from a historical perspective the development of EBG technology has grown from experimentation. The first EBG structures were scaled to sizes appropriate to microwave wavelengths because the submicrometer dimensions of optical structures are difficult and costly to fabricate. Photolithography and micromachining are the most common ways of building optical devices, but the two or three orders of magnitude larger microwave devices can be manufactured using less specialized and perhaps cheaper techniques. These range from the previously mentioned drilling of holes in dielectric blocks to the stacking of printed circuit boards. This review of manufacturing is followed by a brief discussion on some of the experimental apparatus used to measure performance. One advantage of working at lower frequencies is that measurements of both amplitude and phase can be made. This feedback of the complex dielectric response provides valuable information to the theorists and modelers, validating their predictions and facilitating development of computational tools.

The chapter subsequently describes some applications in the microwave and (sub)millimeter-wave frequency range. Already a large part of the research at microwave frequencies is application driven, while the development of submillimeter-wave systems remains technologically more challenging. The simultaneous progress in these two frequency ranges provides verification of

---

[1] Although "photonic" conventionally refers to visible and near-infrared light, the principle of establishing a bandgap applies to electromagnetic waves of all wavelengths. Consequently, there is some discussion within the microwave, millimeter-wave, and submillimeter-wave community about the use of the term "photonic" [5]. The more general alternative title of "electromagnetic bandgap" material or electromagnetic crystal has been proposed and for simplicity the EBG name is adopted here.

scalability of designs and essential insight into possible optical applications of such devices.

Due to the large number of groups that have started working in this field, it is impossible to incorporate a complete list of original and pertinent research in a chapter of this length. The relevant literature includes many books [e.g., 6], special journal issues [e.g., 7–9], and journal articles, and the interested reader should consult them for more details on the many novel configurations that exist. It is noted that the section on applications is an expansion of the contribution made by one of the authors to a previously published review article [10].

9.2   MANUFACTURING

This section concentrates upon the techniques that have been used for the manufacture of 3D EBG structures for frequencies from a few gigahertz up to several terahertz. Expressed in terms of wavelength, this extends from centimeters through the submillimeter region into the mid-infrared. Due to the wide range of frequencies over which the EBG phenomenon has already been observed, the fabrication approaches range from simple mechanical shaping methods at lower frequencies to the sophisticated photolithographic schemes used to form structures for use at near-infrared wavelengths. The EBG crystals themselves have been formed by processing bulk material, either by directly machining a 3D solid or, more usually, by forming 2D layers which are then stacked or by growth of the structure from liquid or powder precursors. The applicability of a particular technique depends upon (a) the scale of the EBG structure, which is inversely proportional to the operational frequency; (b) the material for the finished crystal; and (c) the crystal pattern. As a rule of thumb, the period of the EBG structure is a fraction of the free-space wavelength at the center of the bandgap, and so building structures for operation at frequencies above 300 GHz, that is, for submillimeter and shorter wavelengths, can still be a demanding task. One additional factor which must be considered is the tolerances achievable by the selected process and their relationship to the scale of the structure. The three approaches to manufacture, along with the consideration of acceptable fabrication tolerances, form the basis of this section.

9.2.1   Manufacture of 3D EBGs by Machining from the Solid

The first 3D EBG crystal was machined by drilling sets of parallel holes with angles of $120°$ between each set in a block of Stycast [4] dielectric using a drill press. The resulting Yablonovite structure exhibited a bandgap at a frequency of 14 GHz. This approach is clearly useful for lower frequencies but is limited to circular holes and, perhaps most importantly, is only suitable for processing certain dielectrics (e.g., plastics and composite materials). One more sophisticated extension of this approach used the LIGA (deep X-ray lithography) technique [11] to form a 3D mold in polymethylmethacrylate (PMMA). Collimated high-energy X rays from a synchrotron pass through a shadow mask and alter the solubility of the PMMA. A 3D pattern can thus be achieved by repeated exposures at different angles of incidence. The irradiated polymer is removed using a suitable solvent and a liquid ceramic precursor introduced into the network of holes. This is transformed into a 3D crystal, and the PMMA is removed by pyrolysis. In this way an anti-Yablonovite structure, where holes are replaced by dielectric rods, exhibiting a transmission bandgap centered on 2.4 THz was realized [12]. Like mechanical drilling, the LIGA approach is restricted to parallel-sided holes, which are possibly of limited depth, and the technique is only suitable for use with certain dielectrics.

9.2.2 Manufacture of 3D EBGs by Stacking

The layer-upon-layer stacking method has been used to form structures exhibiting bandgaps from a few gigahertz to near-visible frequencies. One of the first implementations was by Ozbay et al. [13], who stacked alumina rods to form the "woodpile" structure which exhibited a bandgap between 12 and 14 GHz. Woodpiles and their variations are a commonly fabricated EBG structure and are formed from layers of parallel bars with rectangular or, less often, circular cross sections. Bars in adjacent layers are rotated by $90°$, and bars in every second layer are displaced, perpendicular to their long axis, by half a period.

To form woodpiles for higher frequencies, more sophisticated methods of producing entire layers of bars, often surrounded by a supporting frame, have been utilized. One of the first approaches was to utilize the directional etching properties of silicon. Grooves parallel to the {111} planes were etched through a (110)-oriented wafer of high-resistivity silicon using the anisotropic aqueous KOH etchant. This yielded bars 340 μm across with a 1275-μm period, the resulting structure exhibiting an EBG centered on 95 GHz. In an extension of this wet-etching approach, a woodpile with nonorthogonal bars was also produced in silicon with a bandgap frequency around 350 GHz [14]. The possibility of reducing the bandgap width by varying the dimensions of the bars away from the optimum was also investigated in this study.

Mechanical methods have also been used to form the layers for woodpiles. In particular, precision, narrow-kerf, computer-controlled saws designed for the dicing of processed semiconductor wafers have been applied. The usual approach when fabricating EBG layers is to make repeated passes with the narrow diamond saw blade on one side of a silicon wafer to produce a series of bars and grooves. The wafer is then inverted, rotated by $90°$, and a second set of grooves are diced. In this way mechanically robust structures corresponding to one-half of a vertical period of the woodpile structure are produced. This method has been applied to produce woodpiles for center frequencies of 250 GHz [15], 260 GHz, [16], and 500 GHz (see Fig. 9.1 [17]). For still higher frequencies, wet etching and wafer bonding have been used to produce GaAs woodpiles at frequencies above 30 THz, corresponding to a bandgap centered on a wavelength of 7 μm [18]. One noteworthy alignment method adopted in the latter work was the use of laser diffraction to achieve the correct relative position between successive layers of bars. A flexible approach to dicing has also allowed the fabrication of free-standing arrays of pillars [19] and indeed the technique, with appropriate blades, is also suited to the processing of other "hard" dielectrics (e.g., ceramics).

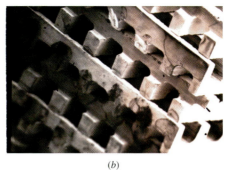

(a) (b)

Figure 9.1 Manufactured woodpile EBG structures for 500 GHz using layer-by-layer approach: (*a*) layers prepared by single-side deep reactive ion etching; (*b*) a closeup of results of mechanical dicing. Bar dimensions are approximately 90 μm × 90 μm in both cases.

An alternative method used to produce silicon EBGs for application at frequencies around 500 GHz involves using deep reactive ion etching (DRIE) (see Fig. 9.1 [20]). Here successive plasma etch and passivation cycles are used to produce nearly parallel sided holes in silicon wafers. The technique is flexible, produces high-aspect-ratio holes to a photolithographically defined pattern, and can be used for double-sided processing of wafers. It is most often applied to silicon, although some other crystalline dielectrics can also be processed [21].

Other EBG structures are also amenable to assembly on a layer-by-layer basis (see Fig. 9.2). Turning first to metallic EBG crystals, a high-pass structure was fabricated by stacking thin stainless steel sheets through which square arrays of circular holes had been cut with near-infrared Nd–YAG laser radiation. Cutoff frequencies of order 10 GHz were observed from holes with diameters of about 1 cm [22]. One novel aspect of the work was the observation that below cutoff resonances were observed if one of the layers in the stack was replaced by a defect layer with larger holes. Metallodielectric EBGs have also been fabricated by drilling triangular arrays of cylindrical holes in Stycast and Teflon sheets and then inserting a metal sphere into each sheet. The 3D structure was then assembled by stacking the layers containing the isolated metal balls to form a face-centered-cubic (FCC) structure [23]. Electromagnetic bandgaps at frequencies up to approximately 30 GHz were realized by using balls with a diameter of about 5 mm.

The use of lasers in creating EBGs by stacking is not restricted to the centimeter wavelength region, for the technique has also been investigated for forming structures operating at submillimeter wavelengths. Rather than using conventional silicon, with $\varepsilon_r \approx 11$, a significantly increased fractional bandgap can be obtained if a higher dielectric constant material is selected. High-permittivity zirconium tin titanate (ZTT) based ceramics, with $\varepsilon_r \approx 36$, are one such possibility [24]. Ceramics are generally very difficult to process by conventional mechanical machining techniques. However, laser ablation, where the absorption of high-intensity light converts the ceramic to plasma, is a particularly valuable

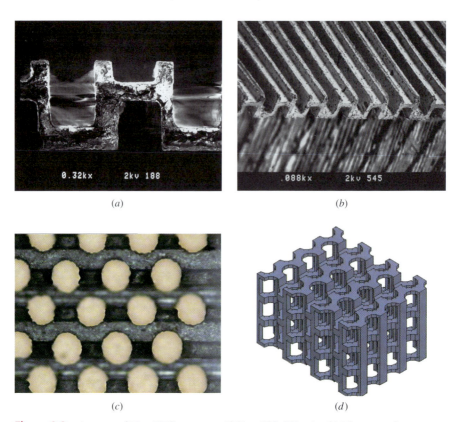

Figure 9.2   Aspects of Fan EBG structure [26] at 500 GHz. (*a*, *b*) Electron micrographs showing edge and oblique views of tiles for Fan structure The horizontal period is 220 μm. (*c*) Same tile after laser drilling with holes of mean diameter 96 μm. (*d*) Artist impression of EBG crystal.

approach for these brittle materials. Care needs to be taken to set up the process parameters, such as wavelength, energy density, and repetition rate; nonetheless, these optimized features can be defined with micrometer-scale precision. The process is best suited to the production of penetrating holes in thin sheets of ceramic, and recent results have shown that closely space hole arrays can been produced for layers of 500-GHz EBG crystals (see Fig. 9.2 [19]).

9.2.3   Manufacture of 3D EBGs by Growth

This section deals with the use of techniques to form a solid 3D EBG structure from either liquid or powder precursors. Application of focused laser light to a thin layer causes either (a) localized heating and hence fusion or (b) photopolymerization. A second layer of precursor is then added on top and the solidification repeated, thereby building the structure on a layer-by-layer basis. This technique has been successfully applied to form a 12-GHz EBG by ultraviolet photopolymerization of a proprietary monomer in a commercial stereolithography machine [25]. The procedure also involved the introduction of an initially fluid, higher dielectric composite material into the voids of the structure by a vacuum casting method. The necessary high-dielectric component of the fluid was provided by powdered calcium titanate. As an alternative approach, focused infrared laser radiation has been used to fuse powdered nylon to make 3D scale models of some EBG structures. This approach also offers the possibility of incorporating high-dielectric fillers in the precursor powder, though it is not known if this has yet been investigated.

9.2.4 Effect of Tolerances in Manufacture of EBGs

The selection of a fabrication process for the realization of a desired EBG structure is determined by both the material to be processed and, equally importantly, the scale of the structure. Every manufacturing approach has associated dimensional limitations, and the successful fabrication of an EBG depends upon how the process tolerances affect the electromagnetic performance of the finished crystal. Thus knowledge of both the tolerances and their effect on the location and width of the bandgap is desirable. It is important to distinguish between random and systematic effects: A surface roughness of 5 µm is unlikely to affect the performance of a submillimeter-wave EBG, but a systematic difference in the structural period by the same amount from the desired value might shift the center frequency by an unacceptable amount.

Some effort has already been expended in trying to understand the effects of fluctuations from the desired value. Initial work focused on the effects of structure sizes, interlayer misalignments, and surface roughness in a 3D structure suggested by Fan et al. [26]. Subsequently the effects of regular displacements of the bars of a woodpile on the bandgap have been investigated [27]. In general, it is found that the bandgap center frequency is reasonably robust to small perturbations, but unsurprisingly the width and depth of the gap are reduced as the perturbations increase. As an example of the tolerance to small fluctuations, systematic changes in any one of the dimensions of a silicon woodpile by up to $\pm 5\%$ are predicted to shift the center frequency by below 3% and the gap width by less than 2% [28]. The sensitivity of the bandgap parameters of other structures to systematic dimensional variations has also been calculated for a range of dielectrics [29].

9.3 EXPERIMENTAL CHARACTERIZATION OF EBG CRYSTALS

The approaches adopted to characterize EBG crystals depend upon the dimensionality of the crystals, their operational frequencies, and their intended applications. Key frequency-dependent EBG performance characteristics include the surface wave attenuation, reflectivity, and transmission, the latter two being complex quantities. Typical dedicated experimental apparatus and methods to determine each of these parameters are described in turn below. For the RF and millimeter-wave frequency range, the apparatus is generally based on commercial vector network equipment which delivers accurate measurements of amplitude and phase at up to several hundred gigahertz. At still higher frequencies, the technique of terahertz time-domain spectroscopy provides the same information.

9.3.1 Surface Wave Characterization

Applications of EBG technology to provide isolation between surface-mounted components (e.g., antennas) demand that the EBG crystal exhibit a high degree of surface wave attenuation. It is thus important to have a precise knowledge of the propagation properties of electromagnetic waves on planar EBG crystals. One method of obtaining this information used two wide-band end-fire antipodal Vivaldi antennas in conjunction with a vector network analyzer [30]. The antennas, which operated from 4 to 20 GHz, were printed on the same dielectric as was used for the planar EBG substrate. Figure 9.3 shows a photograph of the setup using the Vivaldi antenna. The antennas were formed from a symmetrically flared-out slot line, which was fed in turn by a parallel stripline and a microstripline. This setup is very well suited to perform measurements for one specific polarization of the field. Due to the small dielectric thickness, the polarization of the radiated field is predominantly parallel to the plane of the antenna and dielectric, that is, in the transverse electric (TE) mode. There is a very low level of cross-polarization [the transverse magnetic (TM) mode], especially at the low-frequency end of the range.

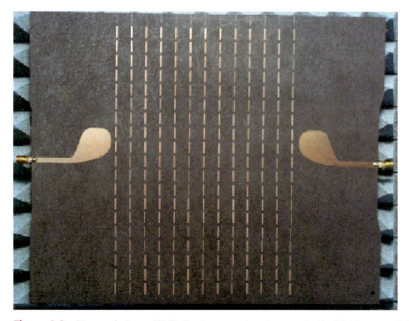

Figure 9.3 Photo of planar EBG dipole array showing one-half of both transmitting and measuring Vivaldi antennas. The other halves of the antennas are situated on the reverse of the dielectric sheet.

A typical measurement of the surface wave propagation along the EBG array is shown in Figure 9.4. In these measurements the forward surface wave transmission response was normalized with respect to the signal from the Vivaldi antennas alone.

Although the above system exhibited very good performance, the general measurement of surface wave propagation is often difficult due to finite sample sizes. As an alternative to planar antennas, small probes have also been used to excite and detect surface waves. For example, horizontally and vertically oriented linear and circular probes were used by Sievenpiper to launch and detect radiation in his characterization of a high-impedance surface [31]. Some edge effects remained in this setup, which led to unwanted scattering and/or standing waves. The effects were strongly reduced by using Saville's dog house arrangement [32]. Here an appropriately curved sample was used to eliminate scattering in one dimension and reduce it in the second. In addition, the curved sample prevented the direct illumination of the receiving probe.

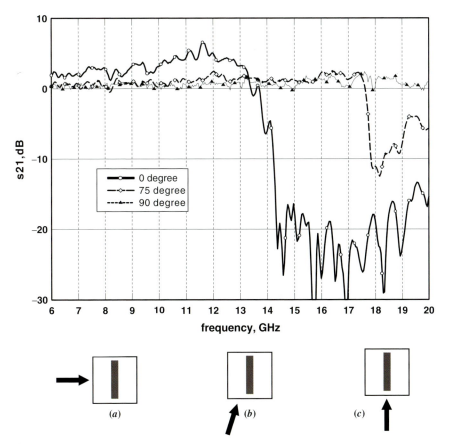

Figure 9.4 Measurement results from dipole array shown in Figure 9.3. The arrows show the direction of incidence with respect to the dipoles: (*a*) 0°, (*b*) 75°, and (*c*) 90°.

9.3.2 Complex Reflectivity Measurements

A double benefit can be derived from the use of metallodielectric EBGs for antenna substrates. Not only are surface waves suppressed, as discussed in the previous section, but also in-phase image currents are induced. The reflection phase–frequency diagram gives information about how the structure reacts to a wave impinging on it. Frequency ranges can be observed over which the configuration behaves similar to a metal plate (or perfect electric conductor). However, a further characteristic feature of metallodielectric EBGs is the existence of a frequency range over which an incident electromagnetic wave does not experience any phase reversal upon reflection. In this range the structure behaves as a perfect magnetic conductor (PMC). The frequency range where the phase reversal is within ±90° from the PMC point is conventionally used to define the bandwidth

Figure 9.5 Photo of setup for measuring EBG reflection phase as function of frequency for different angles of incidence using feedhorns to launch and receive radiation in anechoic chamber.

of the PMC behavior. Despite the significance of the reflection phase diagram, it is generally only shown in the published literature for waves at normal incidence.

One method to make reflection phase measurements of EBGs as a function of frequency for different angles of incidence used feedhorns to launch and receive the radiation in an anechoic chamber (see Fig. 9.5 [33]). For boresight measurements (normal incidence) the reflected signal was received by the transmit horn and was then directed via a circulator to the receiver. For all other angles (bistatic measurements), the illuminating antenna was fixed in position. The EBG sample was rotated through an angle θ degrees while the receiving antenna simultaneously rotated by 2θ. A typical measurement of the reflection phase of the two EBG structures is presented for both TE and TM polarizations in Figures 9.6 and 9.7, respectively. The reflection phase is determined with respect to a reference

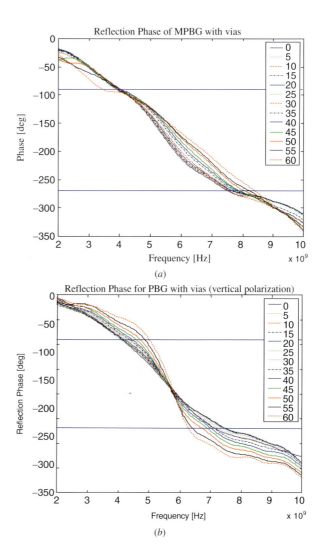

Figure 9.6 Measured phase reflection versus frequency for different angles of incidence for grounded spiral EBG structure [33]: (a) E-plane and (b) H-plane measurements.

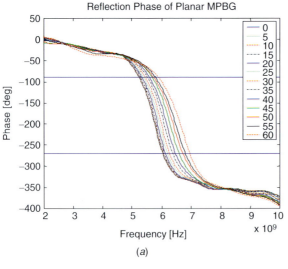

(a)

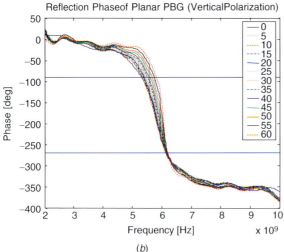

(b)

Figure 9.7   Measured phase reflection versus frequency for different angles of incidence for printed square ring EBG structure [33]: (a) E-plane and (b) H-plane measurements.

aluminum plate; zero degrees means that there is no phase difference between a reflection from the reference metal plate and the EBG sample.

9.3.3   Terahertz Reflection and Transmission Measurements

A full characterization of an EBG structure involves the measurement of its complex reflection and transmission properties over a wide frequency range. For millimeter-wave frequencies and below this can be done by use of vector network equipment as described above. For frequencies between a few hundred gigahertz and several terahertz, the technique of choice is time-domain spectroscopy [34]. In this approach a high-repetition-rate femtosecond laser generates 100-fs pulses of near-infrared radiation. The laser pulses are unequally divided, and the stronger portion excites broadband terahertz radiation from an InGaAs surface field emitter. Parabolic mirrors collimate the terahertz beam, refocus it at the sample position, and then image it onto the detector. This is a dipole antenna with an ultrafast photoconductive sampling switch which is gated by the weaker of the two optical pulses (see Fig. 9.8). Varying the relative delay time of the two optical pulses coherently measures the time-dependent electric field of the propagated pulse and yields both amplitude and phase information following Fourier transformation. The complex transmission of the sample is determined by measuring the terahertz signal with and without sample. The apparatus provides a signal-to-noise level of up to 40 dB, with a usable bandwidth from 150 GHz to above 3 THz. Figure 9.9 indicates the transmission of a silicon woodpile with a bandgap centered on 500 GHz which was measured by this approach [16].

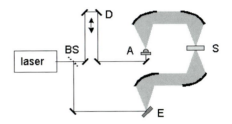

Figure 9.8   The terahertz time-domain spectroscopy setup as used for the transmission experiments: BS, beam splitter; D, optical delay stage; E, InGaAs surface field terahertz emitter; A, photoconductive terahertz detector; S, sample.

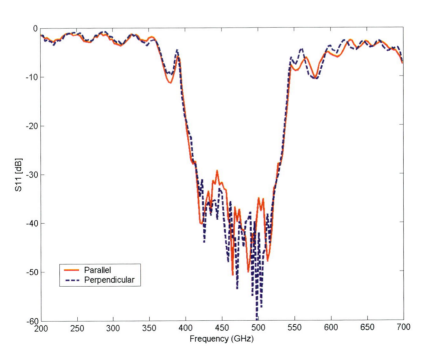

Figure 9.9   Normal-incidence transmission of layer-by-layer silicon woodpile EBG structure measured by terahertz time-domain spectroscopy. The two lines correspond to the incident *E* field polarized parallel and perpendicular to the top bars of the woodpile.

Alternatively the reflectivity of a sample can be measured. In this case care must be taken to substitute accurately a planar high-reflectivity metal mirror for the sample to obtain a reference spectrum.

9.4 CURRENT AND FUTURE APPLICATIONS OF EBG SYSTEMS

A multitude of basic EBG applications now exist, especially within the microwave and low-millimeter-wave region. Examples include reflector systems, electronically scanned phased arrays, high-precision Global Positioning System (GPS), Bluetooth, mobile telephony, and wearable antennas. It would be impossible to treat properly all such uses in the space available, so instead some examples of applications in widely different areas are discussed. It is hoped that the reader can thus gain an appreciation of the scope of the EBG approach.

Current satellite communication systems must provide high-gain links whose coverage is limited to geographical areas with irregular boundaries. Maximum use of the available bandwidth is obtained through frequency reuse and dual-polarization techniques. What must be optimized are not only the functional parameters of the transmitting antenna but also its physical and structural parameters, such as volume, mass, and dimensional stability. One suitable antenna for satellite communication applications is the shaped dual-grid reflector (DGR) antenna. It is composed of two concave shells arranged one behind the other [35]. Typically, a solid graphite back shell, to provide the required stiffness, is used in conjunction with a polarization-sensitive gridded Kevlar front shell. To maintain the structural integrity of the DGR, dielectric stiffener posts are often used as structural reinforcements between the two shells. Although in theory the posts should be RF transparent, in reality they induce significant field disturbances that degrade the side-lobe levels. This can even lead to unacceptable co-channel interference and seriously impair the system performance. An example is shown in Figure 9.10, where 11 dB degradation was observed. Using the simplest form of an EBG stiffener post led to an improvement of 5 dB.

In a typical symmetric reflector antenna configuration, struts will generally be needed to support the feed in the case of a single reflector or the subreflector in a double-reflector system. These struts block the aperture and consequently have an impact on the antenna performance, which is manifested as a reduction of the antenna gain and an increase in the side-lobe level. Conventionally the impact is minimized by optimizing the shape of the struts, but with the advent of EBG materials a new method has been introduced [36]. This method is based on guiding and launching the electromagnetic radiation in preferred directions and reducing the blocking effect to nearly zero in the operational directions.

Electronically scanned phased arrays find their use in many applications. For example, constellations of low Earth orbit satellites can be used for high-data-rate transmission at the Ka band to meet the growing demand for multimedia services. These applications in the Ka band require scanned, dual-polarized, multi-beam antennas with relatively wide bandwidth. Although at lower frequencies

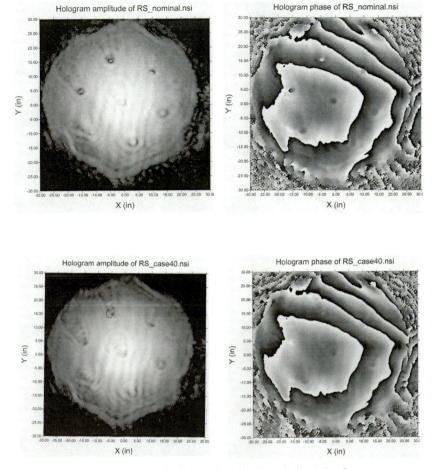

Figure 9.10 Measured field distribution (amplitude and phase) in plane at antenna aperture (from near-field back transformation) in nominal configuration (top) and configuration using EBG stiffeners. The perturbation introduced by the stiffener posts (bottom) is significantly reduced by using EBG stiffeners. Pictures kindly made available by Eric Amyotte, EMS Technologies Canada, Ltd.

mechanically scanned systems are mainly used, the use of active microstrip phased arrays is an attractive Ka band solution. However, the need for bandwidth and scanning increases the undesirable effects of surface waves, effects which may cause scan blindness at specific angles. The suppression of surface waves by incorporating EBG structures has led to improved performance [37,38]. Another area where EBG structures could play a role is in the reduction of the complexity of arrays, for the system gain requirement dictates the number of elements required. If the embedded element gain can be improved, the total element count can be reduced. The EBG "gain enhancement" is nothing special and is based on the same fundamental principles as in any standard reflector system.

The increase in gain is obtained in both cases by enlarging the effective radiating aperture, but EBG technology offers the antenna designer some extra flexibility. An example application of this can be found in overlapped feed clusters [39].

Another microwave application is the provision of high-precision GPS, which can deliver subcentimeter accuracy levels in surveying applications. While software can greatly reduce multipath errors, extra precautions to shield the antenna from unwanted multipath signals and to optimize the antenna's axial ratio and phase center stability are needed to obtain these accuracies. Choke rings provide excellent electrical performance for GPS antennas, but they are usually very large, heavy, and costly. The market demand is clearly for cheaper GPSs with better performance. Making use of the fact that metallodielectric EBG antennas can behave as artificial magnetic conductors, one can design EBG solutions in printed-circuit technology [40], which is much less costly that its conventional counterpart (see Fig. 9.11).

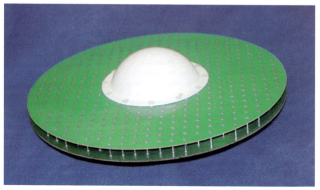

Figure 9.11 Example of GPS antenna made with EBG technology (bottom) in comparison to conventional choke ring approach (top) Pictures kindly made available by William McKinzie, Etenna.

As telecommunications become increasingly wireless, data and voice transmission are bound to become even more common. Attention is now focused on Bluetooth [41], the first implementation of such systems in everyday life. Moreover, for other applications like mobile phones (see Fig. 9.12), more attention is

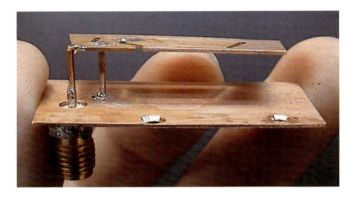

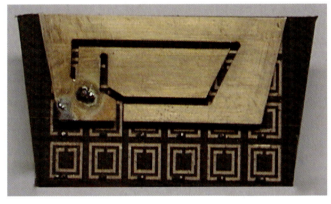

Figure 9.12 Mobile phone antennas fabricated in conventional (top) and EBG technologies (middle and bottom). Pictures kindly made available by Yang Hao, Queen Mary University of London.

being paid to the shielding offered by the antenna and the potential health hazard. Electromagnetic bandgap technology may prove useful in mobile antenna handset designs [42], reducing the radiation level applied to the user's hand and head. Shielding is not only important to reduce possible health concerns. It is also important in multipoint communications. For example, devices placed on the side of a laptop interact with the screen and the case, adversely affecting the maximum bit rate achievable between two computers. Again, EBG materials may play an important role in this area.

Future clothing may have a variety of consumer electronics built into the garments. Wearable antennas have received much interest recently due to the introduction of personal communication technology. Several applications of wearable antennas can be found; for example, radio tagging and miniature remote cameras and eventually the technology may even help parents to pin point their child's position. Antennas play a paramount role in the optimal design of wearable or hand-held units used in these services. Clearly, in designing these antennas, the electromagnetic interaction among the antenna, wearable unit, and human operator is an important factor to be considered. Electromagnetic bandgap technology has been proposed as a design solution [43].

Microwave filtering has also turned out to be an important area where EBG materials may play a significant role [44]. Their broad stop band can be exploited to suppress the spurious passbands that are always present in conventional microstrip filters. The sharp cutoff achievable with EBGs can also be used to improve the roll-off of a low-pass filter. Furthermore, combinations of conventional and EBG materials could lead to very compact structures.

The area of conventional waveguides is another field were hybrid solutions could play an important role. Rectangular waveguides with uniform field distributions are of great interest for applications in quasi-optical power combining. A standard waveguide can be modified by replacing the sidewalls with EBG crystals [45], potentially creating a very efficient waveguiding structure. Additionally, coupled cavity waveguides have recently attracted considerable attention. This concept is predicted to enable very low reflection loss bends to be made in waveguide [46].

9.5 CONCLUSIONS

This chapter has reviewed details of the manufacture, characterization, and application of EBG structures at frequencies ranging from a few gigahertz to over 2 THz. The various approaches to fabrication have been outlined and some of the associated specialized manufacturing approaches have been introduced. Manufacture in itself is not sufficient to advance the EBG area, for careful electromagnetic characterization is required to provide input for the design efforts of modelers and theoreticians. Several approaches to characterization have been presented and explained: These have the common factor of yielding amplitude and phase measurements of the EBG structure under study. Several applications at microwave frequencies have been discussed, and areas where EBG techniques offer significant potential have been indicated. These applications range from improved antennas with reduced side lobes through compact filters to the possibility of an exciting new waveguiding technology.

The space available is not sufficient to treat all approaches to this fascinating and increasing technologically relevant field, but it is hoped that the examples discussed will serve to illustrate the range of approaches and the areas of the spectrum to which EBGs can valuably contribute. The interested reader is encouraged to consult the extensive and continually expanding literature in the area to learn more about the potential and current status of the achievements of EBG technology.

REFERENCES

1. L. Brillouin, *Wave Propagation in Periodic Structures: Electric Filters and Crystal Lattices*, McGraw-Hill, New York, 1946.
2. E. Yablonovitch, "Inhibited spontaneous emission in solid state physics and electronics," *Phys. Rev. Lett.*, Vol. 58, pp. 2059–2062, 1987.
3. S. John, "Strong localization of photons in certain disordered dielectric superlattices," *Phys. Rev. Lett.*, Vol. 58, pp. 2486–2489, 1987.
4. E. Yablonovitch, T. J. Gmitter, and K. M. Leung, "Photonic band structure: The face-centered-cubic case employing nonspherical atoms," *Phys. Rev. Lett.*, Vol. 67, pp. 2295–2298, 1991.
5. A. A. Oliner, "Periodic structures and photonic band-gap terminology: Historical perspectives," in *Proc. 29th European Microwave Conference*, Vol. 3, Munich, Germany, Oct. 1999, pp. 295–298.
6. J. D. Joannopoulos, R. D. Meade, and J. N. Winn, *Photonic Crystals; Moulding the Flow of light*, Princeton University Press, Princeton, NJ, 1995.
7. *IEEE Trans. Microwave Theory Tech.*, Vol. 47, no. 11, 1999, Special Issue on Electromagnetic Crystal Structures, Design, Synthesis, and Applications.
8. *IEEE Trans. Microwave Theory Tech.*, Vol. 49, no. 10, 2001, Special Issue on Microwave and Millimetre Wave Photonics.
9. *IEEE Trans. Antennas Propagation*, Vol. 51, no. 10, 2003, Special Issue on Metamaterials.
10. P. de Maagt, R. Gonzalo, Y. Vardaxoglou, and J-M. Baracco, "Electromagnetic bandgap antennas and components for microwave and (sub)millimeter wave applications," *IEEE Trans. Antennas Propag.*, Vol. 51, pp. 2667–2677, 2003.
11. W. Ehrfeld and H. Lehr, "Deep X-ray lithography for the production of three-dimensional microstructures from metals, polymers and ceramics," *Radiat. Phys. Chem.*, Vol. 45, pp. 349–365, 1995.
12. G. Kiriakidis and N. Katsarakis, "Fabrication of 2-D and 3-D photonic band-gap crystals in the GHz and THz regions," *Mater. Phys. Mech.*, Vol. 1, pp. 20–26, 2000.
13. E. Ozbay, E. Michel, G. Tuttle, R. Biswas, M. Sigalas, and K. M. Ho, "Micromachined millimeter-wave photonic band-gap crystals," *Appl. Phys. Lett.*, Vol. 64, pp. 2059–2061, 1994.
14. E. Ozbay, E. Michel, G. Tutlle, R. Biswas, K. M. Ho, J. Bostak, and D. M. Bloom, "Double-etch geometry for millimeter-wave photonic band crystals," *Appl. Phys. Lett.*, Vol. 65, pp. 1617–1619, 1994.
15. R. Gonzalo, B. Martínez, C. M. Mann, H. Pellemans, P. Haring Bolivar, and P. de Maagt, "A low cost fabrication technique for symmetrical and asymmetrical layer by layer photonic crystal at submillimeter-wave frequencies," *IEEE Trans. Microwave Theory Tech.*, Vol. 50, pp. 2384–2393, 2002.
16. A. Chelnokov, S. Rowson, J.-M. Lourtioz, L. Duvillaret, and J. L. Coutaz, "Terahertz characterisation of mechanically machined 3D photonic crystal," *Electron. Lett.*, Vol. 33, pp. 1981–1983, 1997.

17. R. Gonzalo, I. Ederra, C. M. Mann, and P. de Maagt, "Radiation properties of terahertz dipole antenna mounted on a photonic crystal," *Electron. Lett.*, Vol. 37, pp. 613–614, 2001.

18. S. Noda, N. Yamamoto, H. Kobayashi, M. Okano, and K. Tomoda, "Optical properties of three-dimensional photonic crystals based on III–V semiconductors at infrared to near-infrared wavelengths," *Appl. Phys. Lett.*, Vol. 75, pp. 905–907, 1999.

19. L. Azcona, B. Alderman, D. N. Matheson, P. G. Huggard, B. Martinez, I. Ederra, C. Del Rio, R. Gonzalo, B. de Hon, M. C. van Beurden, L. Marchand, and P. de Maagt, "Micromachined electromagnetic bandgap crystals as antenna substrates for a 500 GHz imaging array," in *Proc. 4th Round Table on Micro/Nano-Technologies for Space*, European Space Agency, Noordwijk, The Netherlands, May 2003, pp. 221–231.

20. S. A. McAuley, H. Ashraf, L. Atabo, A. Chambers, S. Hall, J. Hopkins, and G. Nicholls, "Silicon micromachining using a high density plasma source," *J. Phys. D: Appl. Phys.*, Vol. 34, pp. 2769–2774, 2001.

21. M. E. Ryan, A. C. Camacho, and J. K. Bhardwaj, "High etch rate gallium nitride processing using an inductively coupled plasma source," *Phys. Stat. Sol. (a)*, Vol. 176, pp. 743–746, 1999.

22. N. Katsarakis, E. Chatzitheodoridis, G. Kiriakidis, M. M. Sigalas, C. M. Soukoulis, W. Y. Leung, and G. Tuttle, "Laser-machined layer-by-layer metallic photonic bandgap structures," *Appl. Phys. Lett.*, Vol. 74, pp. 3263–3265, 1999.

23. E. R. Brown and O. B. McMahon, "Large electromagnetic stop bands in metallodielectric photonic crystals," *Appl. Phys. Lett.*, Vol. 67, pp. 2138–2140, 1995.

24. P. Haring Bolivar, M. Brucherseifer, J. Gomez, R. Gonzalo, I. Ederra, A. L. Reynolds, M. Holker, and P. de Maagt, "Measurement of the dielectric constant and loss tangent of high dielectric constant materials at THz frequencies," *IEEE Trans. Microwave Theory Tech.*, Vol. 51, pp. 1062–1066, 2003.

25. T. J. Shepherd, C. R. Brewitt-Taylor, P. Dimond, G. Fixter, A. Laight, P. Lederer, P. J. Roberts, P. R. Tapster, and I. J. Youngs, "3D microwave photonic crystals: Novel fabrication and structures," *Electron Lett.*, Vol. 34, pp. 787–789, 1998.

26. S. Fan, P. R. Villeneuve, and J. D. Joannopoulos, "Theoretical investigation of fabrication-related disorder on the properties of photonic crystals," *J. Appl. Phys.*, Vol. 78, pp. 1416–1418, 1995.

27. A. Chutinan and S. Noda, "Effects of structural fluctuations on the photonic bandgap during fabrication of a photonic crystal: A study of a photonic crystal with a finite number of periods," *J. Opt. Soc. Am.*, Vol. B16, pp. 1398–1402, 1999.

28. L. Azcona, B. Alderman, P. G. Huggard, R. Gonzalo B. Martinez, I. Ederra, C. Del Rio, B. de Hon, M. C. van Beurden, L. Marchand, and P. de Maagt. "EBG technology for imaging arrays at the sub-mm range: Designs, materials and precision micromachining techniques," in *Proceedings of the IEE Seminar Metamaterials for Microwave and (Sub)Millimetre Wave Applications*, DTI, London, Nov. 2003, pp. 1–7.

29. B. Martinez Pascual, "3-D EBG structures at mm-wave frequencies: design, fabrication and measurements," Ph.D. thesis, Public University of Navarra, Pamplona, 2004.

30. Y. L. R. Lee, A. Chauraya, D. S. Lockyer, and J. C. Vardaxoglou, "Dipole and tripole metallodielectric photonic bandgap (MPBG) structures for microwave filter and antenna applications," *IEE Proc. Optoelectron.*, Vol. 147, no. 6, pp. 396–401, 2000.

31. D. Sievenpiper, "High-impedance electromagnetic surfaces," Ph.D. thesis, University of California, Los Angeles, 1999.

32. M. A. Saville, "Investigation of conformal high-impedance ground planes," Ph.D. thesis, AFIT/GE/ENG/00M-17, Air Force Institute of Technology, Wright-Patterson Air Force Base, Ohio, Mar. 2000.

33. J. M. Baracco, M. Paquay, and P. de Maagt, "An electromagnetic bandgap curl antenna for phased array applications," *IEEE Trans. Antennas Propag.*, Special Issue on Artificial Magnetic Conductors, Soft/Hard Surfaces, and other Complex Surfaces, Jan. 2005.

34. M. C. Nuss and J. Orenstein, in *Millimeter and Submillimeter Wave Spectroscopy of Solids*, (G. Gruner, Ed.), Springer-Verlag, Berlin, 1998.

35. R. Caballero, M-J. Martin, E. Ozores, G. Crone, R. Garcia Prieto, and E. Rammos, "Development of shaped dual-gridded reflector antennas," *Preparing for the Future, ESA Publications*, Vol. 8, no. 2, pp. 6–7 1998.

36. P.-S. Kildal, A. Kishk, and A. Tengs, "Reduction of forward scattering from cylindrical objects using hard surfaces," *IEEE Trans. Antennas Propag.*, Vol. 44, pp. 1509–1520, 1996.

37. R. Gonzalo, P. de Maagt, and M. Sorolla, "Enhanced patch antenna performance by suppressing surface waves using photonic band-gap structures," *IEEE Trans. Microwave Theory Tech.*, Vol. 47, pp. 2131–2138, 1999.

38. Z. Iluz, R. Shavit, and R. Bauer, "Microstrip antenna phased array with photonic band gap substrate," *IEEE Trans. Antennas Propag.*, Vol. 52, pp. 1446–1453, 2004.

39. R. Chantalat, T. Monediere, M. Thevenot, B. Jecko, and P. Dumon, "Multibeam reflector antenna with interlaced focal feeds by using a 1-D dielectric EBG resonator," paper presented at the 27th ESA Antenna Technology Workshop on Innovative Periodic Antennas: Electromagnetic Bandgap, Left-handed Materials, Fractal and Frequency Selective Surfaces, Santiago de Compostela, Spain, Mar. 2004.

40. R. Hurtado, W. Klimczak, W. E. McKinzie, and A. Humen, "Artificial magnetic conductor technology reduces weight and size for precision GPS antennas," paper presented at the Navigational National Technical Meeting, San Diego, CA, Jan. 2002.

41. R. Remski, "Modelling photonic bandgap (PBG) structures using Ansoft HFSS7 and optimetrics," presented at the Ansoft International Roadshow (Lecture Series), Aug. Oct. 2000, slides 36–40, Ansoft Corporation, Pittsburgh, PA.

42. R. F. Jimenez Broas, D. F. Sievenpiper, and E. Yablonovitch, "A high-impedance ground plane applied to cellphone handset geometry," *IEEE Trans. Microwave Theory Tech.*, Vol. 49, pp. 1262–1265, 2002.

43. P. Salonen, M. Keskilammi, and L. Sydanheimo, "A low-cost 2.45 GHz photonic band-gap patch antenna for wearable systems," in *Proc. 11th Int. Conf. Antennas Propagation ICAP 2001*, Manchester, UK, Apr. 2001, pp. 719–724.

44. T. Lopetegi, M. A. G. Laso, R. Gonzalo, M. J. Erro, F. Falcone, D. Benito, M. J. Garde, P. de Maagt, and M. Sorolla, "Electromagnetic crystals in microstrip technology," *Opt. Quantum Electron.*, Vol. 34, pp. 279–295, 2002.

45. F-R. Yang, K-P. Ma, Y. Qian, and T. Itoh, "A novel TEM waveguide using uniplanar compact photonic-bandgap (UC-PBG structure)," *IEEE Trans. Microwave Theory Tech.*, Vol. 47, pp. 2092–2098, 1999.

46. U. Peschel, A. Reynolds, B. Arredondo, F. Lederer, P. Roberts, T. Krauss, and P. de Maagt, "Transmission and reflection analysis of functional coupled cavity components," *IEEE J. Quantum Electron.*, Vol. 38, pp. 830–837, 2002.

CHAPTER 10

SUPERPRISM EFFECTS AND EBG ANTENNA APPLICATIONS

Boris Gralak, Stefan Enoch, and Gérard Tayeb

10.1 INTRODUCTION

In this chapter, we will consider some problems using the ability of periodic structures to control the propagation of electromagnetic waves. In this case, we will often consider that the electromagnetic bandgap (EBG) material behaves as a metamaterial. It means that, for our purpose, we will not make use of the ability of the photonic crystal to exhibit an EBG. With this goal, we will not study problems where the photonic crystal acts as a reflector (including substrates [1,2], resonant cavities [3,4], etc.). In this context, the analysis of the phenomena goes through the study of the dispersion curves of the Bloch modes and their related group and phase velocities.

We would first like to point out the pioneering work of R. Zengerle [5] applied to the propagation in periodic planar waveguides. Indeed, many phenomena studied later in the context of photonic crystals were already theoretically investigated and experimentally observed in this reference. The superprism effect using the properties of two-dimensional (2D) and three-dimensional (3D) photonic crystals was studied later experimentally [6–8] and then theoretically [9,10].

Using the same concepts, several other phenomena and applications can be considered, such as the control of emission and its application to directive antennas [11, 12], which will be the subject of Section 10.4, and self-guiding [13, 14].

In Section 10.2, we present a theoretical review of the concepts leading to the understanding of all the refractive phenomena at the boundary between a photonic crystal and a homogeneous medium (including negative refraction, ultrarefraction, etc.). Then we study the superprism effect (Section 10.3). The same "superprism" terminology is used to describe experiments that we have classified in three types, depending on the physical effect involved to obtain highly dispersive components. Finally, in Section 10.4 we focus on directive antennas applications.

Metamaterials: Physics and Engineering Explorations, Edited by N. Engheta and R. W. Ziolkowski

10.2 REFRACTIVE PROPERTIES OF A PIECE OF PHOTONIC CRYSTAL

Throughout this chapter we use the (not necessarily orthonormal) basis $(\mathbf{d}_1, \mathbf{d}_2, \mathbf{d}_3)$: every vector $\mathbf{x}$ in $\mathbb{R}^3$ (or $\mathbf{z}$ in $\mathbb{C}^3$) is represented by its three components x_1, x_2, and x_3 (or z_1, z_2, and z_3). Also, for all complex vector $\mathbf{z}$ in $\mathbb{C}^3$, we use the notation $|\mathbf{z}| = \sqrt{\bar{\mathbf{z}} \cdot \mathbf{z}}$, where $\bar{z}$ is the complex conjugate of z.

10.2.1 General Hypotheses

10.2.1.1 Hypotheses on Electromagnetic Field
The electromagnetic field we consider has a time dependence in $\exp(j\omega t)$ and is represented by the coupled solution $(\mathbf{E}_\omega, \mathbf{H}_\omega)$ of the set of harmonic Maxwell equations

$$\nabla \times \mathbf{E}_\omega = -j\omega\mu_0 \mathbf{H}_\omega \qquad \nabla \times \mathbf{H}_\omega = j\omega\varepsilon \mathbf{E}_\omega \qquad (10.1)$$

where ω is the frequency (real number), μ_0 is the vacuum permeability, and ε is the permittivity depending on the space variable $\mathbf{x}$ in $\mathbb{R}^3$. To describe an electromagnetic field provided by an usual light source, as represented in Figure 10.1, we assume that the electromagnetic energy density

$$\mathscr{E}(\mathbf{x}) = \tfrac{1}{2}[\varepsilon(\mathbf{x})|\mathbf{E}_\omega(\mathbf{x})|^2 + \mu_0|\mathbf{H}_\omega(\mathbf{x})|^2] \qquad (10.2)$$

satisfies for all x_3 in $\mathbb{R}$

$$\int_{\mathbb{R}^2} dx_1\, dx_2 |\mathscr{E}(\mathbf{x})| < \infty \qquad (10.3)$$

This very "weak" hypothesis enables us to consider a large class of electromagnetic fields, including an incident laser beam. If absorption takes place, the quantity $\mathscr{E}$ does not coincide with the electromagnetic energy density [15, 16]. Then, the criterion (10.3) has to be considered as a pure mathematical restriction. Note that we adopted such an "unusual" criterion (10.3) since a pure harmonic field cannot have finite total electromagnetic energy $\int_{\mathbb{R}^3} d\mathbf{x} \mathscr{E}(\mathbf{x})$. Finally, the wavelength associated with this harmonic field is

$$\lambda = \frac{2\pi c}{\omega} \qquad c = (\varepsilon_0\mu_0)^{-1/2} \qquad (10.4)$$

where ε_0 is the vacuum permittivity.

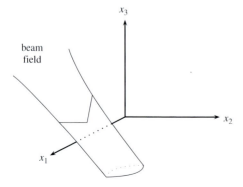

Figure 10.1 Representation of electromagnetic field under consideration. In every horizontal plane (defined by "x_3 is constant"), the electromagnetic energy is finite: It is a beam.

10.2.1.2 Hypotheses on Geometry   In this chapter, we assume the usual physical arguments of optics:

(i) The size of the piece of photonic crystal is large when compared to the wavelength λ of the harmonic electromagnetic field.

(ii) The energy density (10.2) vanishes at the edges of the piece of photonic crystal.

(iii) The radius of curvature of the boundaries delimiting the piece of photonic crystal is large when compared to the wavelength λ.

From (i) we conclude that the evanescent waves play a minor role. Note that here only propagating (or Bloch) waves will be considered, but in the next sections the evanescent waves that can be neglected for the description of the phenomena are fully taken into account in the rigorous numerical calculations. From (ii) we conclude that the photonic crystal can be extended without restriction where the energy density (10.2) is "small enough". Finally, from (iii) we conclude that all the boundaries delimiting the piece of photonic crystal can be considered as plane interfaces.

Then, with these three hypotheses, we can reduce the initial problem to the study of the simple structure represented on the right of Figure 10.2: a plane interface separating a homogeneous medium (with permittivity ε_0 without loss of generality) in the upper half space and a 3D photonic crystal in the lower half space. The plane interface is chosen such that the resulting structure is periodic in two directions; in other words, it is a 2D grating. In practice, this hypothesis is not a restriction since any plane inside a photonic crystal can be approached in this way. From now on, the basis $(\mathbf{d}_1, \mathbf{d}_2, \mathbf{d}_3)$ is chosen such that $(\mathbf{d}_1, \mathbf{d}_2)$ forms a basis of the lattice associated with the 2D grating and such that $(\mathbf{d}_1, \mathbf{d}_2, \mathbf{d}_3)$ corresponds to a basis of the lattice associated with the photonic crystal. Also, its origin can be chosen such that the plane interface is defined by the equation $x_3 = 0$ (see Fig. 10.2). Then the function ε satisfies

$$x_3 \geq 0 \Rightarrow \varepsilon(\mathbf{x}) = \varepsilon_0$$

$$x_3 \leq 0 \Rightarrow \varepsilon(\mathbf{x} + n_i \mathbf{d}_i) = \varepsilon(\mathbf{x}) \qquad n_i \in \mathbb{Z}(n_3 < 0) \qquad i = 1, 2, 3 \qquad (10.5)$$

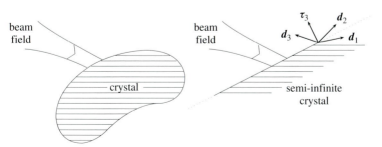

Figure 10.2   Left: initial geometry. Right: simplified geometry.

10.2.2 Rigorous Theory

10.2.2.1 Floquet–Bloch Transform and Decomposition of Initial Problem

After the previous modeling, the resulting structure is periodic with respect to the two variables x_1 and x_2. The basis of the reciprocal lattice associated with this grating is

$$\mathbf{d}_1^\star = 2\pi \frac{\mathbf{d}_2 \times \boldsymbol{\tau}_3}{|\mathbf{d}_1 \times \mathbf{d}_2|} \qquad \mathbf{d}_2^\star = 2\pi \frac{\mathbf{d}_1 \times \boldsymbol{\tau}_3}{|\mathbf{d}_1 \times \mathbf{d}_2|} \qquad \boldsymbol{\tau}_3 = \frac{\mathbf{d}_1 \times \mathbf{d}_2}{|\mathbf{d}_1 \times \mathbf{d}_2|} \qquad (10.6)$$

$\boldsymbol{\tau}_3$ being the vector normal to the plane interface with length $|\boldsymbol{\tau}_3| = 1$. From the finite energy density criterion (10.3) and in order to take advantage of the double periodicity of the structure, it is possible to perform a Floquet–Bloch transform defined by

$$\tilde{\mathbf{F}}_\omega(\mathbf{k}_\parallel, \mathbf{x}) = \frac{|\mathbf{d}_1 \times \mathbf{d}_2|}{(2\pi)^2} \sum_{n_1, n_2 \in \mathbb{Z}} \exp[j\mathbf{k}_\parallel \cdot (n_1 \mathbf{d}_1 + n_2 \mathbf{d}_2)] \mathbf{F}_\omega(\mathbf{x} + n_1 \mathbf{d}_1 + n_2 \mathbf{d}_2)$$

$$(10.7)$$

where $\mathbf{F}_\omega = \mathbf{E}_\omega, \mathbf{H}_\omega$ and $\mathbf{k}_\parallel$ is the tangential component of the wave vector (i.e., the projection of the wave vector $\mathbf{k}$ in the plane interface). After this transform, the set of Maxwell equations (10.1) becomes

$$\nabla \times \tilde{\mathbf{E}}_\omega = -j\omega\mu_0 \tilde{\mathbf{H}}_\omega \qquad \nabla \times \tilde{\mathbf{H}}_\omega = j\omega\varepsilon \tilde{\mathbf{E}}_\omega \qquad (10.8)$$

and the field satisfies the "partial" Bloch boundary conditions: For all $\mathbf{k}_\parallel$ in $\mathbb{R}^2$,

$$\tilde{\mathbf{F}}_\omega(\mathbf{k}_\parallel, \mathbf{x} + n_1 \mathbf{d}_1 + n_2 \mathbf{d}_2) = \exp[-j\mathbf{k}_\parallel \cdot (n_1 \mathbf{d}_1 + n_2 \mathbf{d}_2)] \tilde{\mathbf{F}}_\omega(\mathbf{k}_\parallel, \mathbf{x}) \qquad (10.9)$$

where $\tilde{\mathbf{F}}_\omega = \tilde{\mathbf{E}}_\omega, \tilde{\mathbf{H}}_\omega$, $\mathbf{x}$ is in $\mathbb{R}^3$, and n_1, n_2 are in $\mathbb{Z}$. Note that we denoted these Bloch boundary conditions as partial because they involve two directions while the photonic crystal is a priori 3D.

At this stage, we took advantage of all of the symmetries of the considered structure. This approach leads to the decomposition of the initial problem (10.1) in a collection of independent problems (10.8), indexed by the tangential component $\mathbf{k}_\parallel$ of the wave vector. In other words, the initial electromagnetic field $(\mathbf{E}_\omega, \mathbf{H}_\omega)$ has been decomposed in a collection of independent Bloch components $(\tilde{\mathbf{E}}_\omega, \tilde{\mathbf{H}}_\omega)$ indexed by $\mathbf{k}_\parallel$. From the definition (10.7), we can remark that

$$\tilde{\mathbf{F}}_\omega(\mathbf{k}_\parallel + n_1^\star \mathbf{d}_1^\star + n_2^\star \mathbf{d}_2^\star, \mathbf{x}) = \tilde{\mathbf{F}}_\omega(\mathbf{k}_\parallel, \mathbf{x}) \qquad (10.10)$$

where, $\tilde{\mathbf{F}}_\omega = \tilde{\mathbf{E}}_\omega, \tilde{\mathbf{H}}_\omega$, $\mathbf{x}$ is in $\mathbb{R}^3$, and $n_1^\star, n_2^\star$ are in $\mathbb{Z}$. Then, concerning the coupling at the plane interface, it appears that a wave associated with $\mathbf{k}_\parallel$ is coupled to all the other ones associated with $\mathbf{k}_\parallel + n_1^\star \mathbf{d}_1^\star + n_2^\star \mathbf{d}_2^\star$ since they form a single component. Obviously, they cannot be separated since the problem is already decomposed as much as possible. Also, from the last identity, it is possible to reduce the reciprocal space to $B_\parallel$, the unit cell of the reciprocal lattice associated with the grating, defined by

$$B_\parallel = \left\{ \mathbf{k}_\parallel = k_1^\star \mathbf{d}_1^\star + k_2^\star \mathbf{d}_2^\star | k_1^\star, k_2^\star \in \left[-\tfrac{1}{2}, \tfrac{1}{2} \right] \right\} \qquad (10.11)$$

A very important property of the decomposition (10.7) [17, 18] is that the finite-energy-density criterion (10.3) becomes

$$\int_{\mathbb{R}^2} dx_1\, dx_2 |\mathscr{E}(\mathbf{x})| = \int_{B_\parallel} d\mathbf{k}_\parallel \int_{[0,1]^2} dx_1\, dx_2 |\tilde{\mathscr{E}}(\mathbf{k}_\parallel, \mathbf{x})| < \infty \qquad (10.12)$$

with

$$\tilde{\mathscr{E}}(\mathbf{k}_\parallel, \mathbf{x}) = \tfrac{1}{2}[\varepsilon(\mathbf{x})|\tilde{\mathbf{E}}_\omega(\mathbf{k}_\parallel, \mathbf{x})|^2 + \mu_0|\tilde{\mathbf{H}}_\omega(\mathbf{k}_\parallel, \mathbf{x})|^2] \qquad (10.13)$$

This property is a consequence of the unitarity of the Floquet–Bloch transform, which corresponds to the "Parseval" identity for square integrable functions and the Fourier transform:

$$\int_{\mathbb{R}^2} dx_1\, dx_2 \overline{\mathbf{F}_\omega(\mathbf{x})} \cdot \mathbf{G}_\omega(\mathbf{x}) = \int_{B_\parallel} d\mathbf{k}_\parallel \int_{[0,1]^2} dx_1\, dx_2 \overline{\tilde{\mathbf{F}}_\omega(\mathbf{k}_\parallel, \mathbf{x})} \cdot \tilde{\mathbf{G}}_\omega(\mathbf{k}_\parallel, \mathbf{x})$$

$$(10.14)$$

for all admissible $\mathbf{F}_\omega, \mathbf{G}_\omega$. Also, as the Fourier transform, the Floquet–Bloch transform is invertible. The initial electromagnetic field can be reconstructed from its initial Floquet–Bloch components:

$$\mathbf{F}_\omega(\mathbf{x}) = \int_{B_\parallel} d\mathbf{k}_\parallel \tilde{\mathbf{F}}_\omega(\mathbf{k}_\parallel, \mathbf{x}) \qquad \mathbf{F}_\omega = \mathbf{E}_\omega, \mathbf{H}_\omega \qquad (10.15)$$

This second property is also very important. It tells us that the initial problem (10.1) is equivalent to the decomposed problem (10.8).

10.2.2.2 *Field Coupling at Plane Interface*

In this section we introduce an example of field coupling. We only consider the electric field since the magnetic field can be determined from it. In the upper half space, we have $\varepsilon(\mathbf{x}) = \varepsilon_0$. From the finite-energy-density criterion (10.3), we can use the Fourier transform to obtain the general solution for an incident field as

$$\mathbf{E}_\omega^i(\mathbf{x}) = \int_{\mathbb{R}^2} d\mathbf{k}_\parallel \hat{\mathbf{E}}_\omega^i(\mathbf{k}_\parallel) \exp(-j\mathbf{k}^i \cdot \mathbf{x}) \qquad (10.16)$$

where the wave vector $\mathbf{k}^i = \mathbf{k}_\parallel - \gamma_3(\mathbf{k}_\parallel)\boldsymbol{\tau}_3$ is deduced from the dispersion relation in a homogeneous medium,

$$\omega = c\sqrt{(|\mathbf{k}_\parallel|^2 + \gamma_3^2)} \Rightarrow \begin{cases} \gamma_3(\mathbf{k}_\parallel) = \sqrt{\omega^2/c^2 - |\mathbf{k}_\parallel|^2} & \text{if } \omega/c \geq |\mathbf{k}_\parallel| \\ \gamma_3(\mathbf{k}_\parallel) = -j\sqrt{|\mathbf{k}_\parallel|^2 - \omega^2/c^2} & \text{if } \omega/c \leq |\mathbf{k}_\parallel| \end{cases}$$

$$(10.17)$$

and whose Fourier transform in $x_3 = 0$ forms a square integrable function [to fulfill property (10.3)]:

$$\int_{\mathbb{R}^2} d\mathbf{k}_\parallel |\hat{\mathbf{E}}_\omega^i(\mathbf{k}_\parallel)|^2 < \infty \qquad (10.18)$$

Here, if we apply the Floquet–Bloch transform (10.7) to this incident field, then we obtain after some reasonable calculations

$$\tilde{\mathbf{E}}_\omega^i(\mathbf{k}_\parallel, \mathbf{x}) = \sum_{n_1^\star, n_2^\star \in \mathbb{Z}} \hat{\mathbf{E}}_\omega^i(\mathbf{k}_\parallel + n_1^\star \mathbf{d}_1^\star + n_2^\star \mathbf{d}_2^\star) \exp[-j\mathbf{k}_{n_1^\star, n_2^\star}^i \cdot \mathbf{x}] \qquad (10.19)$$

where $\mathbf{k}^i_{n_1^\star,n_2^\star} = \mathbf{k}_\| + n_1^\star\mathbf{d}_1^\star + n_2^\star\mathbf{d}_2^\star - \gamma_3(\mathbf{k}_\| + n_1^\star\mathbf{d}_1^\star + n_2^\star\mathbf{d}_2^\star)\boldsymbol{\tau}_3$. Now, at this stage, the problem is fully decomposed. As a particular case of the general property (10.10), the right part of Eq. 10.19 shows that the plane wave associated with $\mathbf{k}_\|$ is coupled to all the other ones associated with $\mathbf{k}_\| + n_1^\star\mathbf{d}_1^\star + n_2^\star\mathbf{d}_2^\star$. Note that, in general, the incident field given by (10.16) is chosen such that all the support of its Fourier transform is included in $B_\|$ (in order to obtain a Floquet transform reduced to a single plane wave), and such that the component γ_3 given by (10.17) is always real (there is no evanescent component in the incident field). In the same way, the general solution for a reflected field is

$$\mathbf{E}^r_\omega(\mathbf{x}) = \int_{\mathbb{R}^2} d\mathbf{k}_\| \hat{\mathbf{E}}^r_\omega(\mathbf{k}_\|) \exp(-j\mathbf{k}^r \cdot \mathbf{x}) \qquad (10.20)$$

where the wave vector is now $\mathbf{k}^r = \mathbf{k}_\| + \gamma_3(\mathbf{k}_\|)\boldsymbol{\tau}_3$ and whose Fourier transform $\hat{\mathbf{E}}^r_\omega$ is a square integrable function as well as for the incident field (10.18). Again, we can apply the Floquet–Bloch transform; we then obtain an expression similar to (10.19):

$$\tilde{\mathbf{E}}^r_\omega(\mathbf{k}_\|, \mathbf{x}) = \sum_{n_1^\star,n_2^\star\in\mathbb{Z}} \hat{\mathbf{E}}^r_\omega(\mathbf{k}_\| + n_1^\star\mathbf{d}_2^\star) \exp[-j\mathbf{k}^r_{n_1^\star,n_2^\star} \cdot \mathbf{x}] \qquad (10.21)$$

where $\mathbf{k}^r_{n_1^\star,n_2^\star} = \mathbf{k}_\| + n_1^\star\mathbf{d}_1^\star + n_2^\star\mathbf{d}_2^\star + \gamma_3(\mathbf{k}_\| + n_1^\star\mathbf{d}_1^\star + n_2^\star\mathbf{d}_2^\star)\boldsymbol{\tau}_3$. Finally, in the lower half space, ε is a periodic function of the three space variables. We cannot solve directly the general problem (10.1) as in the upper homogeneous half space. Nonetheless, it is possible to determine solutions $\tilde{\mathbf{E}}^{b,p}_\omega$ of (10.8), which are now Bloch waves; that is in addition to the partial condition (10.9), they satisfy for n_3 in $\mathbb{Z}$

$$\tilde{\mathbf{E}}^{b,p}_\omega(\mathbf{k}_\|, k_3^p, \mathbf{x} + n_3\mathbf{d}_3) = \exp[-j2\pi n_3 k_3^p]\tilde{\mathbf{E}}^{b,p}_\omega(\mathbf{k}_\|, k_3^p, \mathbf{x}) \qquad k_3^p \in \left[-\tfrac{1}{2}, \tfrac{1}{2}\right] \qquad (10.22)$$

where p is in $\{1, \dots, \alpha\}$ (α being the number of modes) and the third Bloch components k_3^p are given by the dispersion relation in the 3D photonic crystal,

$$\omega = f(\mathbf{k}_\|, k_3^p) \qquad (10.23)$$

which has to be determined numerically. From (10.15), the general solution in the lower half space can be written as the packet

$$\mathbf{E}^b_\omega(\mathbf{x}) = \int_{B_\|} d\mathbf{k}_\| \sum_{p=1}^{\alpha} \tilde{\mathbf{E}}^{b,p}_\omega(\mathbf{k}_\|, k_3^p, \mathbf{x}) \qquad (10.24)$$

Now, we are ready to explain how the coupling is working at the plane interface with the example under consideration. The general solution in the upper half space is the superposition of the incident (10.16) and reflected (10.20) fields while it is just the Bloch wave packet (10.24) in the lower half plane. After the Floquet–Bloch decomposition, the field is the superposition of (10.19) and (10.21) above and (10.22) below. Thus, for a fixed $\mathbf{k}_\|$ in $B_\|$ the total electromagnetic field satisfies (10.10). Then it is clear that all of the incident plane waves in (10.19) are coupled to all of the reflected plane waves in (10.21) and to all of the Bloch waves in (10.22) (see Fig. 10.3) with the same tangential component $\mathbf{k}_{\parallel} + n_1^{\star}\mathbf{d}_1^{\star} + n_2^{\star}\mathbf{d}_2^{\star}$ of the wave vector. This is the reason why, from now on, we will associate the field coupling at the plane interface with the "conservation of the tangential component of the wave vector". Note that this conservation is represented by a single vertical red dashed line in Figure 10.3 because, in this

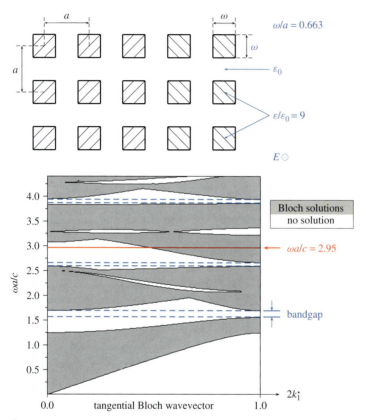

Figure 10.3 Analysis of coupling of field at plane interface separating homogeneous medium and photonic crystal. The considered photonic crystal has a 2D square lattice (edge $a = |\mathbf{d}_1| = |\mathbf{d}_3|$) of square rods (edge $w = 0.663a$) with dielectric constant $\varepsilon/\varepsilon_0 = 9$. The electromagnetic field is s polarized (the electric field is parallel to the rods) and has a normalized frequency $\omega a/c = 2.95$. The bottom left graph shows the dispersion relation in the crystal for all the frequencies. The bottom-right graph shows the equifrequency dispersion relation in the crystal at $\omega a/c = 2.95$ (solid black line) and at a slightly upper one (dashed black line). The top-right graph shows the equifrequency dispersion relation in vacuum at $\omega a/c = 2.95$ (solid black line) and at a slightly upper one (dashed black line). The incident plane wave is coupled with the reflected plane wave and Bloch waves which possess the same tangential component; this conservation is represented by the vertical red dashed line. The group velocity $\mathbf{v}_g$ is perpendicular to the equifrequency dispersion relation and oriented toward the increasing frequencies (this is the reason why we give the curves for two slightly different frequencies).

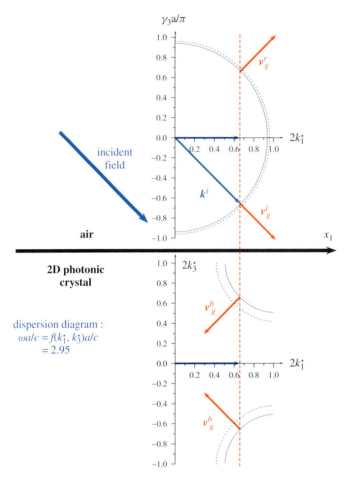

Figure 10.3 (*continued*)

particular case, the period of the crystal a is equal to the period of the grating $|\mathbf{d}_1|$: The unit cell of the reciprocal lattice associated with the grating is then equal to the projection in the interface plane of the Brillouin zone associated with the crystal. In the general case (e.g., $|\mathbf{d}_1| > a$), this conservation has to be represented by a collection of parallel vertical lines separated by $n_1^{\star}\mathbf{d}_1^{\star} + n_2^{\star}\mathbf{d}_2^{\star}$ (e.g., $2\pi/|\mathbf{d}_1|$) inside the Brillouin zone (e.g., $[-\pi/a, \pi/a]$) associated with the crystal; or, equivalently, this conservation can be represented by a single vertical line if the equifrequency dispersion relations are repeated with respect to all of the translations associated with the vectors $n_1^{\star}\mathbf{d}_1^{\star} + n_2^{\star}\mathbf{d}_2^{\star}$.

10.2.2.3 Propagation of Electromagnetic Energy

In this section, to show that the direction of propagation of the electromagnetic energy in a photonic crystal is linked to the group velocity, we do not give the usual proof generally used in the literature. This usual proof is based on the result of Yeh [19]: For a given Bloch wave, the average of the Poynting vector over the unit cell of the Poynting vector is parallel to the group velocity. However, in the case of a wave packet, it is necessary to consider an electromagnetic field with finite total energy (and then a time-dependent field) in order to use this property *via* an identity similar to (10.12). Since our field is purely harmonic and cannot have finite total energy, we cannot use these usual arguments. This is the reason why we prefer to give a proof similar to the one generally used for the Goos–Hänchen effect [20].

Let $\mathbf{x}_\parallel = x_1 \mathbf{d}_1 + x_2 \mathbf{d}_2$ be the projection in the plane interface of the vector $\mathbf{x}$ in $\mathbb{R}^3$. Then, we can define a center of the electromagnetic energy for all x_3 in $\mathbb{R}$ as

$$\mathbf{X}_\parallel(x_3) = \frac{\displaystyle\int_{\mathbb{R}^2} dx_1\, dx_2 \mathbf{x}_\parallel |\mathscr{E}(\mathbf{x})|}{\displaystyle\int_{\mathbb{R}^2} dx_1\, dx_2 |\mathscr{E}(\mathbf{x})|} \tag{10.25}$$

Of course, a proper definition of the numerator in this expression requires a (pure mathematical) restriction $\int_{\mathbb{R}^2} dx_1\, dx_2 |\mathbf{x}_\parallel|\,|\mathscr{E}(\mathbf{x})| < \infty$ which is, in practice, always satisfied. Now we aim to show that the quantity $\mathbf{X}_\parallel(x_3 + 1) - \mathbf{X}_\parallel(x_3)$ is linked to the group velocity in order to find a result similar to the one based on [19].

Using the definition (10.7), it is possible to show that the Floquet–Bloch transform of $x_i \mathbf{F}_\omega$ is

$$\widetilde{x_i \mathbf{F}_\omega} = x_i \tilde{\mathbf{F}}_\omega + \frac{1}{2j\pi} \frac{\partial \tilde{\mathbf{F}}_\omega}{\partial k_i^\star} \qquad k_i^\star = \frac{\mathbf{k}_\parallel \cdot \mathbf{d}_i}{2\pi} \qquad i = 1, 2 \tag{10.26}$$

Using this relationship together with the property of unitarity of the Floquet–Bloch transform (10.14), we obtain for the center of the electromagnetic energy

$$\mathbf{X}_\parallel(x_3) = \frac{\displaystyle\int_{B_\parallel} d\mathbf{k}_\parallel \int_{[0,1]^2} dx_1\, dx_2 \left[\mathbf{x}_\parallel |\tilde{\mathscr{E}}(\mathbf{k}_\parallel, \mathbf{x})| + \frac{1}{2j\pi} \sum_{i=1,2} \Delta_i(\mathbf{k}_\parallel, \mathbf{x}) \mathbf{d}_i \right]}{\displaystyle\int_{B_\parallel} d\mathbf{k}_\parallel \int_{[0,1]^2} dx_1\, dx_2 |\tilde{\mathscr{E}}(\mathbf{k}_\parallel, \mathbf{x})|} \tag{10.27}$$

where, for $i = 1, 2$,

$$\Delta_i(\mathbf{k}_\parallel, \mathbf{x}) = |\varepsilon(\mathbf{x})| \overline{\tilde{\mathbf{E}}_\omega(\mathbf{k}_\parallel, \mathbf{x})} \cdot \frac{\partial \tilde{\mathbf{E}}_\omega(\mathbf{k}_\parallel, \mathbf{x})}{\partial k_i^\star} + |\mu(\mathbf{x})| \overline{\tilde{\mathbf{H}}_\omega(\mathbf{k}_\parallel, \mathbf{x})} \cdot \frac{\partial \tilde{\mathbf{H}}_\omega(\mathbf{k}_\parallel, \mathbf{x})}{\partial k_i^\star} \tag{10.28}$$

For the sake of simplicity, we assume that there is a single Bloch mode inside the structure: $\alpha = 1$ in Eq. 10.22 and we denote by $k_3^\star$ the single third component of the Bloch wave vector k_3^1. Clearly, from its definition (10.13) and the Bloch boundary condition (10.22), the quantity $\tilde{\mathscr{E}}(\mathbf{k}_\parallel, \mathbf{x})$ is periodic with respect to the variable x_3: $\tilde{\mathscr{E}}(\mathbf{k}_\parallel, \mathbf{x}) = \tilde{\mathscr{E}}(\mathbf{k}_\parallel, \mathbf{x} + \mathbf{d}_3)$. Then, only the part containing (10.28) will contribute to $\mathbf{X}_\parallel(x_3 + 1) - \mathbf{X}_\parallel(x_3)$. Again, using the Bloch boundary condition (10.22), it is possible to show that

$$\Delta_i(\mathbf{k}_\parallel, \mathbf{x} + \mathbf{d}_3) - \Delta_i(\mathbf{k}_\parallel, \mathbf{x}) = -2j\pi|\tilde{\mathscr{E}}(\mathbf{k}_\parallel, \mathbf{x})|\frac{\partial k_3^\star}{\partial k_i^\star} \qquad i = 1, 2 \qquad (10.29)$$

Consequently, we obtain

$$\mathbf{X}_\parallel(x_3 + 1) - \mathbf{X}_\parallel(x_3) = -\frac{\int_{B_\parallel} d\mathbf{k}_\parallel \int_{[0,1]^2} dx_1\, dx_2 |\tilde{\mathscr{E}}(\mathbf{x})| \sum_{i=1,2}(\partial k_3^\star/\partial k_i^\star)\mathbf{d}_i}{\int_{B_\parallel} d\mathbf{k}_\parallel \int_{[0,1]^2} dx_1\, dx_2 |\tilde{\mathscr{E}}(\mathbf{x})|} \tag{10.30}$$

Finally, since we have a harmonic field,

$$\omega = f(k_1^\star, k_2^\star, k_3^\star) \Rightarrow d\omega = 0 = df = \frac{\partial f}{\partial k_1^\star}dk_1^\star + \frac{\partial f}{\partial k_2^\star}dk_2^\star + \frac{\partial f}{\partial k_3^\star}dk_3^\star \quad (10.31)$$

we have $\partial k_3^\star/\partial k_i^\star = -(\partial f/\partial k_i^\star)(\partial f/\partial k_3^\star)^{-1}$ for $i = 1, 2$, and then

$$\mathbf{X}_\parallel(x_3 + 1) - \mathbf{X}_\parallel(x_3)$$
$$= \frac{\int_{B_\parallel} d\mathbf{k}_\parallel \int_{[0,1]^2} dx_1\, dx_2 |\tilde{\mathscr{E}}(\mathbf{x})|(\partial f/\partial k_3^\star)^{-1} \sum_{i=1,2}(\partial f/\partial k_i^\star)\mathbf{d}_i}{\int_{B_\parallel} d\mathbf{k}_\parallel \int_{[0,1]^2} dx_1\, dx_2 |\tilde{\mathscr{E}}(\mathbf{x})|} \tag{10.32}$$

If the vector $\mathbf{d}_3$ is added to each side of this equation, we then obtain an expression containing the group velocity $\mathbf{v}_g^b = \sum_{i=1,3}(\partial f/\partial k_i^\star)\mathbf{d}_i$ associated with the Bloch waves under consideration:

$$\mathbf{X}_\parallel(x_3 + 1) + \mathbf{d}_3 - \mathbf{X}_\parallel(x_3)$$
$$= \frac{\int_{B_\parallel} d\mathbf{k}_\parallel \int_{[0,1]^2} dx_1\, dx_2 |\tilde{\mathscr{E}}(\mathbf{x})|(\partial f/\partial k_3^\star)^{-1} \sum_{i=1,3}(\partial f/\partial k_i^\star)\mathbf{d}_i}{\int_{B_\parallel} d\mathbf{k}_\parallel \int_{[0,1]^2} dx_1\, dx_2 |\tilde{\mathscr{E}}(\mathbf{x})|} \tag{10.33}$$

From this expression, we conclude that the center of the electromagnetic energy is following a line whose direction is given by the average of the group velocity. Then, if we consider the example shown in Figure 10.3, it is expected that the electromagnetic energy will propagate with respect to the directions given by vectors $\mathbf{v}_g^b$. Consequently, for this example, negative refraction is expected. In Figure 10.4, we have represented a slab of the photonic crystal described in Figure 10.3, illuminated by an incident Gaussian beam with average incident wave vector $\mathbf{k}^i$ (represented in Fig. 10.3) and corresponding to an incidence angle of $45°$. The direction of propagation of the electromagnetic energy inside the crystal slab shown in Figure 10.4 is exactly the one expected by the theory ($45°$).

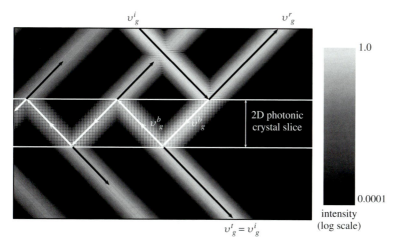

Figure 10.4 Example of negative refraction at two plane interfaces separating vacuum and 2D crystal slice. The parameters are given in the caption of Figure 10.3. The thickness of the crystal slice is $1000a$ and the width of the incident Gaussian beam is $100a$. The maximum value of the incident field intensity is equal to unity.

10.3   SUPERPRISM EFFECT

The purpose of this section is to show how the preceding theoretical tools permit one to design interesting dispersive devices. Indeed, we will see that playing with the conservation of the tangential component of the wave vector at the interface between a homogeneous medium and a photonic crystal enables one to get rapid variations of the light propagation direction when either the angle of incidence or the wavelength varies. These properties are obviously linked to the fact that the dispersion relation can be very sensitive to these parameters under appropriate conditions. Basically, the various studies related in the literature can be classified in three categories, depending on the way the dispersion relation is exploited. Note that, unless otherwise specified, we will consider 2D problems for the sake of simplicity. This assumption means that the structure, as well as the fields, will be independent of the x_2 coordinate. In particular, it means that the k_2 component of the wave vectors vanishes.

10.3.1   Group Velocity Effect

In this first case, the basic idea is to take advantage of a large variation of the group velocity direction in the photonic crystal that appears at some points on the dispersion curves. Such effects have been experimentally studied by Kosaka et al. [7, 8]. We consider a plane interface between a homogeneous medium (usually vacuum) and a photonic crystal, with a harmonic beam impinging on this interface, being incident from the homogeneous medium. Let us now assume that the equifrequency dispersion diagram of the Bloch modes in the photonic crystal is depicted in Figure 10.5. In this 2D problem, the $\mathbf{k}_\parallel$ vector is reduced to

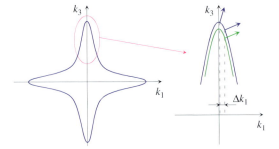

Figure 10.5 Schematic example of equifrequency dispersion diagram (left) and enlargement of interesting part (right). The blue and green curves correspond to two different wavelengths.

its k_1 component. We recall that, from the previous section, each k_1 component of the incident field couples with the Bloch mode with the same value of k_1. Let us consider two different values, k_1 and $k_1 + \Delta k_1$. Since k_1 is linked to both the wavelength and the incidence angle on the photonic crystal boundary, the effect is sensitive to these two parameters. First, let us assume that only the incidence angle changes. On the right part of Figure 10.5, we consider only the blue curve associated with the constant wavelength of interest. The blue arrows represent the direction of the energy flow inside the photonic crystal. It is clear that a small variation Δk_1 can lead to a large variation of the direction. Using the same idea, let us consider now the more practical problem in which the incidence angle is constant but Δk_1 is due to two different wavelengths. The only change is that we have to consider the two equifrequency dispersion diagrams that correspond to each wavelength (in blue and green in Fig. 10.5). Note that since the wavelength variation is small, these two diagrams are generally very close (except in particular situations that will be discussed later). Obviously, from Figure 10.5, the two wavelengths will propagate in different directions and will be spatially split inside the photonic crystal.

The reader interested in a theoretical discussion of the performances of this kind of device will find a detailed study of wavelength sensitivity and resolution in [21].

10.3.2 Phase Velocity Effect

Let us now consider the superprism effect based on phase velocity [22,23]. The situation is depicted in Figure 10.6. An incident monochromatic beam coming from a homogeneous dielectric impinges on the interface $x_3 = 0$ of the photonic crystal. We consider two close incidences of this beam (blue and green arrows). Here, we take advantage of a flat region inside the equifrequency dispersion diagram. Again, the two incidences correspond to k_1 and $k_1 + \Delta k_1$. Due to the flat shape of the dispersion curve, the corresponding variation Δk_3 of the other component of the Bloch wave vector inside the crystal is considerably larger. Let us now consider the interface $x_1 = \text{const}$ where the wave emerges from the crystal. Here, the tangential component that remains constant is k_3. Therefore, the device amplifies the angular variation between the two waves.

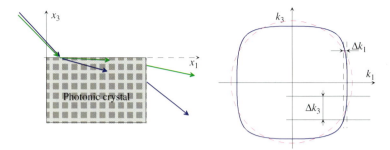

Figure 10.6 Left: schematic description of experiment. Right: equifrequency dispersion diagram of photonic crystal (blue curve) and external medium (pink dashed circle).

Note that for the sake of simplicity we have considered a monochromatic situation with two different incidences. In the same manner, one can consider a given incidence and two different wavelengths. The explanations are slightly more intricate. Indeed, one should consider slightly different dispersion curves for the two wavelengths, as in the previous section. Nonetheless, the final result remains that, as in the situation depicted in Figure 10.6, the two wavelengths will emerge from the crystal with noticeably different deviations.

Here, and in contrast with the group velocity effect, the separation is not due to the propagation inside the crystal itself. In the present case, two interfaces between the photonic crystal and the external medium are involved. Consequently, the group velocity effect gives additional freedom for the design of the structure. More compact structures could be investigated, since the distance of propagation inside the photonic crystal does not directly influence the separation.

10.3.3 Chromatic Dispersion Effect

Here, the basic idea is to take advantage of the fact that the photonic crystals can be highly dispersive [6, 9]. Once more, the graphic construction that gives the energy flow direction is based on the conservation of the tangential component of the wave vector and involves the dispersion curve of the Bloch waves in the photonic crystal. Thus, any rapid variation of the dispersion curve with respect to the wavelength leads to a rapid variation in the direction of propagation. Figure 10.7 shows a schematic representation of the dispersion relation $\omega = f(k_1, k_3)$ for a given Bloch mode. Thus it is represented by a surface. Obviously, in the vicinity of a horizontal tangent to this surface, the equifrequency dispersion curves vary very quickly with respect to the frequency. This case occurs in particular at each band edge.

To illustrate this phenomenon, we will consider a prism such as the one represented in Figure 10.8. The interesting refraction occurs at the slanted interface. Inside the crystal, the incident energy propagates horizontally, that is, with an incidence angle of $45°$ with respect to the slanted interface. This direction is represented by the black arrow in Figure 10.7. For each frequency, the construction expressing the conservation of the tangential wave vector relies on the

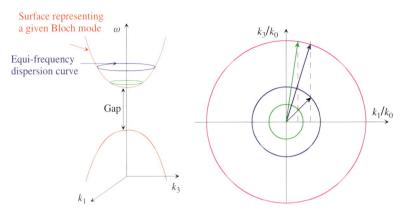

Figure 10.7 Left: schematic representation of dispersion relation $\omega = f(k_1, k_3)$ for given Bloch mode. Right: graphical construction illustrating dispersive refraction.

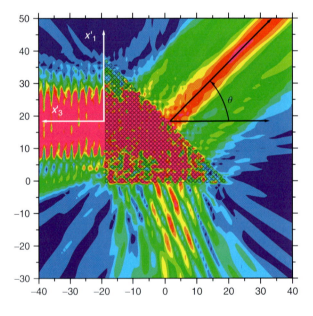

Figure 10.8 Field map of photonic crystal prism illuminated from left-hand side by beam with $\lambda = 1.97a$. On the left-hand side of the crystal the reflected beam interferes with the incident beam and generates a system of fringes.

equifrequency dispersion curve (blue and green curves, each being related to one frequency) and that of the vacuum (pink circle). Finally, the refraction directions in the vacuum are given by the blue and green arrows. Of course, we obtain nothing but the same construction as for a classical prism. But here the dispersion can be much greater. Note that the exact shape of the blue and green curves plays a minor role in the final result.

The photonic crystal in Figure 10.8 is constructed from a set of 465 dielectric rods with optical index equal to 3, lying in vacuum, and with a square lattice of period a. The radius of the rods is $\rho = 0.374a$. It is illuminated by a Gaussian beam whose electric field is parallel to the rods (x_2 axis). The x_2 component of the incident field is given by

$$u^i(x_1', x_3') = \int_{-\infty}^{+\infty} A(k_1') \exp(-jk_1'x_1' + jk_3'x_3') \, dk_1' \tag{10.34}$$

with $k_1' = k_0 \sin \theta^{\text{inc}}$, $k_3'^2 = k_0^2 - k_1'^2$, $k_0 = 2\pi/\lambda$, and with a Gaussian amplitude:

$$A(k_1') = \frac{W}{2\sqrt{\pi}} \exp\left(-\frac{(k_1' - k_0 \sin \theta_0)^2 W^2}{4}\right) \tag{10.35}$$

where θ_0 is the mean incidence of the beam on the left-hand side of the prism. It can be noticed that the parameter W appearing in (10.35) is directly linked to the incident beam width. In the case of Figure 10.8, we have taken $\theta_0 = 0$ and $W = 7.87a$.

Figure 10.8 shows that the beam is going through the hypotenuse with a quasi-normal direction. To precisely define this direction, we plot in Figure 10.9 the scattered intensity at infinity versus the diffraction angle θ (defined in Fig. 10.8). This property is also a consequence of the geometric construction of Figure 10.7, assuming that the equifrequency dispersion curves are small (i.e., close to the origin), which is obviously the case when the wavelengths are chosen near the band edge.

We can remark that the high-index contrast implies that the reflection is important at each interface. It could probably be attenuated by the use of some antireflection structure [24].

In a second step, let us evaluate the dispersion. Changing the wavelength from $\lambda = 1.953a$ to $\lambda = 1.984a$ shifts the maximum diffraction angle (Fig. 10.9) by about $5°$. By the way, since the wavelength is closer from the gap, the transmitted intensity is lower.

It is easy to verify that the dispersion $d\theta/d\lambda$ is much greater with this microprism than with any other classical dispersive device (grating, silica prism). Such microprisms could find interesting applications in the domain of fiber-optic communications and in particular in wavelength multiplexing/demultiplexing.

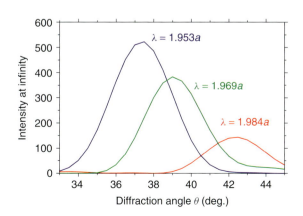

Figure 10.9 Scattered intensity at infinity for different wavelengths.

10.4   ANTENNA APPLICATIONS

From the preceding section, it has been shown that a photonic crystal enables one to couple any propagative wave inside the crystal with a plane wave outside the crystal whose propagation direction can remain close to the normal to the interface provided that the equifrequency dispersion curve remains close to the origin in the reciprocal space. (This is not a necessary condition, as will be illustrated at the end of this section.) Consequently, we can imagine using this property to design directive emitters. We would like to stress that here the antenna properties are not based on a resonant defect inside a photonic crystal, as proposed by several authors [3,4].

The basic ideas that govern the radiation phenomena involved in this section can again be understood using a graphical construction that expresses the conservation of the tangential component of the wave vector at the crossing of the interface between the photonic crystal and the outside homogeneous medium (Fig. 10.10). Let us consider a source embedded in a photonic crystal slab (Fig. 10.10, left). Assuming that the evanescent waves in the crystal play a negligible role, we only consider propagating Bloch modes inside the crystal. Any of these modes is represented by a point on the equifrequency dispersion curve of the crystal, that is, by the red arrow in Figure 10.10. Due to the conservation of the tangential component of the wave vector, this Bloch mode is coupled with the plane wave represented by the blue arrow. Thus, the whole dispersion curve is coupled with the set of directions in the green angular sector.

Let us give an illustration of the design of a directive antenna in the microwave frequencies [12]. The photonic crystal can be considered as a metamaterial that is a composite stack of metallic grids and foam layers. It is backed by a ground plane and excited by a monopole to complete the antenna (Fig. 10.11).

It is well known that such metallic grids behave as a low-frequency filter. It means that there is no propagating solution in the structure for frequencies below the cutoff ω_p (low-frequency bandgap). For our purpose, we are mainly

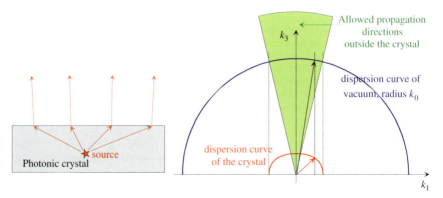

Figure 10.10   Schematic construction representing conservation of tangential component of Bloch wave vectors.

Figure 10.11   Schematic representation of antenna. The six copper grids are separated by foam. The rod in the middle represents the coaxial cable ended with the monopole (in red). The bottom plane is the ground plane.

interested in the small frequency range associated with the transition between the low-frequency bandgap and the allowed propagating solutions. Several studies [25–27] have shown that in this frequency range, where the wavelength is much greater than the period of the periodic media, the metamaterial can be homogenized as a material whose relative permittivity has a behavior governed by a plasma frequency in the microwave domain: $\varepsilon_{\text{eff}} = 1 - \omega_p^2/\omega^2$. Of course, using this expression, one can check that the low frequencies see the metamaterial as a medium with a negative permittivity and, hence, a pure imaginary optical index. Consequently, the only solutions are evanescent waves. But this expression also tells us that for frequencies just a little bit larger than the plasma frequency the relative permittivity stays between 0 and 1, and the same is valid for the optical index. In this case, the metamaterial behaves as a homogeneous material with an effective optical index close to zero. This remarkable property is called ultrarefraction [28].

Another interesting feature of this device is its ability to generate a linearly polarized beam when the current in the exciting source flows parallel to a given direction. This is one of the reasons for our decision to use a monopole to feed the overall EBG antenna. It means that all the currents in the metallic grids also flow parallel to this direction. The consequence is that the electric field inside the metamaterial is also nearly parallel to this direction. Since the grids are made with thin metallic crossed wires, it can be shown that the wires orthogonal to this direction have very little influence on the properties of this antenna. This feature is fundamental for the numerical study, since it proves that a 2D model will allow us to predict the properties of the antenna with good accuracy. After optimization of the parameters in order to match realization constraints

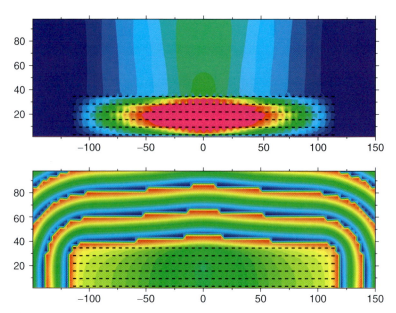

Figure 10.12   Field modulus (top) and lines of equal phase (bottom) of field radiated by 2D antenna.

and to achieve interesting properties around 14 GHz, it appears that convenient parameters for the EBG antenna system are the following: the cross section of the metallic wires is 0.14 mm thick (along the x_3 axis) and 0.71 mm large (along the x_1 axis), $d_1 = 5.8$ mm, and $d_3 = 6.3$ mm. Figure 10.12 shows the modulus and the lines of equal phase of the total field radiated by the structure made with 40×6 of these wires above a ground plane located at $x_3 = 0$. The source is a wire antenna parallel to the x_2 axis and placed in the middle of the metamaterial, with a wavelength $\lambda = 20.7$ mm. The most striking fact is the very slow variation of the phase inside the metamaterial, which is a proof that the effective index in this material is quite low. One can also notice that the phase of the emitted field is nearly constant on planes parallel to the emitting surface. The radiation pattern exhibits a narrow lobe (Fig. 10.13), with a half-power beam width of $2 \times 3.8°$.

The experimental device is made of six crossed grids etched from copper plates. The dimensions are those given above for the 2D case. Figure 10.14 gives the measured radiation pattern at 14.65 GHz. It shows the high directivity of the antenna. The measured half-power beam width is about $2 \times 5°$. The dissymmetry of the radiation pattern in the E plane is probably due to the coaxial cable that feeds the monopole. The cross-polar radiation is not shown on this figure, but it stays low: -23 dB compared to the copolar level in the normal direction (maximum radiation) and about -10 dB for directions far from the normal.

Our approach also allows us to design an emitting device radiating a narrow beam in any direction (and not necessarily toward the normal). To this end, we still need to keep the k_1 values associated with the equifrequency dispersion diagram lying in a small region, but now not centered on the origin. To this aim,

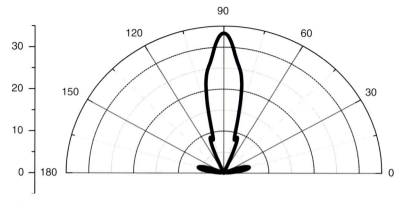

Figure 10.13 Radiation pattern for 2D device of Figure 10.12 (decibel scale). The half-power beam width is $2 \times 3.8°$.

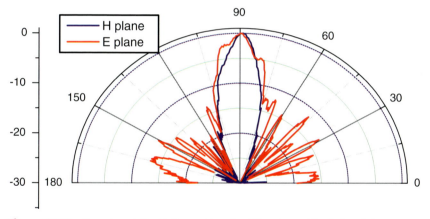

Figure 10.14 Experimental radiation pattern for 3D antenna (decibel scale).

let us consider a dielectric hexagonal crystal made of rods with optical index equal to 2.9, radius $\rho = 0.15a$, where a is the period of the crystal (distance between the centers of two neighbor rods). In the following example, we have taken $a = 4$ using arbitrary units. The dispersion relation has a sixfold symmetry which does not satisfy our requirements. One way to overcome this problem is to break this symmetry. For this purpose, we have chosen to expand the lattice of the crystal in the vertical direction of Figure 10.15: The vertical spacing between two grids is enlarged from $\sqrt{3}a/2 \approx 3.46$ to 3.9. The wavelength $\lambda = 8.01$ is in the gap of the original crystal; it is able to propagate in the expanded direction of the transformed crystal. Then we rotate the expanded crystal clockwise with an angle $\psi = \arctan(3.9/6) \approx 33°$. This choice allows us to obtain a lower row of rods parallel to the ground plane (see Fig. 10.16).

Figure 10.16 shows the field map and the lines of equal phase when this crystal is backed by a ground plane put at the origin of the vertical axis and excited

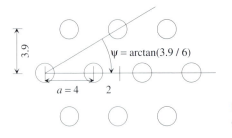

Figure 10.15 Rotation of expanded EBG crystal.

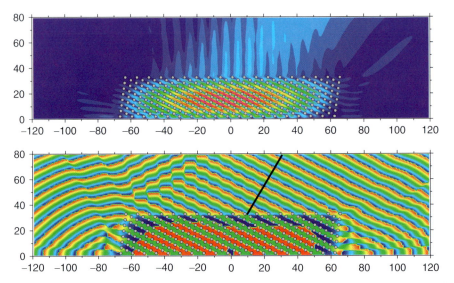

Figure 10.16 Field map (top) and lines of equal phase (bottom) of expanded and rotated EBG crystal backed by ground plane put at origin of vertical axis and excited by wire source.

by a wire source with $\lambda = 8.01$ and located at coordinates (0,4). Note that for this wavelength it can be seen from the dispersion relation (Fig. 10.17) that the energy propagates inside the crystal in all directions (the normal to the equifrequency dispersion diagram). Thus the field fills the entire crystal (Fig. 10.16). In that case the transverse variations of the field are smooth in a direction orthogonal to the desired emitting direction. The Bloch mode propagating inside the crystal enforces an appropriate field phase on the upper boundary of the crystal. This phase is appropriate since it is close to that of a plane wave propagating with an angle $\psi \approx 33°$ with respect to the normal. This example also shows that the location of the wire source is not critical. We see that the field is not particularly large near the source, and the largest values are located in the middle of the crystal. This property is useful, since it could give some freedom to choose the position of the source in order to obtain a convenient input impedance.

Figure 10.18 shows the polar emission diagram for the same structure with the same parameters. As expected, the principal lobe is tilted by an angle of $33°$

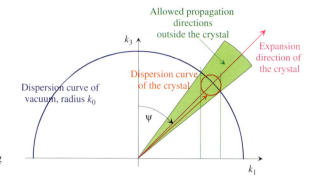

Figure 10.17   Schematic construction associated with off-axis EBG radiating device.

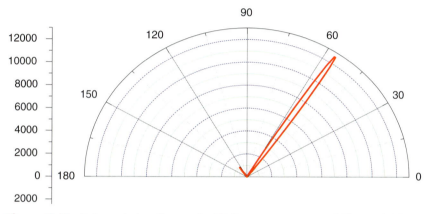

Figure 10.18   Polar emission diagram for EBG antenna structure of Figure 10.16, linear scale.

from the normal. A small part of the energy radiates in the symmetric direction. Assuming again that the outgoing wave has a constant amplitude on the crystal, the half-power beam width should be given by the formula obtained for the field radiated by a screen pierced by an aperture having a width $D \cos 33°$, where $D = 130$ is the lateral size of the crystal, that is, $\pm 0.443\lambda/(D \cos 33°) = \pm 1.9°$. The actual half-power beam width obtained on the emission diagram is $\pm 2.1°$ and thus agrees very well with the model.

10.5   CONCLUSION

We have reviewed several aspects of the propagation of electromagnetic waves in EBG materials. Besides the well-known bandgaps that can arise in periodically structured materials, we have shown how the richness of the dispersion relation of Bloch modes enables one to control the propagation of electromagnetic waves.

The behavior of finite-size EBG structures relies upon theoretical foundations that have been given in the first part of this chapter. Indeed, the study of the propagation of electromagnetic energy is linked to the study of coupling at the interfaces and can be fully understood due to the Floquet–Bloch transform. We have stressed a simple consequence of these theoretical aspects, which can be interpreted graphically using the equifrequency dispersion diagrams. In this way, three different types of superprism effects have been pointed out. The same tools can be used for antenna applications. We have shown that an EBG material can simulate a homogeneous material with a permittivity close to zero (ultrarefractive metamaterial), allowing us to design an innovative directive antenna.

Electromagnetic bandgap materials have paved many new ways for the design of devices, but nowadays only few of them have led to actual applications. No doubt in the near future EBG materials will initiate new technological products, especially in the telecommunication sphere.

REFERENCES

1. E. R. Brown, C. D. Parker, and E. Yablonovitch, "Radiation properties of a planar antenna on a photonic-crystal substrate," *J. Opt. Soc. Am. B*, vol. 10, pp. 404–407, 1993.
2. R. Gonzalo, P. De Maagt, and M. Sorolla, "Enhanced patch-antenna performance by suppressing surface waves using photonic-bandgap substrates," *IEEE Trans. Microwave Theory Tech.*, vol. 47, pp. 2131–2138, 1999.
3. M. Thevenot, C. Cheype, A. Reineix, and B. Jecko, "Directive photonic-bandgap antennas," *IEEE Trans. MTT*, vol. 47, pp. 2115–2122, 1999.
4. R. Biswas, E. Ozbay, B. Temelkuran, M. Bayindir, M. M. Sigalas, and K. M. Ho, "Exceptionally directional sources with photonic-bandgap crystals," *J. Opt. Soc. Am. B*, vol. 18, pp. 1684–1689, 2001.
5. R. Zengerle, "Light propagation in singly and doubly periodic planar waveguides," *J. Mod. Opt.*, vol. 34, pp. 1589–1617, 1987.
6. S-Y. Lin, V. M. Hietala, L. Wang, and E. D. Jones, "Highly dispersive photonic bandgap prism," *Opt. Lett.*, vol. 21, no. 21, pp. 1771–1173, 1996.
7. H. Kosaka, T. Kawashima, A. Tomita, M. Notomi, T. Tamamura, T. Sato, and S. Kawakami, "Superprism phenomena in photonic crystals," *Phys. Rev. B*, vol. 58, no. 16, pp. R10096–R10099, 1998.
8. H. Kosaka, T. Kawashima, A. Tomita, M. Notomi, T. Tamamura, T. Sato, and S. Kawakami, "Photonic crystals for micro lightwave circuits using wavelength-dependent angular beam steering," *Appl. Phys. Lett.*, vol. 74, no. 10, pp. 1370–1372, 1999.
9. B. Gralak, S. Enoch, and G. Tayeb, "Anomalous refractive properties of photonic crystals," *J. Opt. Soc. Am. A*, vol. 17, pp. 1012–1020, 2000.
10. M. Notomi, "Theory of light propagation in strongly modulated photonic crystals: Refractionlike behavior in the vicinity of the photonic band gap," *Phys. Rev. B*, vol. 62, pp. 10696–10705, 2000.
11. S. Enoch, B. Gralak, and G. Tayeb, "Enhanced emission with angular confinement from photonic crystals," *Appl. Phys. Lett.*, vol. 81, no. 9, pp. 1588–1590, 2002.
12. S. Enoch, G. Tayeb, P. Sabouroux, N. Guérin, and P. Vincent, "A metamaterial for directive emission," *Phys. Rev. Lett.*, vol. 89, 213902, 2002.
13. H. Kosaka, T. Kawashima, A. Tomita, M. Notomi, T. Tamamura, T. Sato, and S. Kawakami, "Self-collimating phenomena in photonic crystals," *Appl. Phys. Lett.*, vol. 74, pp. 1212–1214, 1999.

14. D. N. Chigrin, S. Enoch, C. M. Sotomayor Torres, and G. Tayeb, "Self-guiding in two-dimensional photonic crystals," *Opt. Express*, vol. 11, pp. 1203–1211, 2003, http://www.opticsexpress.org/abstract.cfm?URI=OPEX-11-10-1203.

15. A. Tip, "Linear absorptive dielectric," *Phys. Rev. A*, vol. 57, pp. 4818–4841, 1998.

16. A. Tip, "Canonical formalism and quantization for a class of classical fields with application to radiative atomic decay in dielectric," *Phys. Rev. A*, vol. 56, pp. 5022–5041, 1997.

17. M. Reed and B. Simon, *Methods of Modern Mathematical Physics*, Vol. IV: *Analysis of Operators*, Academic, New York, 1978.

18. P. Kuchment, *Floquet Theory for Partial Differential Equations*, Birkhäuser Verlag, Basel, 1993.

19. P. Yeh, "Electromagnetic propagation in birefringent layered media," *J. Opt. Soc. Am.*, vol. 69, pp. 742–756, 1979.

20. J.-P. Hugonin and R. Petit, "Etude générale des déplacements à la réflexion totale," *J. Opt. (Paris)*, vol. 8, pp. 73–87, 1977.

21. T. Baba and T. Matsumoto, "Resolution of photonic crystal superprism," *Appl. Phys. Lett.*, vol. 81, no. 13, pp. 2325–2327, 2002.

22. C. Luo, M. Soljacic, and J. D. Joannopoulos, "Superprism effect based on phase velocities," *Opt. Lett.*, vol. 29, no. 7, pp. 745–747, 2004.

23. T. Matsumoto and T. Baba, "Photonic crystal *k*-vector superprism," *J. Lightwave Technol.*, vol. 22, pp. 917–922, 2004.

24. T. Baba, T. Matsumoto, and M. Echizen, "Finite difference time domain study of high efficiency photonic crystal superprisms," *Opt. Express*, vol. 12, pp. 4608–4613, 2004, http://www.opticsexpress.org/abstract.cfm?URI=OPEX-12-19-4608.

25. D. Felbacq and G. Bouchitté, "Homogenization of a set of parallel fibers," *Waves in Random Media*, vol. 7, pp. 245–256, 1997.

26. J. B. Pendry, A. J. Holden, D. J. Robbins, and W. J. Stewart, "Low frequency plasmons in thin wire structures," *J. Phys. Condens. Matter*, vol. 10, pp. 4785–4809, 1998.

27. G. Guida, D. Maystre, G. Tayeb, and P. Vincent, "Mean-field theory of two-dimensional metallic photonic crystals," *J. Opt. Soc. Am. B*, vol. 15, pp. 2308–2315, 1998.

28. J. P. Dowling and C. M. Bowden, "Anomalous index of refraction in photonic bandgap materials," *J. Mod. Opt.*, vol. 41, pp. 345–351, 1994.

TWO-DIMENSIONAL PLANAR EBG STRUCTURES

CHAPTER 11

REVIEW OF THEORY, FABRICATION, AND APPLICATIONS OF HIGH-IMPEDANCE GROUND PLANES

Dan Sievenpiper

11.1 INTRODUCTION

Engineered electromagnetic surface textures can be used to alter the properties of metal surfaces to perform a variety of functions. For example, specific textures can be designed to change the surface impedance for one or both polarizations, to manipulate the propagation of surface waves, or to control the reflection phase. These surfaces provide a way to design new boundary conditions for building electromagnetic structures, such as for varying the radiation patterns of small antennas. They can also be tuned, enabling electronic control of their electro-magnetic properties. Tunable impedance surfaces can be used as simple steerable reflectors or as steerable leaky-wave antennas.

The simplest example of a textured electromagnetic surface is a metal slab with quarter-wavelength deep corrugations [1–4], as shown in Figure 11.1*a*. This is often described as a soft or hard surface [5] depending on the polarization and direction of propagation. It can be understood by considering the corrugations as quarter-wavelength transmission lines, in which the short circuit at the bottom of each groove is transformed into an open circuit at the top surface. This provides a high-impedance boundary condition for electric fields polarized perpendicular to the grooves and low impedance for parallel electric fields. Soft and hard surfaces are used in various applications, such as manipulating the radiation patterns of horn antennas or controlling the edge diffraction of reflectors. Two-dimensional structures have also been built, such as shorted rectangular waveguide arrays [6] or the inverse structures, often known as pin-bed arrays [7]. These textured surfaces are typically one-quarter-wavelength thick in order to achieve a high-impedance boundary condition.

Metamaterials: Physics and Engineering Explorations, Edited by N. Engheta and R. W. Ziolkowski
Copyright © 2006 the Institute of Electrical and Electronics Engineers, Inc.

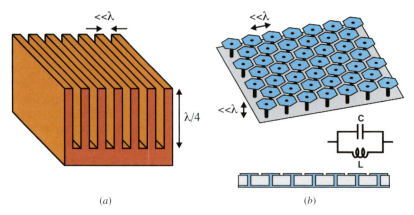

Figure 11.1 (*a*) A traditional corrugated surface consists of a metal slab with narrow quarter-wavelength long slots. The boundary condition at the top surface depends on the polarization of the incoming wave. (*b*) A high-impedance surface is built as a thin two-dimensional lattice of plates attached to a ground plane by metal-plated vias. The plates provide capacitance and inductance, and it has high electromagnetic impedance near its *LC* resonance frequency.

Recently, compact structures have been developed that can also alter the electromagnetic boundary condition of a metal surface but which are much less than one-quarter-wavelength thick [8, 9]. They are typically built as sub-wavelength mushroom-shaped metal protrusions, as shown in Figure 11.1*b*, or overlapping thumbtack-like structures. They can be analyzed as resonant *LC* circuits, and the reduction in thickness is achieved by capacitive loading. These materials provide a high-impedance boundary condition for both polarizations and for all propagation directions. They also reflect with a phase shift of zero, rather than π, as with an electric conductor. They are sometimes known as artificial magnetic conductors because the tangential magnetic field is zero at the surface, rather than the electric field, as with an electric conductor. In addition to their unusual reflection-phase properties, these materials have a surface wave bandgap, within which they do not support bound surface waves of either transverse magnetic (TM) or transverse electric (TE) polarization. They may be considered as a kind of electromagnetic bandgap (EBG) structure or photonic crystal [10, 11] for surface waves [12]. Although bound surface waves are not supported, leaky TE waves can propagate within the bandgap, which can be useful for certain applications.

By incorporating tunable materials or devices into textured surfaces, their capabilities are expanded to include active control of electromagnetic waves. This can be accomplished using mechanical structures such as movable plates or electrical components such as varactor diodes. With a tunable textured surface, one can build devices such as programmable reflectors that can steer or focus a reflected microwave beam [13]. These can provide a low-cost alternative to traditional electrically scanned antennas (ESAs) where phase shifters and complicated feed structures are replaced by a planar array of varactor diodes and a free-space, quasi-optic feed. Despite being low cost, these steerable reflector antennas are ruled out for some applications because they are not entirely planar.

Steerable leaky waves provide an alternative approach to electronic beam steering, without requiring a space feed [14]. The surface is programmed with a periodic impedance function that scatters the surface wave into free space. This steering method allows the scattered radiation to be steered over a wide scan range in both the forward and backward directions. Backward leaky waves can also be understood as resulting from bands of negative dispersion, similar to those in other negative-index materials.

11.2 SURFACE WAVES

By applying a texture to a metal surface, we can alter its surface impedance and thereby change its surface wave properties. The behavior of surface waves on an impedance surface is derived in several electromagnetics textbooks [15]. The derivation proceeds by assuming a surface having an impedance Z_s and a wave that decays exponentially away from a surface with decay constant α, as shown in Figure 11.2. For TM waves, we apply Maxwell's equations to determine the relationship between the surface impedance and the surface wave properties. It can be shown that TM waves occur on an inductive surface, in which the surface impedance is given by the expression

$$Z_s = \frac{j\alpha}{\omega\varepsilon} \tag{11.1}$$

Conversely, TE waves can occur on a capacitive surface, with the following impedance:

$$Z_s = \frac{-j\omega\mu}{\alpha} \tag{11.2}$$

In the above expressions, ε and μ are the permittivity and permeability of the space surrounding the surface, which may be vacuum, and ω is the angular frequency of the wave. We see that TM waves require a positive imaginary impedance, or an inductive surface, while TE waves require a negative imaginary impedance, or a capacitive surface.

Ordinary metals are slightly inductive, due to the skin effect, so they support TM waves. At optical frequencies these are often called surface plasmons [16]. At microwave frequencies, they are simply the ordinary surface currents, and they are only very weakly bound to the surface. A diagram of a TM surface wave is shown in Figure 11.3a. While bare metals do not support TE surface waves, dielectric-coated metals can support TE waves above a cutoff frequency that depends on the

Figure 11.2 A surface wave is a wave that is bound to a surface and decays into the surrounding space.

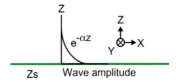

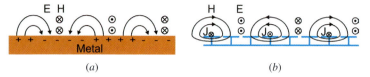

(a)   (b)

Figure 11.3 (*a*) In a TM surface wave, shown here on a flat metal surface, the electric field arcs out of the surface, and the magnetic field is transverse to the surface. (*b*) The fields take the opposite form in a TE surface wave, shown here on a high-impedance surface.

thickness and dielectric constant of the layer. Electromagnetic bandgap structures such as photonic crystals, frequency-selective surfaces, textured surfaces, and other interfaces can also support TE waves if the effective surface impedance is capacitive. The surface impedance of the textured metal surface described in this chapter is characterized by a parallel resonant *LC* circuit. At low frequencies it is inductive and supports TM waves. At high frequencies it is capacitive and supports TE waves, as depicted in Figure 11.3*b*. Near the *LC* resonant frequency, the surface impedance is very high. In this region, waves are not bound to the surface; instead, they radiate readily into the surrounding space as leaky waves.

11.3 HIGH-IMPEDANCE SURFACES

High-impedance surfaces consist of an array of metal protrusions on a flat metal sheet. The protrusions are arranged in a two-dimensional lattice and can be visualized as mushrooms or thumbtacks protruding from the surface. High-impedance surfaces are typically constructed as printed circuit boards, where the bottom side is a solid metal ground plane and the top contains an array of small ($\ll \lambda$) metal patches, as shown in Figure 11.1*b*. The plates are connected to the ground plane by metal-plated vias to form a continuous conductive metal texture. It can be considered as a two-dimensional version of the corrugated ground plane, where the quarter-wavelength resonant corrugations have been folded up into small resonant circuits and distributed on a two-dimensional lattice. For greater capacitance, multilayer circuit boards with overlapping plates can be used.

When the period is small compared to the wavelength of interest, we may analyze the material as an effective medium, with its surface impedance defined by effective lumped-element circuit parameters that are determined by the geometry of the surface texture. A wave impinging on the material causes electric fields to span the narrow gaps between the neighboring metal patches, and this can be described as an effective sheet capacitance *C*. As currents oscillate between the neighboring patches, the conducting paths through the vias and the ground plane provide a sheet inductance *L*. These form a parallel resonant circuit that dictates the electromagnetic behavior of the material, as shown in Figure 11.4. Its surface impedance is given by the expression

$$Z_s = \frac{j\omega L}{1 - \omega^2 LC} \tag{11.3}$$

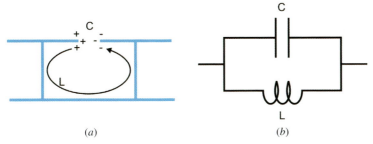

Figure 11.4 (*a*) The capacitance in a high-impedance surface is due to the proximity of the neighboring plates. The inductance comes from the current loops that are formed between the plates and the ground plane through the vias. (*b*) The impedance of the surface can be modeled as a parallel resonant *LC* circuit.

The resonance frequency of the circuit is given by

$$\omega_0 = \frac{1}{\sqrt{LC}} \tag{11.4}$$

Below resonance, the surface is inductive and supports TM waves. Above resonance, the surface is capacitive and supports TE waves. Near ω_0, the surface impedance is much higher than the impedance of free space, and the material does not support bound surface waves.

In addition to its unusual surface wave properties, the high-impedance surface also has unusual reflection-phase properties. In the frequency range where the surface impedance is very high, the tangential magnetic field is small, even with a large electric field along the surface. Such a structure is sometimes described as an artificial magnetic conductor. Because of this unusual boundary condition, the high-impedance surface can function as a new type of ground plane for low-profile antennas. The image currents in the ground plane are in-phase with the antenna current, rather than out of phase, allowing radiating elements to lie directly adjacent to the surface while still radiating efficiently. For example, a dipole lying flat against a high-impedance ground plane is not shorted as it would be on an ordinary metal sheet.

11.4   SURFACE WAVE BANDS

Many of the important properties of the high-impedance surface can be explained using an effective surface impedance model. The surface is assigned an impedance equal to that of a parallel resonant *LC* circuit, as described above. The use of lumped circuit parameters to describe electromagnetic structures is valid as long as the wavelength is much longer than the size of the individual features. The effective surface impedance model can predict the reflection properties and some features of the surface wave band structure, but not the bandgap itself, which by definition must extend to large wave vectors.

The wave vector k is related to the spatial decay constant α and the frequency ω by the dispersion relation

$$k^2 = \mu_0 \varepsilon_0 \omega^2 + \alpha^2 \qquad (11.5)$$

For TM waves we can combine Eq. (11.5) with Eq. (11.1) to find the following expression for k as a function of ω, in which η is the impedance of free space and c is the speed of light in vacuum:

$$k = \frac{\omega}{c}\sqrt{1 - \frac{Z_s^2}{\eta^2}} \qquad (11.6)$$

We can find a similar expression for TE waves by combining Eq. (11.5) with Eq. (11.2):

$$k = \frac{\omega}{c}\sqrt{1 - \frac{\eta^2}{Z_s^2}} \qquad (11.7)$$

Inserting Eq. (11.3) into Eqs. (11.6) and (11.7) we can plot the dispersion diagram for surface waves in the context of the effective surface impedance model. An example of the complete dispersion diagram, calculated using the effective medium model, is shown in Figure 11.5.

Below resonance, TM surface waves are supported. At low frequencies, they lie very near the light line, indicated in Figure 11.5 by the dotted line with a slope equal to the speed of light c. The fields extend many wavelengths beyond the surface, as they do on a flat metal sheet. Near the resonant frequency, the surface waves are tightly bound to the surface and have a very low group velocity. The dispersion curve is bent over away from the light line. In the effective surface impedance limit, there is no Brillouin zone [17] boundary, and the TM dispersion curve approaches the resonance frequency asymptotically. Thus, this approximation does not predict the bandgap.

Above the resonance frequency, the surface is capacitive and TE waves are supported. The lower end of the dispersion curve is close to the light line, and the waves are weakly bound to the surface, extending far into the surrounding space. As the frequency is increased, the curve bends away from the light line and the waves are more tightly bound to the surface. The slope of the dispersion curve

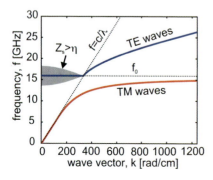

Figure 11.5   The effective surface impedance model can determine many of the properties of the high-impedance surface, including the shape and polarization of the surface wave bands. This is the predicted surface wave dispersion diagram for a surface with sheet capacitance of 0.05 pF and sheet inductance of 2 nH. The surface supports TM waves (red curve) below the resonance frequency and TE waves (blue curve) at higher frequencies. This model does not predict the bandgap, but it does predict a region of radiative loss.

indicates that the waves feel an effective index of refraction that is greater than unity. This is because a significant portion of the electric field is concentrated in the capacitors.

The TE waves that lie to the left of the light line exist as leaky waves that are damped by radiation, which can be modeled as a resistor in parallel with the high-impedance surface. The damping resistance is the impedance of free space, projected onto the surface at the angle of radiation. This blurs the resonance frequency, so the leaky waves actually radiate within a finite bandwidth. Small wave vectors represent radiation perpendicular to the surface, while wave vectors near the light line represent radiation at grazing angles. In place of a bandgap, the effective surface impedance model predicts a frequency band characterized by radiation damping.

In the effective impedance surface model described above, the properties of the textured surface are summarized into a single parameter—the surface impedance. This model correctly predicts the shape and polarization of the surface wave bands and also the reflection phase, to be described later. However, it does not predict the bandgap itself. For a more accurate picture of the surface wave properties, we can use a finite-element numerical model. The metal and dielectric regions are discretized on a grid, and the electric field at all points on the grid is described in terms of an eigenvalue equation, which may be solved numerically. A single unit cell is simulated, and Bloch boundary conditions [18] are used. The calculation yields the allowed frequencies for each wave vector. An example high-impedance surface is shown in Figure 11.6, along with the calculated dispersion diagram. The lowest band is TM, the second band is TE, and both have a similar shape to that predicted by the effective surface impedance model. A bandgap within which the surface does not support bound surface

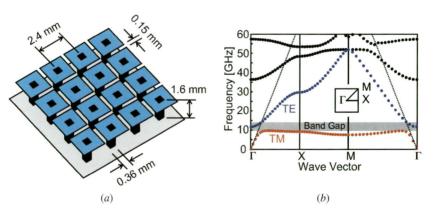

(a)   (b)

Figure 11.6 (a) The complete dispersion diagram can be obtained accurately using numerical methods, and square lattices are often easier to simulate. The substrate (not shown) has a relative dielectric constant of 2.2. (b) The lowest bands are qualitatively similar to that of the effective surface impedance model. The finite-element model also predicts a bandgap where bound surface waves of neither polarization are supported, between the first two bands. It also predicts several higher bands.

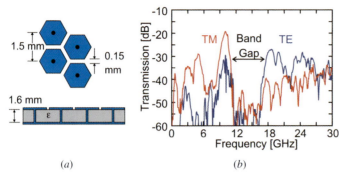

Figure 11.7 (*a*) Measurements were performed on a triangular lattice of hexagons built on a substrate with a relative dielectric constant of 2.2. (*b*) The high-impedance surface supports TM surface waves (red curve) at low frequencies and TE waves (blue curve) at high frequencies. Between these two bands is a gap within which waves of neither polarization are supported.

waves of either polarization extends from the top of the TM band to the point where the TE band crosses the light line. The finite-element model also predicts additional higher order bands that are not predicted by the simple effective surface impedance model.

Surface wave modes can be measured by recording the transmission between a pair of small coaxial probes placed near the surface. Depending on their orientation, the probes will excite surface waves with TM, TE, or both polarizations. An example of a high-impedance surface and the measured surface wave transmission across a 12-cm sample is shown in Figure 11.7 for both TM and TE polarizations. As predicted by both models described above, TM waves are supported at low frequencies and TE waves are supported at high frequencies. The TM and TE bands are separated by a bandgap within which bound surface waves of either polarization are not supported. For comparison, an electric conductor of the same size exhibits nearly flat transmission for TM waves at microwave frequencies, at around −30 dB, and very low transmission for TE waves, at around −60 dB.

11.5 REFLECTION PHASE

The surface impedance defines the boundary condition at the surface for the standing wave formed by incident and reflected waves. For a low-impedance surface, such as an electric conductor, the ratio of electric field to magnetic field is small. The electric field has a node at the surface, and the magnetic field has an antinode. Conversely, for a high-impedance surface, the electric field has an antinode at the surface while the magnetic field has a node. Another term for such a surface is again an artificial magnetic conductor. Using the effective surface impedance model described previously, we can determine the reflection phase for the resonant textured surface described above. For a normally incident wave,

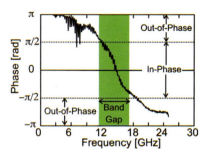

Figure 11.8 The reflection phase was measured for the same surface as shown in Figure 11.7. The phase is zero at the resonance frequency, but it approaches π for frequencies far from the resonance. The phase crosses through $\pi/2$ and $-\pi/2$ near the edges of the surface wave bandgap.

the reflection phase of the surface is given as

$$\Phi = \mathrm{Im}\left[\ln\left(\frac{Z_s - \eta}{Z_s + \eta}\right)\right] \tag{11.8}$$

In the above expression, Z_s is given by Eq. (11.3) and η is the impedance of free space. At very low frequencies, the reflection phase is π, and the structure behaves like a smooth metal surface. At higher frequencies, the reflection phase slopes downward and eventually crosses through zero at the resonance frequency, where it behaves as a magnetic conductor. Above the resonance frequency, the phase returns to $-\pi$. The phase falls within $\pi/2$ and $-\pi/2$ when the magnitude of the surface impedance exceeds the impedance of free space. The behavior of the reflection phase predicted by Eq. (11.8) is identical to the measured result shown in Figure 11.8. This reflection-phase curve was measured using the surface shown in Figure 11.7a. It is worth noting that for a wide range of geometries the edges of the surface wave bandgap occur at the same frequencies where the reflection phase crosses through $\pi/2$ and $-\pi/2$.

11.6   BANDWIDTH

An antenna lying parallel to the textured surface will see the impedance of free space on one side and the impedance of the surface on the other side. Where the textured surface has low impedance, far from the resonance frequency, the antenna current is mirrored by an opposing current in the surface. Since the antenna is shorted out by the nearby conductor, the radiation efficiency is very low. Within the bandgap near resonance, the textured surface has much higher impedance than free space, so the antenna is not shorted out. In this range of frequencies, the radiation efficiency is high.

The textured surface is modeled as an *LC* circuit in parallel with the antenna, and the radiation into free space is modeled as a resistor with a value of the impedance of free space. The amount of power dissipated in the resistor is a measure of the radiation efficiency of the antenna. The maximum radiation efficiency occurs at the *LC* resonance frequency of the ground plane, where the surface reactance is infinite. At very low frequencies or at very high frequencies, currents in the surface cancel the antenna current and the radiated power is reduced. It can be shown that the frequencies where the radiation drops to half of its maximum value occur when the magnitude of the surface impedance is equal to the impedance of free space, as described by the equation

$$|Z_s| = \eta \tag{11.9}$$

In the above expression, η is the impedance of free space and is given by

$$\eta = \sqrt{\frac{\mu_0}{\varepsilon_0}} \tag{11.10}$$

Substituting Z_s from Eq. (11.3) into Eq. (11.9), we can solve for ω:

$$\omega^2 = \frac{1}{LC} + \frac{1}{2\eta^2 C^2} \pm \frac{1}{\eta C}\sqrt{\frac{1}{LC} + \frac{1}{4\eta^2 C^2}} \tag{11.11}$$

The terms in $1/(\eta C)^2$ are typically small compared to the terms in $1/LC$, so we will neglect them. This approximation yields the following equation for the edges of the operating band:

$$\omega = \omega_0 \sqrt{1 \pm \frac{Z_0}{\eta}} \approx \omega_0 \left(1 \pm \frac{1}{2}\frac{Z_0}{\eta}\right) \tag{11.12}$$

In the above expression, Z_0 can be considered as a kind of characteristic impedance of the surface:

$$Z_0 = \sqrt{\frac{L}{C}} \tag{11.13}$$

The two frequencies designated by the plus and minus signs in Eq. (11.12) delimit the range over which an antenna would radiate efficiently on such a surface. The total bandwidth BW is roughly equal to the characteristic impedance of the surface divided by the impedance of free space:

$$BW = \frac{\Delta\omega}{\omega_0} \approx \frac{Z_0}{\eta} = \frac{\sqrt{L/C}}{\sqrt{\mu_0/\varepsilon_0}} \tag{11.14}$$

This is the bandwidth over which the phase of the reflection coefficient falls between $\pi/2$ and $-\pi/2$, and image currents are more in phase than out of phase. As noted in the previous section, this range often coincides with the surface wave bandgap. It also represents the maximum usable bandwidth of a flush-mounted antenna on a resonant surface of this type.

It can be shown that the inductance of the surface L is equal to the product of the permeability μ and the thickness t. Using Eq. (11.4) and substituting for L in Eq. (11.12), we can obtain a more useful expression for the bandwidth of a thin ($t \ll \lambda_0$), nonmagnetic ($\mu = \mu_0$), resonant textured ground plane:

$$BW = \frac{2\pi}{\lambda_0}t \tag{11.15}$$

In the above expression, λ_0 is the free-space wavelength at the resonance frequency. This result is significant because it proves that the bandwidth is determined entirely by the thickness of the surface with respect to the operating wavelength. Note that the dielectric constant of the substrate has no direct effect on the bandwidth, and dielectric loading cannot be used to reduce the thickness, except at the expense of bandwidth. A similar limitation exists for all small antennas, and their bandwidth is determined by a relation analogous to Eq. (11.15) for three-dimensional problems [19–21].

11.7   DESIGN PROCEDURE

The following is a general procedure for designing high-impedance surfaces for a required frequency and bandwidth. For an accurate design, numerical electromagnetic software should be used. A single unit cell can be simulated with minimal computing resources. Electric and magnetic conducting boundaries are used on opposing walls of the unit cell. Simulations of the reflection phase, the geometry, and the materials can be adjusted to provide the desired resonance frequency and bandwidth. However, it is useful to have an intuitive solution, to more rapidly converge on the correct design.

In the two-layer geometry shown in Figure 11.4, the capacitors are formed by the fringing electric fields between adjacent metal patches. For fringing capacitors, the capacitance can be approximated as

$$C_{\text{fringe}} \approx \frac{w(\varepsilon_1 + \varepsilon_2)}{\pi} \cosh^{-1}\left(\frac{a}{g}\right) \tag{11.16}$$

In the above expression, a is the lattice constant, g is the gap between the plates, w is the width of the plates, and ε_1 and ε_2 are the dielectric constants of the substrate and the material surrounding the surface, which may be free space. More accurate expressions for the fringing field capacitance exist [22], but Eq. (11.16) is adequate for first-order designs.

A three-layer design shown in Figure 11.9 achieves a lower resonance frequency for a given thickness by using capacitive loading. In this geometry, parallel-plate capacitors are formed by the top two overlapping layers. The capacitance can be calculated with the well-known equation

$$C_{\text{parallel}} \approx \frac{\varepsilon A}{d} \tag{11.17}$$

In this case, ε is the dielectric constant of the material between the plates, A is the area of the plates, and d is their separation.

In either case, the sheet capacitance is determined by the value of the individual capacitors and a geometric factor F that depends on the choice of lattice:

$$C = C_{\text{individual}} F \tag{11.18}$$

Figure 11.9   Thin high-impedance surface with a low resonance frequency can be built by using greater capacitive loading, such as overlapping plates, as shown in this three-layer structure. For a given resonance frequency, thinner structures have smaller fractional bandwidth.

The geometric factor takes into account the number of capacitors in series or parallel to convert the value of the individual capacitors to the sheet capacitance per square. For a square lattice $F = 1$, for a triangular lattice $F = \sqrt{3}$, and for a hexagonal grid of capacitors $F = 1/\sqrt{3}$. Examples of various lattices for a three-layer design with overlapping capacitors are shown in Figure 11.10.

The inductance of a high-impedance surface is determined entirely by its thickness. This can be understood by considering a solenoid of current that includes two rows of plates and their associated vias. Current flows up one row

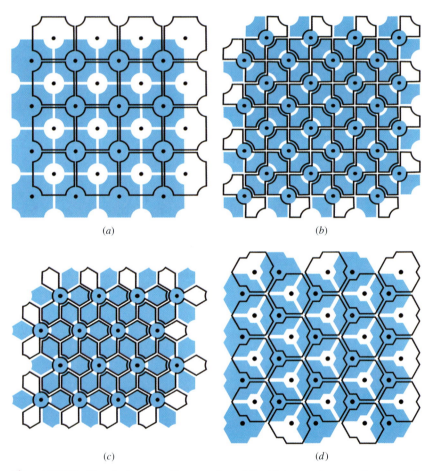

(a) (b)

(c) (d)

Figure 11.10 The electromagnetic properties of the high-impedance surface depend primarily on the surface capacitance and inductance and do not significantly depend on the geometry. Shown here are several three-layer structures, including (a) a square lattice with a completely overlapping layer, (b) a square lattice with two similar layers, (c) a triangular lattice, and (d) a hexagonal lattice, which is another form of the triangular lattice but with two vias per unit cell. In all cases, the shaded regions represent the lower metal layer and the outlined regions represent the upper metal layer. The solid dots represent conductive vias.

of vias, across the capacitors, and down the next set of vias to return through the ground plane. The length and width of the solenoid are canceled to obtain the sheet inductance

$$L = \mu t \qquad (11.19)$$

To design a surface for a desired frequency ω_0 and bandwidth BW, we combine Eqs. (11.4), (11.15), and (11.19). This procedure yields an equation for the required thickness:

$$t = \frac{c\mathrm{BW}}{\omega_0} \qquad (11.20)$$

It also provides an equation for the required sheet capacitance:

$$C = \frac{1}{\omega_0 \eta \mathrm{BW}} \qquad (11.21)$$

Finally, using either Eq. (11.16) or (11.17) together with Eq. (11.18), an appropriate geometry for the capacitors can be found. For the effective surface impedance approximation to be valid, the lattice constant should be small compared to the wavelength, and this often dictates whether a two- or three-layer structure should be used. Note that aside from the effects of the geometric factor F, the choice of lattice and shape and material composition of the capacitors has no effect on the electromagnetic properties of the surface as long as their value and arrangement follow the guidelines given above.

11.8 ANTENNA APPLICATIONS

The high-impedance surface can be used to provide several advantages for antenna applications using either the suppression or enhancement of surface waves or using its unusual reflection phase. Manipulation of surface wave effects can be demonstrated with a simple vertical monopole, shown in Figure 11.11a. It is fabricated by feeding a coaxial cable through a hole in the ground plane. The center conductor is extended through the other side to form a radiating wire, and the outer conductor is shorted to the ground plane.

On a finite metal ground plane, currents generated by the monopole are scattered at the edges of the ground plane. This can be seen as radiation in the backward direction and also as ripples in the forward portion of the radiation pattern because the scattered radiation interferes with the direct radiation from the monopole. Figure 11.11b shows the radiation pattern of a 3-mm monopole on a 5-cm^2 metal ground plane measured at 35 GHz.

If the metal ground plane is replaced with a high-impedance surface designed to resonate near 35 GHz, surface waves are suppressed, and the radiation pattern is changed. While driven currents can exist on any reflective surface, they do not propagate on the high-impedance ground plane. Any induced currents are restricted to a localized region around the antenna and never reach the edges of the ground plane. The absence of radiation from the edges results in a smoother radiation pattern, with less power in the backward direction, as

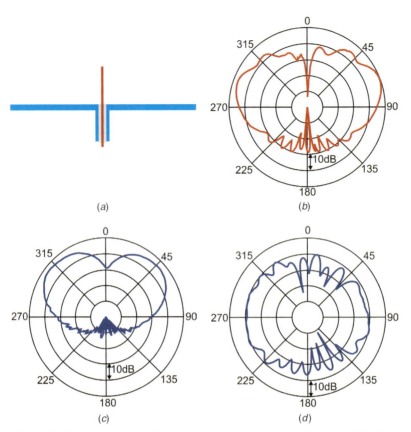

(a)   (b)

(c)   (d)

Figure 11.11   (*a*) A monopole antenna can be built by feeding a coaxial cable through a ground plane. The outer conductor is attached to the ground plane, and the inner conductor is extended to the other side to form the antenna. (*b*) On a flat metal ground plane, the monopole produces the expected radiation pattern. (*c*) On a high-impedance ground plane, at a frequency within the bandgap, the antenna produces a smooth pattern with reduced radiation in the backward direction. (*d*) Outside the bandgap, the antenna produces a complex pattern with significant power in the backward direction.

shown in Figure 11.11*c*. This could be used to reduce effects of nearby objects or discontinuities in the ground plane.

Two additional features are apparent in Figure 11.11*c*. First, the center null is diminished because of asymmetry in the local geometry of the antenna wire and the surrounding metal patches. With more symmetrical construction, the null could be recovered. Second, the received power is lower with the high-impedance ground plane, especially at the horizon. This is because the image currents on the high-impedance ground plane are reversed with respect to their direction on a metal ground plane. For a vertical monopole, this tends to cancel the radiation from the antenna current, particularly along the horizontal directions.

If the antenna is operated outside the bandgap of the high-impedance surface, where surface waves are supported, the radiation pattern is significantly different. Figure 11.11*d* shows the radiation pattern of the same antenna at 26 GHz within the TM surface wave band. The vertical monopole couples strongly into the surface wave modes, and a high density of states at the upper TM band edge also increases the amount of energy in the surface wave modes. Because of the presence of surface waves, the pattern contains many lobes and nulls and a significant amount of power in the backward direction. Such a pattern could be useful for applications requiring nearly omnidirectional radiation in environments where significant shadowing would otherwise occur.

While the vertical monopole illustrates the application of high-impedance surfaces for the suppression or enhancement of surface currents, it does not explain the advantage of the unusual reflection-phase properties. The benefits of an artificial magnetic conductor can be seen by using a horizontal wire antenna, as shown in Figure 11.12*a*. A simple wire antenna is fed through the back of the

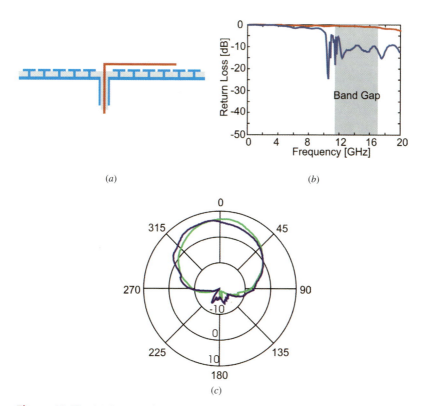

(*a*)

(*b*)

(*c*)

Figure 11.12   (*a*) Low-profile antennas can be built on high-impedance surfaces, such as a horizontal bent-wire antenna that is a small fraction of a wavelength above the surface. (*b*) The measured return loss of the horizontal wire antenna on the high-impedance surface (blue curve) is low within the bandgap. On a smooth metal ground plane (red curve), the antenna is shorted and does not radiate. (*c*) The radiation pattern of the horizontal wire antenna on the high-impedance ground plane is symmetrical, and the *E*-plane pattern (blue curve) is very similar to the *H*-plane pattern (green curve).

surface by a coaxial cable, in a manner similar to the monopole, and it is bent over across the surface. The wire is typically about one-half wavelength long at the resonance frequency of the surface. On a flat metal ground plane, a horizontal wire is shorted out, and most of the power transmitted to the feed is reflected back. However, on the high-impedance surface, a horizontal wire antenna is well matched if operated within the bandgap, as shown by the return loss in Figure 11.12*b*. The radiation pattern in Figure 11.12*c* indicates that the antenna produces significant gain, despite being roughly 1 mm above the ground plane. This is because the reflection phase of the surface is zero, rather than π, as with an ordinary conductor. Thus, currents in the high-impedance surface reinforce the currents in the wire, instead of canceling them as a smooth metal surface does. This effect can be used to build a variety of low-profile antennas that can lie directly adjacent to the artificial magnetic ground plane, such as antennas with various polarizations, including circular, as well as various directive radiation patterns.

11.9   TUNABLE IMPEDANCE SURFACES

The resonance frequency and the reflection phase of a high-impedance surface can be tuned by changing the effective capacitance, inductance, or both. However, without magnetically active materials, the inductance is determined entirely by the thickness of the surface and is difficult to tune. On the other hand, the capacitance can be controlled by changing the geometry and arrangement of the metal plates or by adding tunable lumped capacitors. Because the reflection phase is determined by the frequency of the incoming wave with respect to the resonance frequency, such a surface can perform as a distributed phase shifter. As the resonance frequency is swept from low to high values, the curve in Figure 11.8 is shifted from left to right, so the reflection phase at any fixed frequency varies from $-\pi$ to π.

An electrically tunable impedance surface can be built by connecting neighboring cells with varactor diodes. Changing the bias voltage on the diodes adjusts the capacitance and tunes the resonance frequency. To supply the required voltage to all of the varactors, we alternately bias half of the cells and ground the other half in a checkerboard pattern, as shown in Figure 11.13. At the center of each biased cell, a metal via passes through a hole in the ground plane and connects to a control line located on a separate circuit layer on the back of the surface. The varactors are oriented in opposite directions in each alternate row, so that when a positive voltage is applied to the control lines, all the diodes are reverse biased. By individually addressing each cell, the reflection phase can be programmed as a function of position across the surface.

The reflection phase for various bias conditions is shown in Figure 11.14. As the voltage across the varactors is increased, the capacitance decreases, and the resonance frequency increases. For a fixed frequency, the reflection phase increases with bias voltage. For frequencies within the tuning range, nearly any

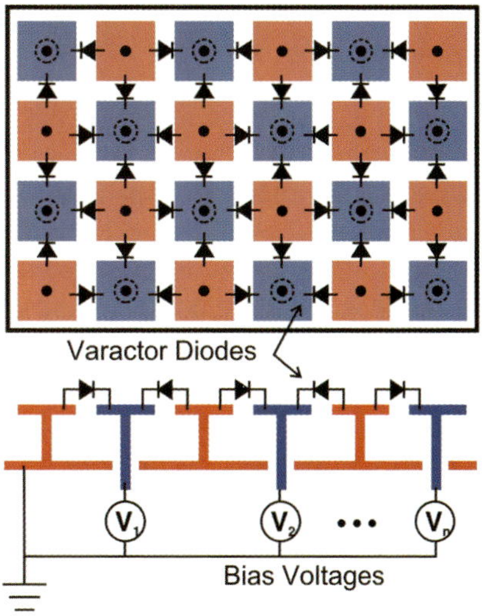

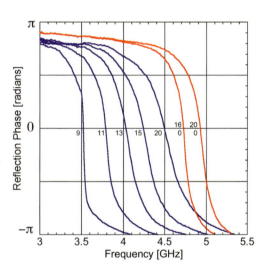

Figure 11.13 A tunable impedance surface consists of a high-impedance surface in which adjacent cells have been connected by varactor diodes, which have voltage-tunable capacitance. Half of the vias are grounded, but the other half are attached to a voltage control network on the back of the surface. The grounded and biased plates are arranged in a checkerboard pattern.

Figure 11.14 The reflection phase of the surface can be tuned electronically by varying the bias voltage on the varactors. Numbers by each curve represent voltages. The two red curves are for alternating voltages on every other row. For frequencies within the tuning range, nearly any reflection phase can be created by the appropriate choice of bias voltage.

reflection phase can be obtained by choosing the correct bias voltage. A series of measured data relating the reflection phase to frequency and voltage forms the basis of a calibration table that can be used to steer a reflected beam at any frequency within the tuning range.

11.10   REFLECTIVE-BEAM STEERING

If the reflection phase is programmed as a function of position across the surface, it can be used for beam steering. A linear phase gradient $\partial\phi(x, y)/\partial x$ will reflect a normally incident microwave beam to an angle θ that depends on the magnitude of the gradient:

$$\theta = 2\tan^{-1}\left(\frac{\lambda}{2\pi}\frac{\partial\phi(x, y)}{\partial x}\right) \qquad (11.22)$$

Other phase functions can be used for other tasks, such as a parabolic phase function for focusing. These concepts have been demonstrated previously using arrays of various resonant elements ranging from dipoles to patches, and beamforming structures employing this technique are commonly known as reflectarrays [23–28]. Tunable reflectarrays using varactor diodes, and related devices known as grid arrays [29, 30] have also been built. The tunable impedance surface has the advantage, when compared to other kinds of tunable reflectarrays, that the bias lines do not interfere with the microwave fields on the front side and two-dimensional steering is possible.

To create an electronically steerable reflector, the tunable impedance surface is illuminated with a microwave beam and a phase gradient is created electronically, as shown in Figure 11.15. To steer the beam into a particular angle, we calculate the required reflection phase gradient, as described by Eq. (11.18); select a frequency; and then calculate the corresponding voltages for each bias line based on a previously measured calibration table. The radiation patterns for several sets of control voltages corresponding to several beam-steering angles are shown in Figure 11.16. Since each of the cells is individually addressable through the bias lines in the back, the surface can steer in two dimensions. For this example, the surface is about 3.75 wavelengths square and operates at about 4.5 GHz. The surface can steer a reflected beam over $\pm40°$ for both polarizations. Wider steering angles would be possible with a larger surface.

Limitations on the varactor tuning range lead to limitations on the achievable phase range. Using a surface with a steeper phase curve can mitigate this

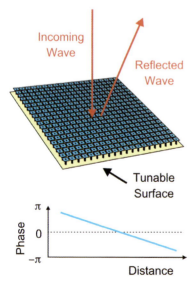

Figure 11.15   The tunable surface can be used as an electronic beam-steering reflector by programming the surface to have a reflection-phase gradient. A reflected microwave beam will be steered to an angle that depends on the phase gradient. This can serve as a simple electronically scanned antenna.

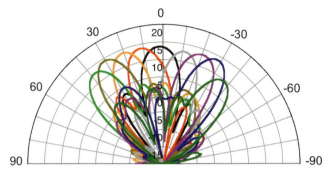

Figure 11.16 The beam can be steered over a range of ±40°, and greater steering would be possible with a larger structure. Each colored curve represents the radiation pattern for different beam-steering angles. The side lobes, which overlap at the center of the plot, are roughly 10 to 15 dB below the main beams.

problem, but at the expense of bandwidth. There is ultimately a trade-off between intrinsic surface bandwidth, varactor tuning range, and allowable phase error. These three parameters affect the side-lobe levels and usable bandwidth of the antenna.

11.11 LEAKY-WAVE BEAM STEERING

Despite being simple and low cost, steerable reflectors based on tunable impedance surfaces are ruled out for some applications because they require a free-space feed and thus are not entirely planar. An alternative is to use a leaky-wave design [31–34], where a surface wave is excited directly in the surface and then radiates energy into the surrounding space as it propagates. This method involves programming the surface with a periodic impedance function that scatters the surface wave into free space. The period of the surface impedance can be varied to change the phase-matching condition between the surface wave and the space wave and thus steer the radiated wave. The beam can be electronically steered over a wide range in both the forward and backward directions. The decay rate of the surface waves can also be controlled independently of the beam angle to allow adjustment of the aperture profile.

To build a steerable leaky-wave antenna, a feed structure is integrated into the tunable surface, such as the flared notch antenna shown in Figure 11.17. It can be as close as a small fraction of a wavelength from the surface, but it should not be close enough to detune the capacitance between the plates below it. A flared notch antenna will generate TE waves or a wire antenna can be used for TM waves.

A periodic pattern of voltages is applied to the tunable surface to create a periodic surface impedance function. When waves propagate across the surface, they are scattered by the nonuniform surface impedance. The scattered energy radiates at an angle determined by the wave vector of the surface wave and the periodicity of the surface impedance. The radiation angle may be determined by

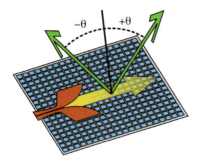

Figure 11.17 The tunable impedance surface (shown in blue) can be used as an electronically steerable leaky-wave antenna by incorporating a conformal feed, such as a flared notch antenna (orange). The surface wave (yellow) propagates away from the antenna, but the radiation (green) can propagate in either the forward or backward direction, depending on the phase-matching condition at the surface.

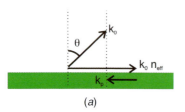
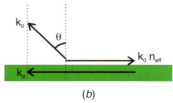

(a)   (b)

Figure 11.18 The direction of radiation is determined by phase matching at the surface. The tangential component of the wave vector of the space wave must match the difference between that of the surface wave and that of the periodic surface impedance. (a) Forward leaky waves are generated when the surface impedance has a period that is greater than the wavelength of the surface waves, corresponding to a shorter wave vector. (b) Backward leaky waves are generated when the period of the surface impedance is shorter than the wavelength of the surface wave.

assuming that a wave launched into the surface feels an effective refractive index of n_{eff}. Its wave vector is $k_0 n_{\text{eff}}$, where $k_0 = 2\pi/\lambda$ is the free-space wave vector. The surface impedance has period p, corresponding to a wave vector $k_p = 2\pi/p$. The scattered radiation in free space must have a total wave vector of k_0, and phase matching requires that it have a component parallel to the surface that is equal to the sum of the wave vectors of the surface wave and the surface impedance function. As illustrated in Figure 11.18, the radiation is scattered into the forward direction if $k_p < k_0 n_{\text{eff}}$ and it is scattered backward if $k_p > k_0 n_{\text{eff}}$. In general, the radiation angle is given by the expression

$$\theta = \sin^{-1}\left[\frac{k_0 n_{\text{eff}} - k_p}{k_0}\right] \tag{11.23}$$

For backward leaky waves, the energy still travels outward from the feed, so its group velocity is in the forward direction but its phase velocity, which determines the radiation angle, is in the backward direction. Leaky-wave structures capable of backward or broadside radiation have been studied extensively [35–37], but tunable impedance surfaces are novel because they can be electronically reconfigured to steer continuously from the forward to the backward direction at a single frequency. Figure 11.19 shows examples of radiation in both the forward and backward directions.

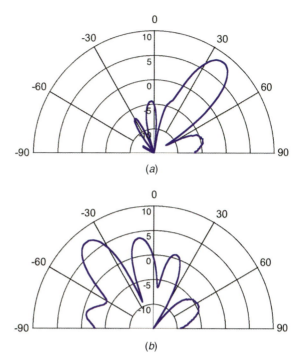

Figure 11.19 The surface can be configured for either forward or backward leaky-wave radiation. This shows examples of the leaky-wave radiation patterns when the surface is programmed for (*a*) forward radiation and (*b*) backward radiation.

11.12 BACKWARD BANDS

The existence of backward leaky waves can be analyzed in terms of backward bands, similar to those produced in other kinds of metamaterials [38–41]. In this section, we explore the properties of backward bands on textured surfaces and study their behavior through reflection measurements. Consider a tunable impedance surface in which alternate rows of plates are biased at two different voltages, thus creating rows of alternating capacitance values. For a TE wave, the electric field is transverse to the direction of propagation, so it sees alternating capacitance as it propagates from row to row. The effective lattice period is doubled, and the Brillouin zone is halved, as shown in Figure 11.20. The upper half of the TE band is folded into a reduced Brillouin zone, labeled BZ′. In the upper part of the TE band, the sign of the phase velocity ω/k is opposite to that of the group velocity $d\omega/dk$, so we may describe this as a backward band. The group velocity corresponds to the direction of energy propagation along the surface, which is always outward from the feed. The phase velocity, which determines the direction of radiation, progresses backward toward the feed.

Using mode analysis, the direction and relative strength of the electric field in the capacitors can be determined for various points on the band diagram. Groups of small arrows in Figure 11.14 illustrate the electric field in four adjacent rows. At the bottom of the TE band, the electric field is parallel throughout the entire surface. In the mode at the top of the backward band, the fields are

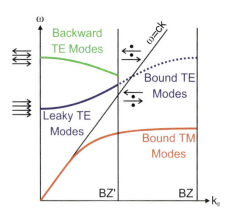

Figure 11.20   Backward leaky-wave radiation can be understood in terms of backward bands. If the surface is tuned so that every other row has alternate voltages, the TE waves will see a surface with a period that is twice as large. The Brillouin zone will be reduced by half, and the upper portion of the TE band will be folded into the reduced zone, labeled BZ′. The phase velocity and group velocity in this band will have opposite signs, corresponding to a backward wave. The fields in each row can be deduced by mode analysis for points at the edges of the bands.

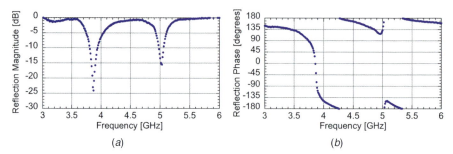

(a)   (b)

Figure 11.21   The presence of a backward band, as depicted in Figure 11.20, can be measured from the normal-incidence reflection properties of the surface. Modes at zero wave vector are visible as (a) dips in the magnitude and (b) corresponding curves in the phase. The second mode corresponds to the top of the backward band. It disappears if a uniform voltage is applied to the entire surface.

antiparallel in each adjacent row of capacitors. Because the capacitors have alternating values on every other row, the period of this mode matches that of the surface, with alternating capacitance values, and it lies at $k = 0$. For modes that occur at the edge of the folded zone, one-half wavelength fits in each period of two capacitors. Thus, the field is zero in alternate rows of capacitors and antiparallel in every other alternate row.

It is possible to detect the presence of the backward band using reflection measurements. Modes at $k = 0$ are standing waves that support a finite tangential electric field at the surface, and they can be identified by frequencies where the reflection phase is zero and by decreased reflectivity due to losses in the varactor diodes. Figure 11.21 shows the reflection magnitude and phase when adjacent rows were biased at 10 and 20 V. When two different voltages are applied to alternate rows, two modes are visible, corresponding to the lower edge of the forward TE band and the upper edge of the backward TE band. The presence of the second mode is experimental evidence of the backward band. It is not present when a uniform voltage is applied to all of the varactors.

11.13 SUMMARY

Subwavelength textures can be applied to metal surfaces to change their electromagnetic properties. These include corrugated structures to produce hard or soft boundary conditions and a variety of two-dimensional structures. Thin coatings containing lattices of grounded metal plates can behave as a high-impedance surface and can be analyzed using a simple lumped-circuit-parameter model. These surfaces have two important properties: (1) they suppress the propagation of surface waves within a bandgap and (2) they provide a reflection phase of zero at the resonance frequency. The bandwidth of these properties is related to the thickness of the surface. High-impedance surfaces can be used for a variety of antenna applications, such as to suppress scattering of surface waves by nearby structures or to build various kinds of low-profile antennas.

Electronically tunable impedance surfaces can be built by incorporating varactor diodes into the lattice. These can be used as electronically steerable reflectors for low-cost beam-steering applications. They can also be used as steerable leaky-wave antennas by incorporating a conformal feed. Leaky waves can be steered over a wide range of angles in both the forward and backward directions. Backward leaky waves can be understood in terms of backward bands, the presence of which can be measured directly. These backward bands are similar to those produced by other means in various other kinds of metamaterials.

REFERENCES

1. L. Brillouin, "Wave guides for slow waves," *J. Appl. Phys.*, vol. 19, no. 11, pp. 1023–1041, Nov. 1948.
2. W. Rotman, "A study of single-surface corrugated guides," *Proc. IRE*, vol. 39, pp. 952–959, Aug. 1951.
3. R. Elliot, "On the theory of corrugated plane surfaces," *IRE Trans. Antennas Propag.*, vol. 2, pp. 71–81, Apr. 1954.
4. A. Harvey, "Periodic and guiding structures at microwave frequencies," *IRE Trans.*, vol. 8, pp. 30–61, June 1959.
5. P.-S. Kildal, "Artificially soft and hard surfaces in electromagnetics," *IEEE Trans. Antennas Propag.*, vol. 38, pp. 1537–1544, June 1990.
6. S. Lee and W. Jones, "Surface waves on two-dimensional corrugated surfaces," *Radio Sci.*, vol. 6, pp. 811–818, Aug. 1971.
7. R. King, D. Thiel, and K. Park, "The synthesis of surface reactance using an artificial dielectric," *IEEE Trans. Antennas Propag.*, vol. 31, pp. 471–476, May 1983.
8. D. Sievenpiper, "High-impedance electromagnetic surfaces," Ph.D. dissertation, Department of Electrical Engineering, University of California, Los Angeles, CA, 1999.
9. D. Sievenpiper, L. Zhang, R. Broas, N. Alexopolous, and E. Yablonovitch, "High-impedance electromagnetic surfaces with a forbidden frequency band," *IEEE Trans. Microwave Theory Tech.*, vol. 47, pp. 2059–2074, Nov. 1999.
10. E. Yablonovitch, "Inhibited spontaneous emission in solid-state physics and electronics," *Phys. Rev. Lett.*, vol. 58, pp. 2059–2062, May 1987.
11. J. Joannopoulos, R. Meade, and J. Winn, *Photonic Crystals: Molding the Flow of Light*, Princeton University Press, Princeton, NJ, 1995.
12. W. Barnes, T. Priest, S. Kitson, and J. Sambles, "Physical origin of photonic energy gaps in the propagation of surface plasmons on gratings," *Phys. Rev. B*, vol. 54, pp. 6227–6244, Sept. 1996.

13. D. Sievenpiper, J. Schaffner, H. J. Song, R. Loo, and G. Tangonan, "Two-dimensional beam steering reflector using an electrically tunable impedance surface," *IEEE Trans. Antennas Propag.*, vol. 51, pp. 2713–2722, Oct. 2003.

14. D. Sievenpiper, "Forward and backward leaky wave radiation with large effective aperture from an electronically tunable textured surface," *IEEE Trans. Antennas Propag.*, vol. 53, pp. 236–247, Jan. 2005.

15. S. Ramo, J. Whinnery, and T. Van Duzer, *Fields and Waves in Communication Electronics*, 2nd ed., Wiley, New York, 1984.

16. H. Raether, *Surface Plasmons on Smooth and Rough Surfaces and on Gratings*, Springer-Verlag, New York, 1988.

17. L. Brillouin, *Wave Propagation in Periodic Structures*, McGraw-Hill, New York, 1946.

18. N. Ashcroft and N. Mermin, *Solid State Physics*, Saunders College Publishing, Orlando, FL, 1976.

19. H. Wheeler, "Fundamental limitations of small antennas," *Proc. IRE*, vol. 35, pp. 1479–1484, Dec. 1947.

20. L. Chu, "Physical limitations of omni-directional antennas," *J. Appl. Phys.*, vol. 19, pp. 1163–1175, Dec. 1948.

21. J. McLean, "A re-examination of the fundamental limits on the radiation Q of electrically small antennas," *IEEE Trans. Antennas Propag.*, vol. 44, pp. 672–675, May 1996.

22. S. Tretyakov and C. Simovski, "Dynamic model of artificial reactive impedance surfaces," *J. Electromagnetic Waves Appl.*, vol. 17, pp. 131–145, 2003.

23. D. G. Berry, R. G. Malech, and W. A. Kennedy, "The reflectarray antenna," *IEEE Trans. Antennas Propag.*, vol. 11, pp. 645–651, Nov. 1963.

24. R. D. Javor, X.-D. Wu, and K. Chang, "Design and performance of a microstrip reflectarray antenna," *IEEE Trans. Antennas Propag.*, vol. 43, no. 9, pp. 932–939, Nov. 1995.

25. D. C. Chang and M. C. Huang, "Multiple polarization microstrip reflectarray antenna with high efficiency and low cross-polarization," *IEEE Trans. Antennas Propag.*, vol. 43, pp. 829–834, Aug. 1995.

26. D. M. Pozar, S. D. Targonski, and H. D. Syrigos, "Design of millimeter wave microstrip reflectarrays," *IEEE Trans. Antennas Propag.*, vol. 45, no. 2, pp. 187–296, Feb. 1997.

27. M. E. Bialkowski and H. J. Song, "Investigations into a power combining structure using a reflect array of dual-feed aperture coupled microstrip patch antennas," *IEEE Trans. Antennas Propag.*, vol. 50, pp. 841–849, June 2002.

28. R. Waterhouse and N. Shuley, "Scan performance of infinite arrays of microstrip patch elements loaded with varactor diodes," *IEEE Trans. Antennas Propag.*, vol. 41, pp. 1273–1280, Sept. 1993.

29. W. Lam, C. Jou, H. Chen, K. Stolt, N. Luhmann, and D. Rutledge, "Millimeter wave diode grid phase shifters," *IEEE Trans. Microwave Theory Tech.*, vol. 36, pp. 902–907, May 1988.

30. L. B. Sjogren, H. X. Liu, X. Qin, C. W. Domier, and N. C. Luhmann, "Phased array operation of a diode grid impedance surface," *IEEE Trans. Antennas Propag.*, vol. 42, pp. 565–572, Apr. 1994.

31. G. Broussaud, "Un nouveau type d'antenne de structure plane," *Ann. Radioelectricite*, vol. 11, pp. 70–88, Jan. 1956.

32. J. W. Lee, J. J. Eom, K. H. Park, and W. J. Chun, "TM-wave radiation from grooves in a dielectric-covered ground plane," *IEEE Trans. Antennas Propag.*, vol. 49, no. 1, pp. 104–105, Jan. 2001.

33. C.-N. Hu and C.-K. C. Tzuang, "Analysis and design of large leaky-mode array employing the coupled-mode approach," *IEEE Trans. Microwave Theory Tech.*, vol. 49, pp. 629–636, Apr. 2001.

34. P. W. Chen, C. S. Lee, and V. Nalbandian, "Planar double-layer leaky wave microstrip antenna," *IEEE Trans. Antennas Propag.*, vol. 50, pp. 832–835, June 2002.

35. T. Tamer and F. Kou, "Varieties of leaky waves and their excitation along multilayered structures," *IEEE J. Quantum Electron.*, vol. 22, pp. 544–551, Apr. 1986.

36. M. Guglielmi and D. Jackson, "Broadside radiation from periodic leaky-wave antennas," *IEEE Trans. Antennas Propag.*, vol. 41, pp. 31–37, Jan. 1993.

37. S.-G. Mao and M.-Y. Chen, "Propagation characteristics of finite-width conductor-backed coplanar waveguides with periodic electromagnetic bandgap cells," *IEEE Trans. Microwave Theory Tech.*, vol. 50, pp. 2624–2628, Nov. 2002.

38. D. Smith, W. Padilla, D. Vier, S. Nemat-Nasser, and S. Schultz, "Composite medium with simultaneously negative permeability and permittivity," *Phys. Rev. Lett.*, vol. 84, pp. 4184–4187, May 2000.

39. D. Smith and N. Kroll, "Negative refractive index in left-handed materials," *Phys. Rev. Lett.*, vol. 85, pp. 2933–2936, Oct. 2000.

40. G. Eleftheriades, A. Iyer, and P. Kremer, "Planar negative refractive index media using periodically loaded *L-C* transmission lines," *IEEE Trans. Microwave Theory Tech.*, vol. 50, pp. 2702–2712, Dec. 2002.

41. L. Liu, C. Caloz, and T. Itoh, "Dominant mode leaky-wave antenna with backfire to endfire scanning capability," *Electron. Lett.*, vol. 38, pp. 1414–1416, Nov. 2002.

CHAPTER 12

DEVELOPMENT OF COMPLEX ARTIFICIAL GROUND PLANES IN ANTENNA ENGINEERING

Yahya Rahmat-Samii and Fan Yang

12.1   INTRODUCTION

Novel artificial electromagnetic materials, such as photonic crystals [1], electromagnetic bandgap (EBG) structures [2, 3], and double-negative (DNG) materials [4–6], have attracted increasing attention in the electromagnetics community. These structures are broadly classified as metamaterials [7] and are typically realized by periodic dielectric substrates and metallization patterns [8]. Metamaterials exhibit novel electromagnetic features that may not occur in nature, and they have led to a wide range of applications in the electromagnetics area.

The periodic metamaterials can be classified into two groups: (1) three-dimensional volumetric structures and (2) two-dimensional surface designs. This chapter focuses on the latter, which possesses the advantages of low profile, light weight, and low fabrication cost and hence is desirable in wireless communication systems.

The artificial surfaces have been investigated over many years, and representative examples include the frequency-selective surfaces (FSSs) [9], artificial soft and hard surfaces [10], and micromachined substrates [11, 12]. Recently, planar EBG surfaces [13, 14] have been proposed which exhibit distinctive electromagnetic properties with respect to incident electromagnetic waves:

1. When the incident wave is a surface wave ($k_x^2 + k_y^2 \geq k_0^2$), the analyzed structures show a frequency bandgap through which the surface wave cannot propagate for any incident angles and polarization states.

2. When the incident wave is a plane wave ($k_x^2 + k_y^2 < k_0^2$), the reflection coefficient of the analyzed structures is +1 at a certain frequency, which resembles a perfect magnetic conductor (PMC) that does not exist in nature.

In the above equations, k_x and k_y are the wavenumbers in the horizontal directions while k_0 is the free-space wavenumber. Various applications have been presented based on the above properties. For example, the surface wave bandgap has been utilized to enhance antenna gain, decrease back radiation, and reduce mutual

Metamaterials: Physics and Engineering Explorations, Edited by N. Engheta and R. W. Ziolkowski
Copyright © 2006 the Institute of Electrical and Electronics Engineers, Inc.

TABLE 12.1   Comparison of Conventional PEC and Artificial Ground Planes in Antenna Designs

Options	Efficiency	Low Profile
J ↑ —— PEC ↑	🙂	🙁
J → —— PEC ←	🙁	🙂
J → —— Artificial ground plane	🙂	🙂

coupling of microstrip antennas and arrays [15–17]. The in-phase reflection coefficient has been used to design transverse electromagnetic (TEM) waveguide [18] and low-profile wire and slot antennas [19, 20].

To illustrate the favorable features of the novel artificial ground plane in antenna engineering, Table 12.1 compares it with the traditional perfect electric conductor (PEC) ground plane in wire antenna designs. When an electric current is perpendicular to a PEC ground plane, its image current has the same direction and reinforces the radiation from the original current. Thus, this antenna has good radiation efficiency. However, the antenna height is relatively large because of the vertical placement of the current. To realize a low-profile design that is always desired in modern wireless communication systems, one may put a wire antenna horizontally close to the ground plane. However, the problem with this structure is the poor radiation efficiency because the image current from the PEC ground plane, which has an opposite flowing direction, cancels the radiation from the original current. The novel artificial ground plane, as will be discussed in this chapter, is capable of providing a constructive image current even with a low-profile configuration, resulting in good radiation efficiency. It overcomes the difficulty of the PEC ground plane to realize both low-profile and high-efficiency design goals. Therefore, the artificial ground plane has a great potential for antenna applications.

This chapter covers the analysis, design, and applications of complex artificial ground planes. Although we concentrate on a mushroomlike EBG structure [14], the approaches developed here are widely applicable for other types of artificial surfaces. The chapter starts with the finite-difference time-domain (FDTD) analysis of artificial ground planes. The bandgap properties of the surface are illustrated, and both the dispersion diagram and scattering features are presented. After the establishment of analysis methods, various artificial ground-plane designs will be presented, including EBG parametric study and polarization-dependent EBG designs. The important role of the conducting vias in the mushroomlike structure will also be discussed. Special emphasis will be given to the applications of the artificial ground planes in antenna engineering. A wealth of antenna examples will be introduced, ranging from wire antennas to microstrip antennas, from linearly polarized antennas to circularly polarized antennas, and from the conventional antenna structures to novel surface wave antenna concepts and reconfigurable antenna designs.

This chapter summarizes in a comprehensive and unified fashion the authors' latest work on the development of complex artificial ground planes in antenna engineering. We strongly believe that it will stimulate discussion and new avenues of research in this area.

12.2 FDTD ANALYSIS OF COMPLEX ARTIFICIAL GROUND PLANES

12.2.1 Bandgap Characterizations of an EBG Structure

A mushroomlike EBG structure was proposed in [14]. It consists of four parts: a ground plane, a dielectric substrate, periodic patches, and connecting vias, as shown in Figure 12.1. This structure exhibits a distinct stop band for surface wave propagation. The operation mechanism of this EBG structure can be explained by an LC filter array: the inductance L results from the current flowing through the vias, and the capacitance C is due to the gap effect between the adjacent patches. Some empirical formulas for the inductance L and the capacitance C are presented in [21, 22].

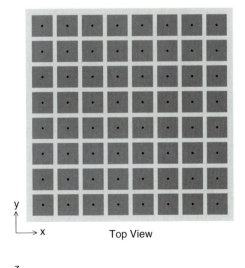

Figure 12.1 Geometry of mushroomlike EBG structure. From: F. Yang, A. Aminian, and Y. Rahmat-Samii, "A novel surface wave antenna design using a thin periodically loaded ground plane," *Microwave and Optical Technology Letters*, vol. 47, pp. 240–245, Nov. 2005, copyright © 2005 by John Wiley & Sons, Inc.

Top View

Cross View

To comprehensively understand the bandgap properties, the FDTD method [23, 24] is used to analyze the EBG structures. An EBG structure with the following parameters is analyzed:

$$W = 0.12\lambda_{12 \text{ GHz}} \qquad g = 0.02\lambda_{12 \text{ GHz}} \qquad h = 0.04\lambda_{12 \text{ GHz}} \qquad \varepsilon_r = 2.20$$

$$(12.1)$$

where W is the patch width, g is the gap width, h is the substrate thickness, and ε_r is the dielectric constant of the substrate. The free-space wavelength at 12 GHz, $\lambda_{12 \text{ GHz}}$, is used as a reference length for the EBG structure. The vias' radius is $0.005\lambda_{12 \text{ GHz}}$. The FDTD method is used to simulate the field distributions of a vertical infinitesimal dipole source surrounded by the EBG structure. For comparison purposes, a conventional case is also analyzed. This conventional (CONV) case consists of a PEC ground plane and a dielectric substrate with the same thickness and permittivity as the EBG case. The electromagnetic fields of these two cases are calculated and compared. Since the EBG structure can suppress the surface waves in a certain bandgap, the electromagnetic fields outside the EBG structure should be weaker than that of the CONV case.

To visualize the bandgap effect, the near field distributions of the EBG case and the CONV case are graphically presented. Figure 12.2 plots the near fields of both cases at 12 GHz, which is inside the bandgap. The field level is normalized to 1 W delivered power and is shown on a decibel scale. The field level outside the EBG structure is around 10 dB. In contrast, the field level of the CONV case is around 20 dB. The difference of field levels is due to the existence of the EBG structure, which suppresses the propagation of surface waves so that the field level in the EBG case is much lower than in the CONV

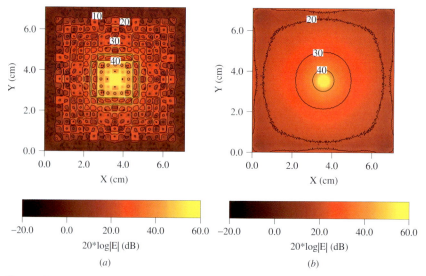

Figure 12.2 Near-field distributions at 12 GHz, which is inside the bandgap: (*a*) EBG case; (*b*) CONV case. The outside field of the EBG case is about 10 dB lower than that of the CONV case. From [17], copyright © 2003 by the Institute of Electrical and Electronics Engineers, Inc.

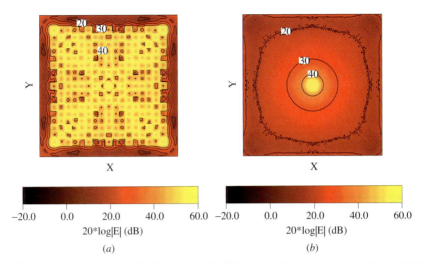

Figure 12.3 Near-field distributions at 10 GHz, which is outside the bandgap: (*a*) EBG case; (*b*) CONV case. The outside field of the EBG case has a similar level as that of the CONV case. From [17], copyright © 2003 by the Institute of Electrical and Electronics Engineers, Inc.

case. In contrast, Figure 12.3 plots the near fields of both cases at 10 GHz, which is outside the bandgap. The field distribution of the CONV case remains similar to its distribution at 12 GHz. However, the field value outside the EBG structure is increased to around 20 dB, which is similar to that of the CONV case. This means that although there are some interactions between the dipole source and the EBG structure, the field can still propagate through the EBG structure. From this comparison it is clear that surface waves can be suppressed by the EBG structure in a certain frequency band.

12.2.2   Modal Diagram and Scattering Analysis of EBG Structure

In the previous section, the FDTD method was used to analyze a finite EBG structure to illustrate its surface wave suppression behavior. With the utilization of periodic boundary conditions (PBCs), the FDTD method can also be used to analyze an infinite EBG structure. In this implementation, a single unit of the EBG structure with PBCs on four sides is simulated to model an infinite periodic structure. Both the modal diagram and scattering coefficient are calculated, and the bandgap frequency can be identified.

The dimensions of the analyzed EBG structure are

$$W = 0.10\lambda \qquad g = 0.02\lambda \qquad h = 0.04\lambda \qquad \varepsilon_r = 2.94 \qquad (12.2)$$

The vias' radius in the EBG structure is 0.005λ. The free-space wavelength at 4 GHz, $\lambda = 75$ mm, is used as a reference length to define the physical dimensions of the EBG structure. These parameters are selected for antenna applications that will be discussed in the following sections, and they are readily scaled to other frequencies of interests.

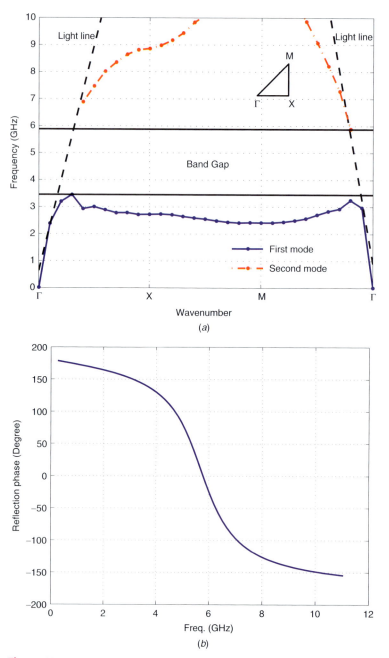

Figure 12.4 Analysis of EBG structure using FDTD method with PBCs: (*a*) $\omega-\beta$ diagram. From: F. Yang, A. Aminian, and Y. Rahmat-Samii, "A novel surface wave antenna design using a thin periodically loaded ground plane," *Microwave and Optical Technology Letters*, vol. 47, pp. 240–245, Nov. 2005, copyright © 2005 by John Wiley & Sons, Inc. (*b*) reflection phase. The EBG structure exhibits a surface wave bandgap in (*a*) and an in-phase reflection coefficient for plane-wave incidence in (*b*).

Figure 12.4*a* shows the $\omega-\beta$ modal diagram of the EBG structure characterized using a spectral FDTD method [25]. The vertical axis shows the frequency and the horizontal axis represents the values of the transverse wavenumbers (k_x, k_y). In the spectral FDTD method, each simulation outputs the frequencies of the surface wave modes for a given combination of wavenumbers (k_x, k_y). The simulation is repeated for 30 combinations of k_x and k_y in the Brillion zone, and the corresponding frequencies of surface waves are extracted and plotted in Figure 12.4*a*. Thus, each point in the modal diagram represents a certain surface wave mode. It is observed that no surface waves exist in the frequency range from 3.5 to 5.9 GHz. Thus, this frequency region is defined as a surface wave bandgap.

Besides the surface wave bandgap feature, the EBG structure also exhibits interesting plane-wave scattering behavior corresponding to the angle of incidence. Figure 12.4*b* shows the reflection phase curve for a normally incident plane wave. The reflection phase is defined as the phase of the reflected electric field normalized to the phase of the incident electric field at the reflecting surface. It is known that a PEC has an $180°$ reflection phase and a PMC has a $0°$ reflection phase. In contrast, the reflection phase of the EBG surface decreases continuously from $180°$ to $-180°$ as frequency increases. For example, the EBG surface exhibits a $90°$ reflection phase around 4.6 GHz and a $0°$ reflection phase around 5.8 GHz. It is worthwhile to point out that the reflection phase varies with incident angles and polarization states. By adjusting the setup of the incident waves [26], one can obtain the reflection phase for arbitrary incident angles and polarizations [27].

12.3 VARIOUS COMPLEX ARTIFICIAL GROUND-PLANE DESIGNS

12.3.1 Parametric Study of EBG Ground Plane

The mushroomlike EBG structure shows interesting behaviors with respect to surface waves and plane waves. These behaviors are frequency dependent, which is mainly determined by EBG parameters, namely, patch width W, gap width g, substrate permittivity ε_r, and substrate thickness h. In this section, some parametric studies are carried out to establish some engineering design guidelines.

The patch width plays an important role in determining the frequency behavior of the EBG structure. To study the effect of the EBG patch width, the gap width, substrate permittivity, and substrate thickness are kept constant while the patch width varies. Initial EBG parameters are

$$W = 0.12\lambda_{12 \text{ GHz}} \qquad g = 0.02\lambda_{12 \text{ GHz}} \qquad h = 0.04\lambda_{12 \text{ GHz}} \qquad \varepsilon_r = 2.20$$
(12.3)

The vias' radius is $0.005\lambda_{12 \text{ GHz}}$. The patch width then varies from $0.04\lambda_{12 \text{ GHz}}$ to $0.20\lambda_{12 \text{ GHz}}$. It is worthwhile to point out that these parameters can be easily scaled for operations at other frequencies. The reflection phases of the EBG surfaces with different patch widths are plotted in Figure 12.5*a* to characterize

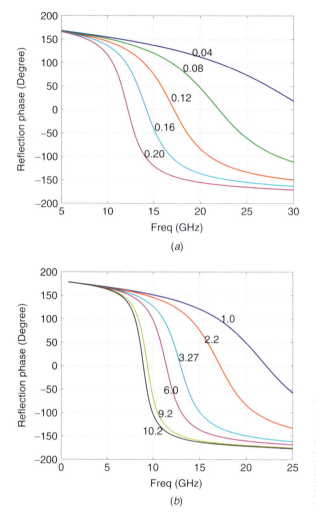

Figure 12.5 EBG parametric studies: effects of variations in (a) patch width W and (b) substrate relative permittivity ε_r. From [20], copyright © 2003 by the Institute of Electrical and Electronics Engineers, Inc.

the patch width effect. It is observed that when the patch width is increased, the frequency at which the in-phase reflection occurs decreases. At the same time, the slope of the reflection-phase curve becomes steep, which implies a narrow frequency band.

The relative permittivity ε_r, also called the dielectric constant of a substrate, is another effective parameter that determines the frequency behavior of the EBG structure. Some commonly used commercial materials such as RT/Duroid substrates and transfer matrix method (TMM) substrates have been investigated as well as air. The EBG structure analyzed here has the same parameters as given in (12.3) except that the permittivity is varied. The reflection phases of the EBG surfaces with various permittivities are plotted in Figure 12.5b. It is observed that when air is used as the substrate, the in-phase reflection frequency of the EBG surface has its highest value and the corresponding largest bandwidth. When the relative permittivity is increased, the frequency band decreases, as does the bandwidth. Therefore, when the RT/Duroid 6010 substrate ($\varepsilon_r = 10.2$) is used, the in-phase reflection frequency has its lowest value and the corresponding bandwidth also becomes very narrow. Therefore, the relative permittivity of the substrate of the EBG structure has an effect on its in-phase reflection behavior similar to that of the patch width but provides an additional degree of freedom to tune the EBG design. In practical applications, if one would like to design a compact EBG structure at a given frequency, increasing the permittivity is a feasible approach.

The gap width and substrate thickness also affect the operational frequency band of an EBG structure. For example, when the gap width is increased, the operational frequency will increase. In contrast, when the substrate thickness is increased, the operational frequency will decrease. Detailed data can be found in [20].

12.3.2   Polarization-Dependent EBG (PDEBG) Surface Designs

The mushroomlike EBG structure discussed in the previous sections uses square units so that its reflection phase for normal incidence is independent of the polarization of the incident plane wave. When the unit geometry is modified, EBG structures with polarization-dependent reflection phases can be obtained [28].

Figure 12.6a shows a PDEBG design using rectangular patch units. The reflection phase of the EBG surface becomes dependent on the X or Y polarization state of the incident plane wave. Figure 12.6b depicts reflection phases of the rectangular-patch EBG surface compared to a square-patch EBG surface. The dimensions of the EBG surfaces are given in the caption of Figure 12.6. When the incident plane wave is Y polarized, the rectangular-patch EBG surface has the same reflection phase as the square-patch EBG surface because the patch widths are the same. For the X-polarized incident plane wave, it is the patch length that determines the reflection phase. Since X-directed patch length is longer than the width, the reflection-phase curve shifts down to lower frequencies. It is noticed that near 3 GHz, the EBG surface shows a $-90°$ reflection phase for the X-polarized wave and a $+90°$ reflection phase for the Y-polarized wave. Thus, the phase difference between orthogonal polarizations is $180°$. This feature will be used in a circularly polarized dipole antenna design in the following section.

Another approach to realizing the polarization-dependent feature is to adjust the vias' location. When the vias are offset from the center of the patch, different reflection phases are obtained with respect to the polarizations of the incident plane wave. As shown in Figure 12.7a, consider the EBG structure in which the vias are offset along the X direction while they are still centered along the Y direction. Therefore, the reflection phase for the Y-polarized wave remains unchanged and the reflection phase for the X-polarized wave changes with the vias' position, as shown in Figure 12.7b. When the vias are located in the center of the patch, only one in-phase reflection frequency is observed. Once the vias are offset, in-phase reflection frequencies appear with one higher than the original frequency and the other lower. The different in-phase reflection frequencies

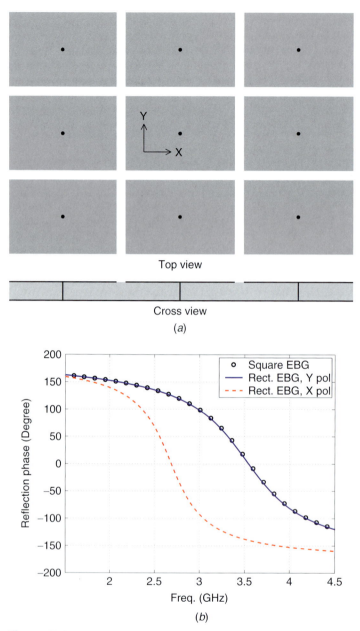

Figure 12.6 (*a*) Rectangular patch EBG surface ($0.24\lambda_{3\text{ GHz}} \times 0.16\lambda_{3\text{ GHz}}$).
(*b*) Reflection phases resulting from different polarizations of incident wave. A square-patch EBG surface ($0.16\lambda_{3\text{ GHz}} \times 0.16\lambda_{3\text{ GHz}}$) is used as a reference. The gap width is $0.02\lambda_{3\text{ GHz}}$ and the via radius is $0.0025\lambda_{3\text{ GHz}}$. The substrate thickness is $0.04\lambda_{3\text{ GHz}}$ and the dielectric constant is 2.20. From [28], copyright © 2004 by John Wiley & Sons, Inc.

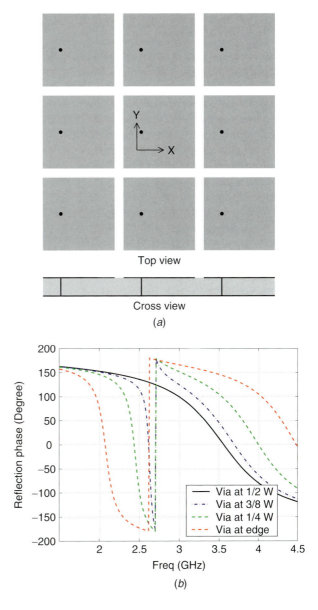

Figure 12.7 (*a*) EBG surface with offset vias. (*b*) Reflection phases with different via positions. From [28], copyright © 2004 by John Wiley & Sons, Inc.

correspond to the different widths of the left and right sides of the patch with respect to the via. The left part is narrower and thus is responsible for the higher frequency value. The right part is wider and thus is responsible for the lower frequency value. When the vias are closer to the patch edges, the separation of the two in-phase reflection frequencies increases because the width difference between the two sides of the patch becomes larger.

12.3.3 Characterizations of Grounded Slab Loaded with Periodic Patches

In the previous section it was shown that the vias' position has a significant effect on the in-phase reflection feature of the mushroomlike EBG structure. This section discusses the vias' effect from another viewpoint: existence of the vias. In particular, the properties of another complex artificial ground plane will be discussed: a grounded slab loaded with periodic patches, which is similar to the mushroomlike EBG structure except that the vertical vias are removed.

Figure 12.8 presents the dispersion diagram and reflection phase of the patch-loaded grounded slab. The dimensions of this artificial ground plane are the same as those given in Eq. (12.2). It is clear from Figure 12.8*a* that when the vertical vias are removed, the surface wave bandgap disappears. Therefore, the surface waves can exist over the entire frequency band.

However, removing the vias has little effect on the in-phase reflection features when the plane wave is normally incident. According to Figure 12.8*b* both the EBG surface (with vias) and the patch-loaded grounded slab (no vias) have very similar reflection-phase characteristics. The distinct surface wave and plane-wave features of the patch-loaded grounded slab have led to a novel surface wave antenna design, which will be discussed in the antenna application section.

Another point observed from Figure 12.8 is that the surface wave bandgap and in-phase reflection feature are not necessarily associated with each other. Therefore, one needs to be cautious when analyzing a complex artificial surface. The in-phase reflection coefficient does not guarantee the existence of the surface wave bandgap.

12.4 APPLICATIONS OF ARTIFICIAL GROUND PLANES IN ANTENNA ENGINEERING

12.4.1 Enhanced Performance of Microstrip Antennas and Arrays

Microstrip antennas are widely used in wireless communication due to their advantages, including a low-profile configuration, light weight, low fabrication cost, and conformability with radio frequency (RF) circuitry. In typical microstrip antenna designs, unwanted surface waves are excited in the substrate. These surface waves in turn degrade the antenna performance by decreasing the antenna gain, increasing the back lobe and the mutual coupling. This problem becomes more severe when high-dielectric-constant materials are used to design compact microstrip antennas. Several methods have been proposed recently to solve this problem. One approach suggested is to manipulate the antenna substrate using micromachining techniques [11, 12, 29]. A reduced surface wave (RSW) microstrip antenna structure is presented in [30] that does not excite surface waves. The EBG metamaterials [15–17] have also been utilized in microstrip antenna designs to reduce the surface wave effect. In this section, the mushroomlike EBG structure is integrated into several microstrip antenna designs. The antenna performance is compared to normal microstrip antenna designs to appreciate the benefits of using such EBG structures.

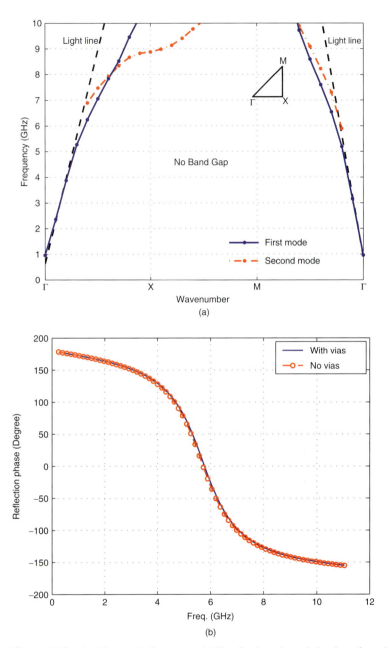

Figure 12.8 (*a*) The $\omega-\beta$ diagram and (*b*) reflection-phase behavior of patch-loaded dielectric slab that is similar to mushroomlike EBG structure except that vias are removed. As a result, the bandgap disappears while the reflection phase for a normally incident wave remains the same. From: F. Yang, A. Aminian, and Y. Rahmat-Samii, "A novel surface wave antenna design using a thin periodically loaded ground plane," *Microwave and Optical Technology Letters*, vol. 47, pp. 240–245, Nov. 2005, copyright © 2005 by John Wiley & Sons, Inc.

(*a*)   (*b*)

Figure 12.9   Photos of two microstrip patch antenna with enhanced radiation performance: (*a*) antenna surrounded by mushroomlike EBG structure; (*b*) antenna on steplike substrate.

First, the application of the mushroomlike EBG structure in a single microstrip antenna element [31] is analyzed. Figure 12.9*a* shows a patch antenna surrounded by such an EBG structure. The basic idea is to properly design the EBG structure so that the resonant frequency of the patch antenna falls inside the bandgap of the EBG structure; hence the surface waves that could propagate along the substrate will be inhibited. It should be emphasized that the EBG structure is very compact because a high dielectric constant and a thick substrate are employed. For comparison purposes, a microstrip patch antenna on an inhomogeneous substrate is also designed and examined, as plotted in Figure 12.9*b*. The idea is to use a thick substrate immediately under the patch, which helps to maintain its compact size and broad bandwidth, but then to use a thin substrate around the patch, which reduces the generation of surface waves. The substrate thus has a stepped shape.

To validate the above design concepts, four antennas were fabricated on a RT/Duroid 6010 ($\varepsilon_r = 10.2$) substrate with a finite ground plane that was 52 mm × 52 mm ($1\lambda \times 1\lambda$ at 5.8 GHz) in size. To establish references, two normal patch antennas were built on 1.27- and 2.54-mm-thick substrates, respectively. The steplike structure stacked two 1.27-mm-thick substrates under the patch and the distance from the patch edge to the step was 10 mm. The EBG structure was built on the 2.54-mm-thick substrate and its patch size was 2.5 mm × 2.5 mm with a 0.5-mm gap width. Four rows of EBG patches were used to suppress the surface waves.

Figure 12.10*a* depicts the measured S_{11} results of these four antennas. All the four patches were tuned to resonate at the frequency 5.8 GHz. It is noticed that the patch on the thin substrate has the narrowest impedance bandwidth, only 1 percent, while the other three have similar bandwidths, about 3 to 4 percent,

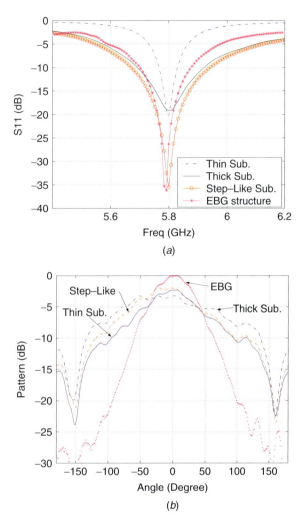

Figure 12.10 Experimental results of four microstrip antenna designs: (*a*) return loss results; (*b*) E-plane radiation patterns.

because of the thicker substrates. Figure 12.10*b* presents the *E*-plane radiation patterns of these antennas. The antenna on the thick substrate has the lowest front radiation while its back radiation is the largest. The steplike structure has a radiation performance that is similar to the antenna on the thin substrate. The radiation performance of the EBG structure is the best: Its front radiation is the highest, 3.24 dB higher than the thick case, and its back radiation is the lowest, more than 15 dB lower than the other cases. In summary, the EBG structure increases the gain and reduces the back lobe of a microstrip antenna while maintaining a similar impedance bandwidth.

After the successful implementation of the EBG structure in a single microstrip antenna element environment, the integration of the EBG structure with microstrip antenna arrays was explored with the objective of reducing the mutual coupling between elements in the array [17]. Two pairs of microstrip

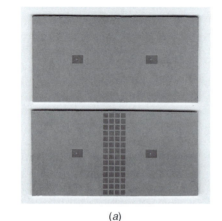

(a)

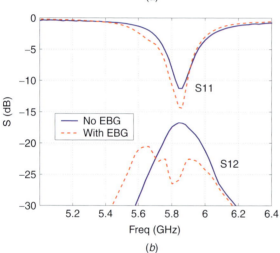

(b)

Figure 12.11 Experiments of microstrip antenna arrays with and without EBG structures. (*a*) Photographs of normal microstrip antenna array (above) and design that integrates EBG structure and antennas (below). (*b*) Measured scattering coefficients of microstrip antenna arrays. An 8-dB-mutual-coupling reduction is observed at the resonant frequency when the EBG structure is used. From [17], copyright © 2003 by the Institute of Electrical and Electronics Engineers, Inc.

antennas were fabricated on Roger RT/Duroid 6010 substrates. The relative permittivity of the substrate was 10.2, and the substrate thickness is 1.92 mm (75 mils). Figure 12.11*a* shows a photograph of the fabricated antennas with and without the EBG structure. The antenna's size was 6.8 mm × 5 mm, and they were fabricated on a finite ground plane whose size was 100 mm × 50 mm. The patch distance was 38.5 mm ($0.75\lambda_{5.8\,\text{GHz}}$). The EBG patch width was 3 mm and the gap width was 0.5 mm.

The measured results are shown in Figure 12.11*b*. It is observed that both antennas resonate at 5.86 GHz with a return loss better than −10 dB. For the antenna array without the EBG structure, the mutual coupling at 5.86 GHz was −16.8 dB. In comparison, the mutual coupling of the antennas with the EBG structure was only −24.6 dB. An 8-dB reduction of mutual coupling was achieved at the resonant frequency. This mutual coupling design is potentially useful for a variety of array applications, such as eliminating the blind angles in radar systems.

12.4.2 Dipole Antenna on EBG Ground Plane: Low-Profile Design

In the previous section, the surface wave bandgap feature of the EBG structure was utilized to improve the performance of microstrip antennas and arrays. As is known, the EBG structure has another important feature: the in-phase reflection coefficient for plane waves. This section will discuss how the in-phase reflection feature is used to design low-profile wire antennas [20]. The low-profile design usually refers to the antenna structures whose overall height is less than one-tenth of the operating wavelength, which is desirable in many wireless communication systems.

12.4.2.1 *Comparison of PEC, PMC, and EBG Ground Planes*   Let us begin
with a comparison of different ground planes in low-profile wire antenna design, for example, a PEC, a PMC, and an EBG surface. As shown in Figure 12.12, a dipole antenna is horizontally positioned near a ground plane to obtain a low-profile configuration. The PEC, PMC, and EBG surfaces are each used as the ground plane to compare their capabilities for low-profile designs [19].

The dipole length is $0.40\lambda_{12\text{ GHz}}$ and its radius is $0.005\lambda_{12\text{ GHz}}$, while $\lambda_{12\text{ GHz}}$, the free-space wavelength at 12 GHz, is used as a reference length to define the physical dimensions of various EBG and antenna structures studied in this section. A finite ground plane, having a $1\lambda_{12\text{ GHz}} \times 1\lambda_{12\text{ GHz}}$ size, is used in the analysis. The EBG structure has the following parameters:

$$W = 0.12\lambda_{12\text{ GHz}} \qquad g = 0.02\lambda_{12\text{ GHz}} \qquad h = 0.04\lambda_{12\text{ GHz}}$$
$$r = 0.005\lambda_{12\text{ GHz}} \qquad \varepsilon_r = 2.20 \tag{12.4}$$

The height of the dipole over the top surface of the EBG ground plane is $0.02\lambda_{12\text{ GHz}}$. Thus, the overall height of the dipole antenna from the bottom conductor of the EBG structure is $0.06\lambda_{12\text{ GHz}}$. The dipole height on the PEC and PMC ground plane is then set to $0.06\lambda_{12\text{ GHz}}$ so that all three cases have the same overall height.

Figure 12.13 compares the FDTD simulated return loss of dipole antennas over the PEC, PMC, and EBG ground planes. The input impedance in each case is matched to a 50-Ω transmission line. With the PEC surface as the ground plane, the return loss of the dipole is only -3.5 dB. This large value is obtained because

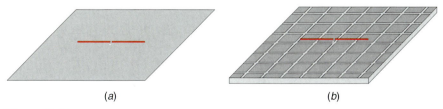

(a)   (b)

Figure 12.12   Dipole antennas over (*a*) PEC or PMC ground plane and (*b*) EBG ground plane. From [20], copyright © 2003 by the Institute of Electrical and Electronics Engineers, Inc.

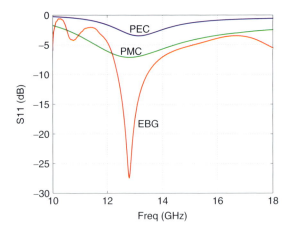

Figure 12.13 FDTD simulated return loss results of dipole antenna over PEC, PMC, and EBG ground planes. The dipole length is $0.40\lambda_{12 \text{ GHz}}$ and the overall antenna height is $0.06\lambda_{12 \text{ GHz}}$. From [20], copyright © 2003 by the Institute of Electrical and Electronics Engineers, Inc.

TABLE 12.2 Comparison of PEC, PMC, and EBG Ground Planes for Low-Profile Dipole Antenna Designs

Ground Plane	Reflection Phase	Return Loss (dB)	Comments
PEC	180°	−3.5	Reverse image
PMC	0°	−7.2	Mutual coupling
EBG	Varies from 180° to −180° with frequency	−27	Where is the suitable frequency band?

the PEC surface has a 180° reflection phase, so that the direction of the image current is opposite to that of the original dipole (see Table 12.1). The radiations from the image current and the original dipole cancel each other, resulting in a very poor return loss.

When the PMC surface, which has a reflection phase of 0°, is used as the ground plane, the dipole has a return loss of −7.2 dB. Despite the image being in phase with the actual source, this value is large because a strong mutual coupling occurs between the image current and the dipole due to their close proximity. This changes the input impedance of the dipole. Therefore, the antenna cannot be directly matched well to a 50-Ω transmission line. In addition, the PMC surface is an ideal surface that does not exist in nature.

The best return loss of −27 dB is achieved by the dipole antenna over the EBG ground plane. The reflection phase of the EBG surface varies with frequency from 180° to −180°. In a certain frequency band, the EBG surface successfully serves as the ground plane to achieve a low-profile dipole antenna, so that the dipole can radiate efficiently. From this comparison it is found that the EBG surface is a good ground-plane candidate for low-profile wire antenna designs.

Table 12.2 summarizes the comparisons of the PEC, PMC, and EBG ground planes for low-profile dipole antenna designs. An important question arises from the comparison: What is the suitable frequency band of the EBG structure that is used as a ground plane for a low-profile wire antenna?

12.4.2.2 Operational Frequency Band of EBG Structure

For an EBG structure, various frequency band definitions have been proposed. For example, a frequency bandgap was defined in [14] using the dispersion diagram. However, this definition only refers to the surface waves that propagate in the horizontal plane. In low-profile wire antenna applications, such a bandgap definition is not applicable because complicated interactions occur between the antenna and the EBG surface, and electromagnetic waves are not restricted to propagate in the horizontal plane. Thus, to ensure that resulting designs will meet the criteria for low-profile antenna applications, an operational frequency band of an EBG surface is defined as the frequency region inside which a low-profile wire antenna radiates efficiently with a good return loss and acceptable radiation patterns.

To search for this operational frequency band, the parameters of the EBG surface are fixed and the length of the horizontal dipole is varied to resonate at different frequencies. Since properties of the EBG surface such as the reflection phase change with frequency, the return loss and the radiation pattern of the dipole will change as well. By observing the return loss and radiation patterns of the dipole at different frequencies, one can find a useful operational frequency band of the EBG surface for low-profile wire antenna designs.

From a computational efficiency viewpoint, it would be interesting to know if one could directly use the reflection-phase feature of the EBG structure to identify the operational frequency band. To this end, a plane-wave model is established in the FDTD method and the reflection phase of the EBG surface is evaluated, as discussed in Section 12.2.2. The simulation results obtained from the low-profile dipole model and plane-wave model are compared to each other to establish a methodology as how to use the reflection-phase curve to identify the useful frequency band of the EBG/antenna structure.

The EBG surface analyzed here has the same parameters as given in (12.4). Figure 12.14a shows the return loss results of a dipole with its length varying from $0.26\lambda_{12 \text{ GHz}}$ to $0.60\lambda_{12 \text{ GHz}}$. The radius of the dipole remains $0.005\lambda_{12 \text{ GHz}}$. It is observed that the dipole shows a return loss better than -10 dB from 11.5 to 16.6 GHz. Figure 12.14b shows the reflection-phase results obtained from the plane-wave model. In contrast to the $180°$ reflection phase of a PEC surface and $0°$ reflection phase of a PMC surface, if one chooses the $90° \pm 45°$ reflection phases as the criterion for the EBG surface, one obtains a frequency region from 11.3 to 16 GHz, which is nearly the same frequency region as obtained in the dipole model.

From this comparison, it is revealed that the reflection-phase feature of an EBG surface may be used to identify the operational frequency band for low-profile wire antenna applications. The operational frequency band where a low-profile wire antenna obtains a good return loss is the frequency region inside which the EBG surface shows a reflection phase in the range $90° \pm 45°$.

The radiation patterns of dipole antennas on the EBG surface are also calculated to verify the radiation efficiency. Figure 12.15 displays both the E-

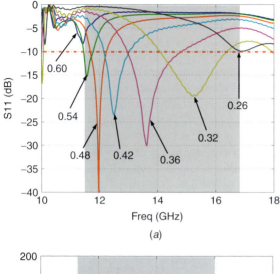

(a)

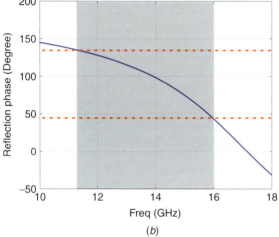

(b)

Figure 12.14 Comparison of frequency band results from two FDTD models: (a) return loss results when dipole length is varied from $0.26\lambda_{12\ GHz}$ to $0.60\lambda_{12\ GHz}$; (b) reflection phase of EBG surface. From [20], copyright © 2003 by the Institute of Electrical and Electronics Engineers, Inc.

and H-plane patterns of three dipole antennas at their resonant frequencies: (1) $0.48\lambda_{12\ GHz}$ dipole, which resonates at 12 GHz; (2) $0.36\lambda_{12\ GHz}$ dipole, which resonates at 13.6 GHz; and (3) $0.32\lambda_{12\ GHz}$ dipole, which resonates at 15.3 GHz. It is observed that all three dipoles radiate efficiently and have directivities around 8 dB.

The quadratic reflection-phase criterion ($90° \pm 45°$) has been further verified by several EBG cases with different parameters, as analyzed in [20]. In addition, this criterion has been demonstrated by experimental results of a low-profile curl antenna over the EBG ground plane [32]. In conclusion, the quadratic reflection phase ($90° \pm 45°$) of the EBG structure, different from the PEC and PMC surfaces, answers the question raised by the results in Table 12.2. It appears to provide a useful guideline to design EBG ground planes for low-profile wire antenna designs.

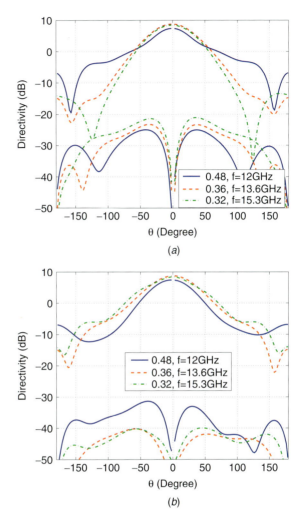

Figure 12.15 Radiation patterns of three dipoles at their resonant frequencies: (*a*) *E*-plane pattern; (*b*) *H*-plane pattern. The radiation patterns show that dipoles radiate efficiently throughout the frequency band in Figure 12.14. From [20], copyright © 2003 by the Institute of Electrical and Electronics Engineers, Inc.

12.4.3   Novel Surface Wave Antenna Design for Wireless Communications

In the previous section, it was revealed that the quadratic reflection phase of an EBG ground plane enables one to obtain a good return loss for a half-wavelength dipole with a low-profile configuration. Recall that a patch-loaded grounded slab was investigated in Section 12.3.3 that had the same reflection phase as the corresponding mushroomlike EBG surface but had no surface wave bandgap due to the removal of the vertical vias. Therefore, it is interesting to examine the performance of a horizontal dipole near this type of artificial ground plane.

12.4.3.1 Antenna Performance   Figure 12.16*a* shows the geometry of a dipole antenna near a patch-loaded grounded slab. The dipole is fed by a 50-Ω coaxial cable. One arm of the dipole is connected to the center conductor of the

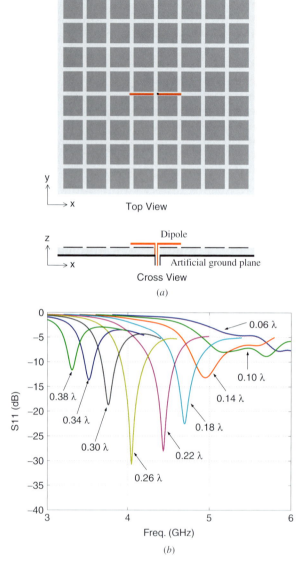

Figure 12.16 Horizontal dipole near patch-loaded grounded slab: (*a*) antenna geometry; (*b*) FDTD simulated return loss for dipoles with different lengths. From: F. Yang, A. Aminian, and Y. Rahmat-Samii, "A novel surface wave antenna design using a thin periodically loaded ground plane," *Microwave and Optical Technology Letters*, vol. 47, pp. 240–245, Nov. 2005, copyright © 2005 by John Wiley & Sons, Inc.

cable, and the other arm is connected to the outside conductor of the cable, which is soldered to the lower PEC of the complex surface. It is important to point out that no balun is used in the feed structure.

The FDTD method is used to simulate the behavior of this radiating structure. The dimensions of the artificial ground plane are the same as given in Eq. (12.2) in Section 12.2.2, and the electromagnetic properties of this surface, that is, dispersion diagram and reflection-phase curve, are depicted in Figure 12.8. The dipole with a radius of $0.005\lambda_{4 \text{ GHz}}$ is positioned $0.02\lambda_{4 \text{ GHz}}$ above the top surface of the artificial ground plane in order to get a low-profile configuration. The return loss results of the dipoles with different lengths are presented in Figure 12.16b. The dipole length varies from $0.06\lambda_{4\ GHz}$ to $0.38\lambda_{4\ GHz}$. The reader is reminded that when the length of the dipole is increased, its resonant frequency decreases. The frequency range inside which the antenna can obtain a good return loss ($S_{11} < -10$ dB) is close to the frequency region (3.7 to 5.3 GHz) where the artificial surface exhibits a reflection phase in the range $90° \pm 45°$ (see Figure 12.8). This observation agrees with the conclusion drawn in the previous section.

When the dipole length is $0.26\lambda_{4\ GHz}$, the antenna achieves a good return loss of -30 dB at 4.05 GHz with a 7.1 percent impedance bandwidth. It is worthwhile to point out that the length of the dipole is much smaller than the half wavelength of the operating frequency. As a comparison, when a dipole is located near an EBG ground plane and resonates at the same frequency, the length of the dipole is close to a half wavelength.

Figure 12.17a shows the radiation patterns of the dipole antenna illustrated in Figure 12.18 at its resonant frequency of 4.05 GHz. Several interesting observations can be made from this figure. First, the antenna shows a small radiation power in the broadside direction ($\theta = 0°$). This is different from the dipole antenna on an EBG ground plane whose main beam points to the broadside direction, as shown in Figure 12.15. The main beam of this antenna points to the $\theta = 50°$ direction with a directivity of 5 dB. Second, E_θ is the copolarized field in both the xz ($\varphi = 0°$) and yz planes ($\varphi = 90°$). Thus, both the xz and yz planes are E planes. In contrast, if the dipole is near an EBG ground plane, the xz plane is the E plane but the yz plane is the H plane. Therefore, this antenna has an entirely different radiation pattern, in contrast to a horizontal dipole on an EBG ground plane.

12.4.3.2 Radiation Mechanism

To understand the different behaviors of a dipole antenna over the EBG ground plane and over the patch-loaded grounded slab, the radiation mechanisms of these two antennas are examined. When a dipole is positioned near an EBG ground plane, no surface wave can be excited because the EBG structure has a bandgap for surface waves. The radiation is contributed by the dipole; thus the dipole length is close to its half wavelength for resonance. The radiation of the dipole dictates the antenna beam direction and polarizations.

In contrast, for the patch-loaded grounded slab, since the vertical vias are removed, the bandgap disappears and the surface wave can propagate along the ground plane. When a dipole is positioned near such a patch-loaded grounded slab, it excites strong surface waves. The dipole works more like a transducer rather than a radiator. Therefore, the optimal length of the dipole is not necessarily equal to a half wavelength. It is also noticed that the surface waves are dominated by the transverse magnetic TM_z mode (z axis being normal to the ground plane), and the electric field is vertically polarized. When the TM surface waves diffract at the boundary of the ground plane, the radiation pattern is determined. For example, since the diffractions at the edge are hard boundary

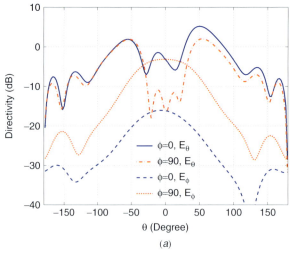

(a)

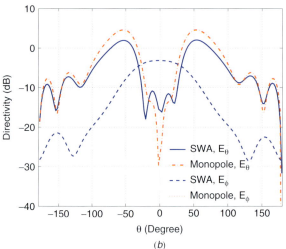

(b)

Figure 12.17 (a) Radiation patterns of antenna in Figure 12.16. (b) Comparison of radiation pattern of surface wave antenna (SWA) and conventional vertical monopole antenna. From: F. Yang, A. Aminian, and Y. Rahmat-Samii, "A novel surface wave antenna design using a thin periodically loaded ground plane," *Microwave and Optical Technology Letters*, vol. 47, pp. 240–245, Nov. 2005, copyright © 2005 by John Wiley & Sons, Inc.

diffractions, two diffracted rays from opposite edges will cancel in the broadside direction, resulting in a radiation null. The vertically polarized surface waves also cause the radiation field to be polarized in the θ direction. Thus, a monopole-type radiation pattern is generated. Therefore, this antenna can be identified as a SWA due to its radiation mechanism.

The attractive feature of the SWA design is the low-profile configuration. As shown in Figure 12.17b, the SWA has a similar radiation pattern as a vertical monopole antenna. However, the height of the horizontal dipole over the artificial ground plane is only $0.02\lambda_{4\text{ GHz}}$, whereas the height of the vertical monopole antenna is $0.22\lambda_{4\text{ GHz}}$. Thus, the dipole height is less than 10 percent of the monopole antenna. Therefore, this low-profile SWA design has a promising potential in wireless communication systems such as vehicle radio systems.

12.4.4   Low-Profile Circularly Polarized Antennas: Curl and Dipole Designs

In the previous sections, complex artificial ground planes were utilized in linearly polarized antenna designs. This section focuses on their applications in circularly polarized (CP) antenna designs that are desired in many communication systems such as Global Positioning System (GPS) and satellite links.

12.4.4.1   *Curl Antenna on EBG Ground Plane*   The first approach to design a low-profile CP antenna begins with a curl antenna. The curl antenna was proposed as a simple radiator to generate a circular polarization pattern [33]. However, it does not function well when it is placed close to a conventional PEC ground plane because of the reversed image current. To improve the radiation efficiency, an EBG ground plane was used to replace the PEC ground plane [34], as shown in Figure 12.18.

The EBG surface was built on a 2-mm-thick RT/Duroid 5880 ($\varepsilon_r = 2.20$) substrate. The EBG patch size was 6 mm × 6 mm and the gap between patches was 1 mm wide. The patches were connected to the ground plane by vias in the center of the patches. The EBG structure was truncated at 52 mm × 52 mm (about $1.20\lambda \times 1.20\lambda$ at 7 GHz). The parameters of the curl antenna were

$$R = 5.5 \text{ mm} \qquad L = 5 \text{ mm} \qquad h = 3 \text{ mm} \tag{12.5}$$

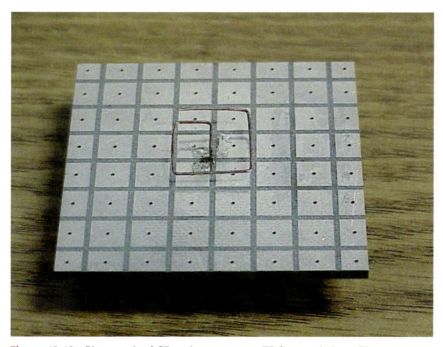

Figure 12.18   Photograph of CP curl antenna over EBG ground plane. The antenna height is only 0.07λ. From [32], copyright © 2001 by John Wiley & Sons, Inc.

where R is the distance between the center to the first edge of the square curl, L is the excessive length over one round of the curl, and h is the curl height above the EBG ground plane. The distance from the center to the edge of the square curl increases 0.5 mm with every turn.

Figure 12.19a compares the return loss of the curl antenna over the EBG and PEC ground planes. Since the height, 3 mm, above the ground plane was only 0.07λ, the curl antenna over the PEC ground plane was not matched well. In contrast, the curl over the EBG ground plane showed a good match in the frequency range from 6 to 8.5 GHz due to its in-phase reflection feature. Figure 12.19b shows the measured antenna AR at broad side versus frequency. The curl antenna over the EBG surface achieved a good AR of 0.9 dB at 7.18 GHz. It should be pointed out that according to [20] the reflection phase of this EBG structure

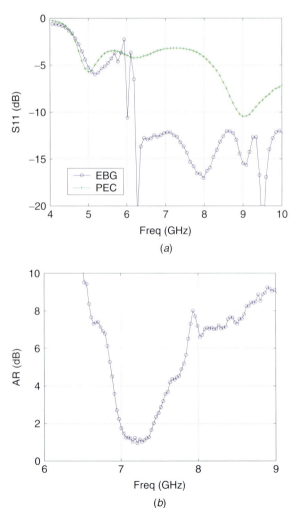

Figure 12.19 Measured results of low-profile CP curl antenna: (a) return loss results; (b) axial ratio (AR) at broadside direction. From [32], copyright © 2001 by John Wiley & Sons, Inc.

at 7.18 GHz is around $90°$. An 8.4 percent CP bandwidth (AR $<$ 3 dB) was obtained for this design.

12.4.4.2 Single-Dipole Antenna Radiating CP Waves

The second approach that was studied to realize a low-profile CP design utilized linearly polarized antenna elements, such as dipole antennas, instead of the CP elements. Reference [35] uses two perpendicular dipole antennas with quadrature feeding phases to realize the desired CP pattern. Here, a novel antenna design is presented that radiates CP waves with only a single dipole element. The unique idea in this design is to use an artificial surface with polarization-dependent reflection phases [36] as the ground plane for a horizontal dipole. The artificial ground plane operates like a meander line polarizer [37] to convert the linear polarization to circular polarization while achieving a low-profile configuration of the overall antenna structure.

Figure 12.20a shows the geometry of a horizontal dipole near a typical ground plane. The height of the dipole over the ground plane is very small compared to the operating wavelength, such as 0.02λ. The dipole is oriented along the $\phi = 45°$ direction. The total radiation field in the broadside direction can be approximated by the summation of the field directly radiated from the dipole and the field reflected from the ground plane:

$$\mathbf{E} = \mathbf{E}^d + \mathbf{E}^r = \tfrac{1}{2}E_0(\hat{x} \cdot e^{-jkz} + \hat{y} \cdot e^{-jkz}) + \tfrac{1}{2}E_0(\hat{x} \cdot e^{-jkz-2jkd+j\theta_x} \\ + \hat{y} \cdot e^{-jkz-2jkd+j\theta_y}) \qquad (12.6)$$

where E_0 denotes the magnitude of electrical fields, d is the height of the dipole over the ground plane, θ_x is the reflection phase of the ground plane for an x-polarized incident wave, and θ_y is its value for a y-polarized incident wave. The time-harmonic variation is represented by $e^{j\omega t}$. When the dipole is located very close to the ground plane, the phase value $2kd$ is very close to zero.

If the ground plane is a PEC, then the reflection phases $\theta_x = \theta_y = 180°$. Thus, the total radiating field becomes zero in (12.6) because the reflected field cancels the directly radiating field. When a PMC is used as the ground plane, the reflection phases $\theta_x = \theta_y = 0$. Then, $\mathbf{E} = E_0 e^{-jkz}(\hat{x} + \hat{y})$. The dipole still radiates linearly polarized waves. In contrast, when an artificial surface with polarization-dependent reflection phases, namely $\theta_x = 90°$ and $\theta_y = -90°$, is used as the ground plane, the radiation field becomes

$$\mathbf{E} = \tfrac{1}{2}E_0 e^{-jkz}[(\hat{x} + \hat{y}) + j(\hat{x} - \hat{y})] \qquad (12.7)$$

The reflected field becomes perpendicular to the directly radiating field with a $90°$ phase difference. Therefore, a right-hand circularly polarized (RHCP) wave is obtained.

To verify the antenna concept, an artificial ground plane with polarization-dependent reflection phases was designed and fabricated, and a dipole antenna was then mounted on this surface, as shown in Figure 12.20b. The periodic artificial ground plane consisted of rectangular patch units, as discussed in Section 12.3.2. A RT/Duroid 6002 high-frequency laminate ($\varepsilon_r = 2.94 \pm 0.04$) with 6.10 mm thickness was used as the substrate. The width of the patch was

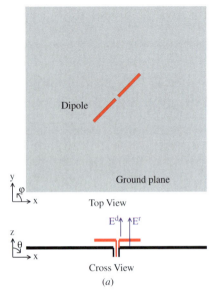

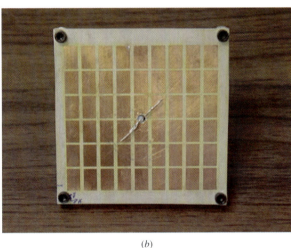

(b)

Figure 12.20 Low-profile CP dipole antenna design using artificial ground plane: (a) dipole antenna oriented along $\phi = 45°$ direction near ground plane; (b) photograph of tilted dipole near artificial ground plane. From: F. Yang and Y. Rahmat-Samii, "A low profile single dipole antenna radiating circularly polarized waves," *IEEE Trans. Antennas Propagat.*, vol. 53, pp. 3083–3086, Sept. 2005, copyright © 2005 by the Institute of Electrical and Electronics Engineers, Inc.

8 mm and its length was 13 mm. The gap width between adjacent patches was 2 mm along the x direction and 1 mm along the y direction. A finite ground plane with a 100×100 mm size was used, which included 9×6 rectangular patches. A 45° oriented dipole was positioned in the center of the ground plane. The length of the dipole was 34 mm, the height was 3 mm, and the radius was 0.34 mm. The antenna was fed by a 50-Ω coaxial cable.

The measured return loss result is shown in Figure 12.21a. It was better than -10 dB in the frequency range of 3.25 to 4.14 GHz. The AR of the antenna at the broadside direction was measured, as plotted in Figure 12.21b. An AR of

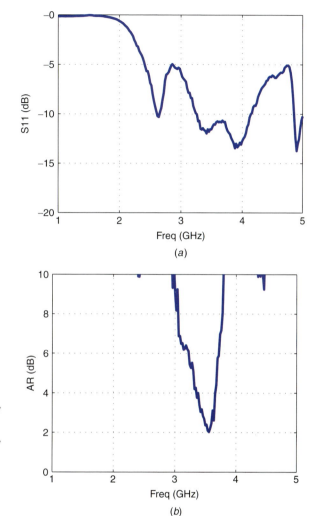

Figure 12.21 Measured results of low-profile CP dipole antenna: (*a*) return loss result; (*b*) axial ratio at broadside direction. From: F. Yang and Y. Rahmat-Samii, "A low profile single dipole antenna radiating circularly polarized waves," *IEEE Trans. Antennas Propagat.*, vol. 53, pp. 3083–3086, Sept. 2005, copyright © 2005 by the Institute of Electrical and Electronics Engineers, Inc.

2 dB was obtained at 3.56 GHz. The 3-dB AR bandwidth was 200 MHz (3.45 to 3.65 GHz, 5.6 percent).

12.4.5   Reconfigurable Wire Antenna with Radiation Pattern Diversity

Reconfigurable antennas are desirable for modern wireless communication systems because they can provide more functionalities than classic antenna designs [38–40]. They are capable of fulfilling various requirements of wireless communication systems by reconfiguring their radiation performance such as their operating frequencies [41,42], polarizations [43,44], radiation patterns [45–47], or combinations [48–50]. The reconfigurability is realized through different approaches, including mechanical tuning [51], semiconductor devices such as diodes or varactors [52], and microelectromechanical system (MEMS) actuators [53, 54]. Compared to traditional designs, reconfigurable antennas have several advantages, including compact antenna volumes and low cosite interference.

It is worthwhile to point out here that the CP dipole antenna structure in the previous section can be used to realize reconfigurable polarization by adjusting the orientation of the dipole. For example, when the dipole is oriented along the x or y direction, a linear polarization (LP) is obtained. A left-hand circular polarization (LHCP) can also be achieved if the dipole is oriented along the $\phi = 135°$ direction.

This section focuses on the radiation pattern reconfigurability, which has been used to realize beam scanning in radar systems, to avoid noise sources, and to direct signals toward intended users in wireless communication networks. Although pattern reconfigurability has been well developed in *large* antenna systems such as phased arrays and reflector antennas, it is still a challenging area for *small* antenna elements that are widely used in personal communication devices. The goal of this section is to present a reconfigurable wire antenna on the EBG ground plane that can provide radiation pattern diversity.

We begin by investigating a bent monopole antenna near an EBG ground plane [55–58], as shown in Figure 12.22a. The parameters of the EBG ground plane are as follows:

$$W = 7.5 \text{ mm} \qquad g = 1.5 \text{ mm} \qquad h = 3 \text{ mm}$$
$$\varepsilon_r = 2.94 \qquad r = 0.375 \text{ mm} \qquad (12.8)$$

These parameters guarantee that the EBG ground plane provides a suitable operation frequency at around 4 to 5 GHz for wire antennas. The size of the ground plane is 75 mm ×75 mm, including 8 × 8 EBG cells. To maintain the low-profile advantage, the height of the bent monopole over the ground plane is set to 1.5 mm. The bent monopole, which is made of a 1.5-mm-width strip, is located in the center of the ground plane and the feeding probe is connected to one end of the strip. When the strip length is 39 mm, the antenna resonates at 4.40 GHz with a good return loss of −15 dB and a 13.7 percent bandwidth [59]. An interesting feature of this antenna structure is observed in its radiation patterns, shown in Figure 12.22b. A tilted beam pointing to $\theta = -36°$ is obtained in the xz plane. This is different from the performance of a dipole antenna whose beam points into the broadside direction ($\theta = 0°$).

The tilted beam of the bent monopole structure has led to a novel antenna design with reconfigurable radiation patterns: One can switch the direction of the antenna beam by controlling the orientation of the bent monopole. A reconfigurable wire antenna was designed, as shown in Figure 12.23a. The same EBG ground plane was used in this structure and a feeding probe was located at the center of the ground plane. The probe was connected to two metal strips through two switches. When the left switch was on and the right switch was off, the probe had an electrical connection to the left strip and the bent monopole was oriented along the $-x$ direction. When the left switch was off and the right switch was on, the probe had an electrical connection to the right strip and the bent monopole was oriented along the $+x$ direction. As a result, the direction

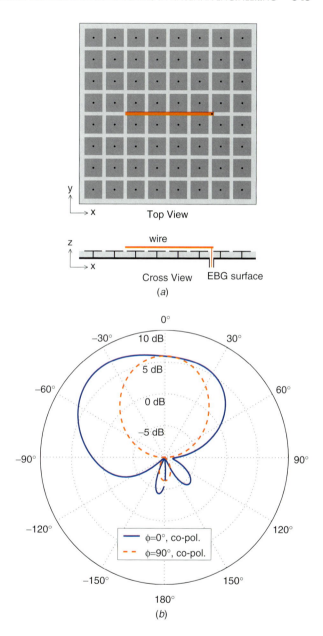

Figure 12.22 Bent monopole antenna on EBG ground plane: (*a*) antenna geometry; (*b*) radiation patterns. A tilted beam in the *xz* plane ($\varphi = 0°$) is observed at the $-36°$ direction with a directivity of 7.5 dB.

of the antenna beam can be switched and, hence, the diversity in the radiation pattern can be obtained.

A reconfigurable antenna prototype was built to demonstrate the switchable beam operation, and Figure 12.23*b* shows a photograph of the fabricated antenna. A RT/Duroid 6002 high-frequency laminate ($\varepsilon_r = 2.94 \pm 0.04$) with 120 mils (3.048 mm) thickness was used as the substrate and the ground-plane dimensions

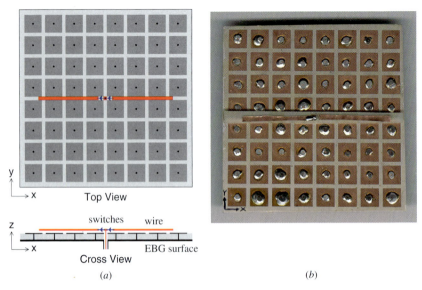

Figure 12.23 Reconfigurable bent monopole design over EBG ground plane: (*a*) antenna geometry; (*b*) photograph of fabricated antenna.

were the same as those given in Eq. (12.8). The length of the strip was tuned to 30 mm to obtain the similar resonant frequency as the single bent monopole antenna. The adjustment of the dipole length was due to the coupling effect from the parasitic strip. The feeding probe was alternatively soldered to the left and right metal strips to represent the operational mode of the $-x$ and $+x$ oriented bent monopoles. The validity of this on/off representation of the switch status has been verified in [42].

The measured return loss result of the antenna is shown in Figure 12.24*a* and compared to the FDTD simulation result. The antenna resonated at 4.40 GHz with a good return loss of -20 dB. The bandwidth of the antenna ($S_{11} < -10$ dB) was 8.2 percent. It is worthwhile to point out that regardless of the $+x$ or $-x$ orientation of the bent monopole the return loss of the antenna remained the same. The radiation patterns of the antenna were measured at the resonant frequency 4.40 GHz, as shown in Figure 12.24*b*. As expected from the FDTD simulation, a switchable antenna beam was observed in the xz plane (E plane). When the bent monopole was $-x$ oriented, the antenna beam pointed to $\theta = 26°$ with a gain of 6.5 dB. When the bent monopole was $+x$ oriented, the antenna beam was switched to $\theta = 26°$ with the same gain. The measured results demonstrate the concept of this radiation pattern reconfigurable antenna.

The reconfigurable concept has also been extended to realize two-dimensional beam switching, as shown in Figure 12.25. Four strips are connected to the center probe through four switches. By controlling the switches, the bent monopole can orient along the $-x$, $+x$, $-y$, $+y$ directions, respectively, resulting in four different antenna beams. Therefore, the antenna beam can be switched not only in the xz plane but also in the yz plane.

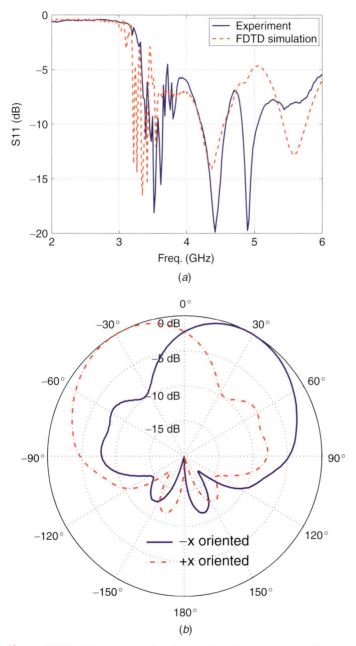

Figure 12.24 Measured results of reconfigurable bent monopole antenna: (*a*) return loss results; (*b*) normalized *E*-plane radiation patterns at 4.40 GHz. Switchable beams between $\pm 26°$ are observed.

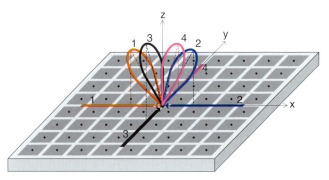

Figure 12.25   Reconfigurable antenna scheme with two-dimensional radiation pattern diversity.

12.5   SUMMARY

The complex artificial ground planes exhibit interesting electromagnetic properties such as a surface wave bandgap and an in-phase reflection coefficient. The analysis methods and various artificial ground-plane designs were discussed in this chapter. Furthermore, extensive examples were provided to demonstrate the potential usefulness of artificial ground planes in antenna applications. It was demonstrated that artificial ground planes not only improve the performance of conventional antenna designs, including wire and microstrip antennas, but also lead to novel radiator concepts such as the surface wave and reconfigurable antennas described here. Given the evident and tangible progress in the artificial ground-plane research, it is clear that the complex artificial ground planes can be considered as a new paradigm in antenna engineering and may provide antenna engineers with many future opportunities.

REFERENCES

1. J. D. Joannopoulos, R. D. Meade, and J. N. Winn, *Photonic Crystals*, Princeton University Press, Princeton, NJ, 1995.
2. *IEEE Trans. Microwave Theory Tech.*, Special Issue on Electromagnetic Crystal Structures, Designs, Synthesis, and Applications, vol. 47, no. 11, Nov. 2003.
3. Y. Rahmat-Samii and H. Mosallaei, "Electromagnetic band-gap structures: Classification, characterization and applications," *Proc. 11th Int. Conf. Antennas and Propagation*, Manchester, U.K., Apr. 17–20, 2001, pp. 560–564.
4. V. G. Veselago, "The electrodynamics of substances with simultaneous negative values of ε and μ," *Sov. Phys. Usp.*, vol. 10, no. 4, pp. 509–514, 1966.
5. J. B. Pendry, "Negative refraction makes a perfect lens," *Phys. Rev. Lett.*, vol. 85, no. 18, pp. 3966–3969, Oct. 2000.
6. R. A. Shelby, D. R. Smith, and S. Schultz, "Experimental verification of a negative refractive index of refraction," *Science*, vol. 292, pp. 77–79, Apr. 2002.
7. *IEEE Trans. Antennas Propag.*, Special Issue on Meta-Materials, vol. 51, no. 10, Oct. 2003.

8. A. S. Barlevy and Y. Rahmat-Samii, "Characterization of electromagnetic band-gaps composed of multiple periodic tripods with interconnecting vias: Concept, analysis, and design," *IEEE Trans. Antennas Propag.*, vol. 49, no. 3, pp. 343–353, Mar. 2001.

9. B. A. Munk, *Frequency Selective Surfaces, Theory and Design*, Wiley, New York, 2000.

10. P.-S. Kildal, "Artificial soft and hard surfaces in electromagnetics," *IEEE Trans. Antennas Propag.*, vol. 38, no. 10, pp. 1537–1544, Oct. 1990.

11. I. Papapolymerous, R. F. Drayton, and L. P. B. Katehi, "Micromachined patch antennas," *IEEE Trans. Antennas Propag.*, vol. 46, pp. 275–283, Feb. 1998.

12. J. S. Colburn and Y. Rahmat-Samii, "Patch antennas on externally perforated high dielectric constant substrates," *IEEE Trans. Antennas Propag.*, vol. 47, pp. 1785–1794, Dec. 1999.

13. F.-R. Yang, K.-P. Ma, Y. Qian, and T. Itoh, "A uniplanar compact photonic-bandgap (UC-PBG) structures and its applications for microwave circuit," *IEEE Trans. Microwave Theory Tech.*, vol. 47, no. 8, pp. 1509–1514, Aug. 1999.

14. D. Sievenpiper, L. Zhang, R. F. J. Broas, N. G. Alexopoulos, and E. Yablonovitch, "High-impedance electromagnetic surfaces with a forbidden frequency band," *IEEE Trans. Microwave Theory Tech.*, vol. 47, no. 11, pp. 2059–2074, Nov. 1999.

15. R. Gonzalo, P. de Maagt, and M. Sorolla, "Enhanced patch antenna performance by suppressing surface waves using photonic-bandgap substrates," *IEEE Trans. Microwave Theory Tech.*, vol. 47, pp. 2131–2138, Nov. 1999.

16. R. Cocciolo, F. R. Yang, K. P. Ma, and T. Itoh, "Aperture coupled patch antenna on UC-PBG substrate," *IEEE Trans. Microwave Theory Tech.*, vol. 47, pp. 2123–2130, Nov. 1999.

17. F. Yang and Y. Rahmat-Samii, "Microstrip antennas integrated with electromagnetic band-gap (EBG) structures: A low mutual coupling design for array applications," *IEEE Trans. Antennas Propag.*, vol. 51, no. 10, pp. 2936–2946, Oct. 2003.

18. F.-R. Yang, K.-P. Ma, Y. Qian, and T. Itoh, "A novel TEM waveguide using uniplanar compact photonic-bandgap (UC-PBG) structure," *IEEE Trans. Microwave Theory Tech.*, vol. 47, no. 11, pp. 2092–2098, Nov. 1999.

19. Z. Li and Y. Rahmat-Samii, "PBG, PMC and PEC ground planes: A case study for dipole antenna," *2000 IEEE APS Int. Symp. Dig.*, vol. 4, pp. 2258–2261, Salt Lake City, UT, July 16–21, 2000.

20. F. Yang and Y. Rahmat-Samii, "Reflection phase characterizations of the EBG ground plane for low profile wire antenna applications," *IEEE Trans. Antennas Propag.*, vol. 51, pp. 2691–2703, Oct. 2003.

21. D. F. Sievenpiper, "High Impedance electromagnetic surfaces", Ph.D. dissertation, Electrical Engineering Department, University of California, Los Angeles, 1999.

22. M. Rahman and M. A. Stuchly, "Transmission line-periodic circuit representation of planar microwave photonic bandgap structures," *Microwave Opt. Technol. Lett.*, vol. 30, no. 1, pp. 15–19, July 2001.

23. A. Taflove and S. Hagness, *Computational Electromagnetics: The Finite-Difference Time-Domain Method*, 2nd ed., Artech House, Boston, MA, 2000.

24. M. A. Jensen and Y. Rahmat-Samii, "Performance analysis of antennas for hand-held transceiver using FDTD," *IEEE Trans. Antennas Propag.*, vol. 42, pp. 1106–1113, Aug. 1994.

25. A. Aminian and Y. Rahmat-Samii, "Spectral FDTD: A novel computational technique for the analysis of periodic structures," paper presented at the 2004 IEEE APS International Symposium, Monterey, CA, June 20–26, 2004.

26. H. Mosallaei and Y. Rahmat-Samii, "Periodic bandgap and effective dielectric materials in electromagnetics: Characterization and applications in nanocavities and waveguides," *IEEE Trans. Antennas Propag.*, vol. 51, no. 3, pp. 549–563, Mar. 2003.

27. A. Aminian, F. Yang, and Y. Rahmat-Samii, "In-phase reflection and EM wave suppression characteristics of electromagnetic band gap ground planes," *2003 IEEE AP-S Dig.*, vol. 4, pp. 430–433, June 2003.

28. F. Yang and Y. Rahmat-Samii, "Polarization dependent electromagnetic band gap (PDEBG) structures: Designs and applications," *Microwave Opt. Technol. Lett.*, vol. 41, no. 6, pp. 439–444, July 2004.

29. G. P. Gauthier, A. Courtay, and G. H. Rebeiz, "Microstrip antennas on synthesized low dielectric-constant substrate," *IEEE Trans. Antennas Propag.*, vol. 45, no. 8, pp. 1310–1314, Aug. 1997.

30. D. R. Jackson, J. T. Wlliams, A. K. Bhattacharyya, R. L. Smith, S. J. Buchheit, and S. A. Long, "Microstrip patch designs that do not excite surface waves," *IEEE Trans. Antennas Propag.*, vol. 41, no. 8, pp. 1026–1037, Aug. 1993.

31. F. Yang, C.-S. Kim, and Y. Rahmat-Samii, "Step-like structure and EBG structure to improve the performance of patch antennas on high dielectric substrate," *2001 IEEE AP-S Dig.*, vol. 2, pp. 482–485, July 2001.

32. F. Yang and Y. Rahmat-Samii, "A low profile circularly polarized curl antenna over electromagnetic band-gap (EBG) surface," *Microwave Opt. Technol. Lett.*, vol. 31, no. 3, pp. 165–168, 2001.

33. H. Nakano, S. Okuzawa, K. Ohishi, H. Mimaki, and J. Yamauchi, "A curl antenna," *IEEE Trans. Antennas Propag.*, vol. 41, no. 11, 1570–1575, Nov. 1993.

34. F. Yang and Y. Rahmat-Samii, "A low profile circularly polarized curl antenna over electromagnetic band-gap (EBG) surface," *Microwave Opt. Technol. Lett.*, vol. 31, no. 3, pp. 165–168, 2001.

35. W. E. Mckinzie III and R. Fahr, "A low profile polarization diversity antenna built on an artificial magnetic conductor," *2002 IEEE AP-S Int. Symp. Dig.*, vol. 1, pp. 762–765, June 2002.

36. F. Yang and Y. Rahmat-Samii, "Polarization dependent electromagnetic band-gap surfaces: Characterization, designs, and applications," *2003 IEEE AP-S Dig.*, vol. 3, pp. 339–342, June 2003.

37. B. A. Munk, *Finite Antenna Arrays and FSS*, Wiley, New York, 2003.

38. Y. Qian and T. Itoh, "Progress in active integrated antennas and their applications," *IEEE Trans. Microwave Theory Tech.*, vol. 46, no. 11, pp. 1891–1900, Nov. 1998.

39. J. T. Bernhard, "Reconfigurable antennas and apertures: State-of-the-art and future outlook," *Proc. SPIE Conf. Smart Electronics, MEMS, BioMEMS, Nanotechnology*, vol. 5055, pp. 1–9, Mar. 2003.

40. F. Yang and Y. Rahmat-Samii, "Patch antennas with switchable slots (PASS) in wireless communications: Concepts, designs, and applications," *IEEE Antennas Propag. Mag.*, vol. 47, no. 2, pp. 13–29, Apr. 2005.

41. D. H. Shaubert, F. G. Farrar, A. Sindoris, and S. T. Hayes, "Microstrip antennas with frequency agility and polarization diversity," *IEEE Trans. Antennas Propag.*, vol. 29, no. 1, pp. 118–123, Jan. 1981.

42. F. Yang and Y. Rahmat-Samii, "Patch antenna with switchable slot (PASS): Dual frequency operation," *Microwave Opt. Technol. Lett.*, vol. 31, no. 3, pp. 165–168, Nov. 2001.

43. M. Boti, L. Dussopt, and J. M. Laheurte, "Circularly polarized antenna with switchable polarization sense," *Electron. Lett.*, vol. 36, no. 18, pp. 1518–1519, Aug. 2000.

44. F. Yang and Y. Rahmat-Samii, "A CP polarization diversity patch antenna using switchable slots," *IEEE Microwave Wireless Components Lett.*, vol. 12, no. 3, pp. 96–98, Mar. 2002.

45. K. Chang, M. Li, T.-Y. Yun, and C. T. Rodenbeck, "Novel low cost beam steering technique," *IEEE Trans. Antennas Propag.*, vol. 50, no. 5, pp. 618–627, May 2002.

46. K. W. Lee and R. G. Rojas, "Novel scheme for design of adaptive printed antenna element," *IEEE AP-S Dig.*, vol. 3, pp. 1252–1255, July 2000.

47. D. Sievenpiper, J. Schaffner, R. Loo, G. Tangonan, S. Ontiveros, and R. Harold, "A tunable impedance surface performing as a reconfigurable beam steering reflector," *IEEE Trans. Antennas Propag.*, vol. 50, no. 3, pp. 384–390, Mar. 2003.

48. J. Sor, C.-C. Chang, Y. Qian, and T. Itoh, "A reconfigurable leaky-wave/patch microstrip aperture for phased-array applications," *IEEE Trans. Microwave Theory Tech.*, vol. 50, pp. 1877–1884, Aug. 2002.

49. G. H. Huff, J. Feng, S. Zhang, and T. Bernhard, "A novel radiation pattern and frequency reconfigurable single turn square spiral microstrip antenna," *IEEE Microwave Wireless Components Lett.*, vol. 13, no. 2, pp. 57–59, Feb. 2003.

50. N. Jin, F. Yang, and Y. Rahmat-Samii, "A novel reconfigurable patch antenna with both frequency and polarization diversities for wireless communications," *2004 IEEE AP-S Int. Symp. Dig.*, vol. 2, pp. 1796–1799, Monterey, CA, June 20–26, 2004.

51. J. T. Bernhard, E. Kiely, and G. Washington, "A smart mechanically-actuated two-layer electromagnetically coupled microstrip antenna with variable frequency, bandwidth, and antenna gain," *IEEE Trans. Antennas Propag.*, vol. 49, no. 4, pp. 597–601, Apr. 2001.

52. K. L. Virga and Y. Rahmat-Samii, "Low profile enhanced-bandwidth PIFA antennas for wireless communications packaging," *IEEE Trans. Microwave Theory Tech.*, vol. 45, no. 10, pp. 1879–1888, Oct. 1997.

53. E. R. Brown, "RF-MEMS switches for reconfigurable integrated circuits," *IEEE Trans. Microwave Theory Tech.*, vol. 46, no. 11, pp. 1868–1880, Nov. 1998.

54. R. Simon, D. Chun, and L. Katehi, "Reconfigurable array antenna using micro-electro-mechanical system (MEMS) actuators," *IEEE AP-S Int. Symp. Dig.*, vol. 3, pp. 674–677, July 2001.

55. D. Sievenpiper, R. Broas, and E. Yablonovitch, "Antennas on high-impedance ground planes," *1999 IEEE MTT-S Int. Symp. Dig.*, vol. 3, pp. 1245–1248, June 1999.

56. D. Sievenpiper, H.-P. Hsu, J. Schaffner, G. Tangonan, R. Garcia, and S. Ontiveros, "Low-profile, four-sector diversity antenna on high-impedance ground plane," *Electron. Lett.*, vol. 36, no. 16, pp. 1343–1345, Aug. 2000.

57. S. Clavijo, R. E. Diaz, and W. E. McKinzie III, "Design methodology for Sievenpiper high-impedance surfaces: An artificial magnetic conductor for positive gain electrically small antennas," *IEEE Trans. Antennas Propag.*, vol. 51, no. 10, pp. 2678–2690, Oct. 2003.

58. S. Rogers, J. Marsh, W. McKinzie, and J. Scott, "An AMC-based 802.11a/b antenna for laptop computers," *2003 IEEE AP-S Int. Symp. Dig.*, vol. 2, pp. 10–13, June 2003.

59. F. Yang and Y. Rahmat-Samii, "Bent monopole antennas on EBG ground plane with reconfigurable radiation patterns," *2004 IEEE AP-S Int. Symp. Dig.*, vol. 2, pp. 1819–1822, Monterey, CA, June 20–26, 2004.

CHAPTER 13

FSS-BASED EBG SURFACES

Stefano Maci and Alessio Cucini

13.1 INTRODUCTION

Among the variety of electromagnetic bandgap (EBG) structures which have been developed in recent years, an interesting role is played by frequency-selective surfaces (FSSs) printed on grounded stratified dielectric media. This type of structure creates a particular class of so-called artificial or metamaterial surfaces, whose applications are intended to create artificial magnetic conductors (AMCs) [1,2] or surfaces which exhibit "soft" and/or "hard" equivalent boundary conditions [3]. Further applications are concerned with obtaining a stop band for surface wave (SW) propagation along the interface. Consequent benefits are the suppression of SW coupling [4], the reduction of diffraction lobes, the improvement of planar antenna efficiency, the development of compact antennas [5] and resonators [6], and the suppression of parallel-plate waveguide modes [7].

A simple and intuitive way to design an FSS-based metamaterial resorts to a model based on equivalent *LC* circuits based on quasi-static concepts. The *LC* model describes the dominant physics at low frequencies. For increasing frequencies, the *LC* parameters are increasingly dependent on the phasing and on the wave polarization, and the simple, quasi-static concepts are increasingly insufficient. This introduces the need for a more rigorous network derivation which extracts from the full-wave analysis the minimum number of wavenumber-dependent parameters for an analytical synthesis of the surface impedance. Along this line, this chapter illustrates first an integral equation method for the analysis of FSS-based metamaterials and then a general process to rigorously derive equivalent networks for designing reflection and dispersion properties.

When the emphasis is on the reflection properties of the metamaterial surface, the structure response is conventionally presented in terms of the phase of the reflection coefficient versus frequency for various plane-wave incidence angles and for both transverse electric (TE) and transverse magnetic (TM) incident polarizations. The goal is to obtain a phase–frequency curve which is as flat as possible around the zero-degree phase frequency value and invariant with respect to the incidence angle. However, when the emphasis is on the EBG properties, the analysis leads to the dispersion diagram, and the aim is to enlarge the SW frequency stop band or reduce the dependence on the SW direction of propagation along the surface. All these design purposes are pursued by acting

Metamaterials: Physics and Engineering Explorations, Edited by N. Engheta and R. W. Ziolkowski
Copyright © 2006 the Institute of Electrical and Electronics Engineers, Inc.

on the shape and stratification of the printed elements [8] as well as on via groundings [1,9]. The lowering of the resonance frequencies is indeed obtained by increasing the capacitance with a coupled-FSS approach [10].

The two above-mentioned methods used to investigate the properties of the artificial surface (i.e., the phases of the reflection coefficient and the dispersion diagram of the SW modes) actually focus on two different physical aspects of the problem, which also implies the use of different analysis schemes. The analysis of the reflection coefficient is based on the expansion of the field in terms of Floquet waves (FWs) centered around a wavenumber imposed by the excitation, that is, known a priori. This allows us to approach the problem in a very efficient way by resorting to integral equations and to a relevant method-of-moments (MoM) solution. However, when analyzing the dispersion properties, the modal wavenumber is the unknown of the problem and may be found through the numerical solution of a resonance equation. This implies the annulment of the MoM matrix determinant and hence at least of one of the eigenvalues. As a consequence, there are theoretical and numerical problems associated with this latter approach. First, the size of the MoM matrix depends on the actual wavenumber, which dictates to a large extent the scale of variation of the current. Second, the search for zeros is extremely sensitive to the initial guess. For these reasons, emphasis will be given here to the full-wave-based dispersion analysis by suggesting alternative ways to proceed.

13.1.1 Quasi-Static Admittance Models

As is well known, an intuitive model for the problem of plane-wave scattering from an FSS is given in terms of a quasi-static reactive LC impedance network placed in a TE or TM vertical transmission line (Fig. 13.1). Several authors have treated the periodic surface problem via equivalent LC circuits to characterize either the reflection or the dispersion properties of metamaterial surfaces [11,12].

In the absence of losses, the descriptive impedance is purely reactive. For printed dipoles, at a frequency lower than their resonant frequency, the FSS is capacitive. This capacitance may resonate in parallel with the inductance provided

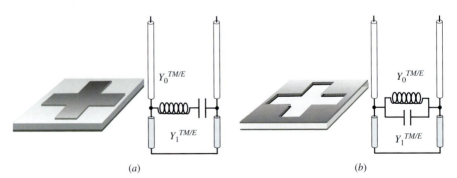

Figure 13.1 Low-frequency equivalent transmission lines for FSS-based surface: (a) series LC circuit for patch-type FSS; (b) parallel LC circuit for aperture-type FSS. The transmission line parameters are those derived by the dominant Floquet mode for the TE and TM polarization.

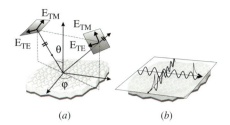

Figure 13.2   An FSS-based EBG surface and relevant plane-wave excitation: (*a*) propagating incident and reflected plane wave ($k_x^2 + k_y^2 < \omega^2/c^2$); (*b*) evanescent plane wave ($k_x^2 + k_y^2 > \omega^2/c^2$) with attenuation along z.

by the piece of short-circuited transmission line, thus leading to the equivalent magnetic properties. Nearby in frequency, an EBG, that is, a frequency band in which no surface wave propagation will occur, can be present. Analogous considerations apply to the aperture-type FSS (Fig. 13.2*b*); the latter exhibits an inductive behavior at a lower frequency and can be described by a parallel *LC* circuit.

The equivalent transmission line for planar periodic surfaces can be oriented in a *horizontal* instead of a *vertical* direction. This leads to a cascade of periodic *LC* cells (see, e.g., [13, 14], relevant to the case of negative-refractive-index structures). This horizontal transmission line model can substantially be derived from a vertical transmission line by changing the wavenumber via a standard analytical transformation.

The simple circuits described in Figure 13.1 contain the essential physics to give a qualitative justification to the basic aspects of the magnetic conducting properties. Actually, they are rigorous in the quasi-static limit. However, the lack of important aspects, such as the quantification of the wavenumber dependence and the coupling between the TE and TM polarizations, requires a more accurate generalization of the model. This generalization will be substantiated next on the basis of a full-wave MoM analysis.

13.1.2   Chapter Outline

This chapter is organized as follows. Section 13.2 provides a brief overview of the spectral domain FW-based MoM for both patch-type and aperture-type FSSs. Although the content of this section is rather standard [15, 16], it is essential to emphasize the basic concepts which prepare the subsequent innovative parts. Section 13.3 starts with the definition of an FW-based network and the relevant admittance matrix at the accessible modal ports and proceeds with the definition of a dispersion equation based on the accessible modes, with emphasis on numerical advantages and disadvantages compared to a more standard approach. Section 13.4 presents the "pole–zero matching" method for the characterization of the FSS equivalent circuit. The basic ideas of this method have been discussed in a recent paper by the authors [17]. Here, the method is generalized by defining the poles and zeros of the equivalent FSS admittance from the visible to the nonvisible region, avoiding nonrigorous extrapolation processes. The pole–zero matching method invokes the *LC* properties of the admittance associated to the FSS. However, it does not specify the inductance and capacitance of the network, but rather identifies the essential parameters, which synthesize all the necessary physical information of the artificial surface and exhibit a weak variation against the wavenumber. These parameters are the resonant frequencies (i.e., poles and zeros) of the FSS equivalent admittance. Their appropriate use, based on Foster's theorem, leads to an accurate analytical approximation of the surface response on a large-frequency region of the dispersion diagram. Explicit representations of inductances and capacitances of the admittance network can be easily obtained by a conventional network synthesis that will not be explicitly treated here. The conclusions of this chapter are given in Section 13.5.

13.2 MOM SOLUTION

Let us consider an infinite planar FSS consisting of patches printed on a grounded slab or of apertures etched on the upper wall of a parallel-plate dielectric-filled waveguide. The entire structure is assumed to be without losses. A rectangular (x, y, z) reference system is assumed with the z axis orthogonal to the FSS and the origin at the FSS level. The periodicities of the FSS are d_x and d_y along x and y, respectively. An incident, either TE or TM, plane wave is assumed to illuminate the structure, with zero phase at the origin of the reference system. The plane wave imposes a phasing k_x and k_y in the principal directions (we will refer here to a rectangular lattice for reasons of simplicity, but the present formulation is valid for a nonorthogonal lattice as well). The impressed plane wave could be either propagating (Fig. 13.1a, $k_x^2 + k_y^2 < \omega^2/c^2$) or evanescent (Fig. 13.1b, $k_x^2 + k_y^2 > \omega^2/c^2$) along z. In the latter case, the incident field is considered as a planar impressed phased distribution of the TE or TM field placed at the FSS level with attenuation in the z direction.

13.2.1 Patch-Type FSS (Electric Current Approach)

Let us first consider a patch-type FSS. The numerical computation of the equivalent currents at the interface of the planar periodic structure is performed via a numerical solution of the electric field integral equation (EFIE) by using a spectral periodic MoM approach. More than discussing the numerical implication of the MoM scheme, our objective is to construct an appropriate and simple admittance matrix through the MoM matrix to characterize the FSS surface.

Due to the periodicity of the problem, the analysis can be reduced to that of a single periodic cell, with phase shift boundary conditions applied to the ideal vertical walls. By applying the equivalence theorem (Fig. 13.1), an electric current distribution is assumed on the region of the metallic patches, radiating with the Green's function (GF) of the grounded slab. By imposing the boundary conditions on the surface of the metallic patches, the EFIE is derived as follows:

$$\mathbf{E}_S(\mathbf{J}) + \mathbf{E}_{imp} = 0 \qquad (13.1)$$

where $\mathbf{E}_S$ is the field radiated by the currents $\mathbf{J}$ induced on the patches, and $\mathbf{E}_{imp} = \mathbf{E}_{inc} + \mathbf{E}_{ref}$ is the impressed field at the interface (in the absence of printed elements), which is given by the sum of the incident ($\mathbf{E}_{\mathrm{inc}}$) and reflected ($\mathbf{E}_{\mathrm{ref}}$) fields. From here on, the bold characters indicate vectors and the carets indicate unit vectors. As suggested by Tascone and Orta [16, pp. 221–238], the equivalent currents $\mathbf{J}$ are expressed in terms of basis functions $\mathbf{f}_n(\mathbf{r}_t)$,

$$\mathbf{J}(\mathbf{r_t}) = \sum_{n=1}^{N} I_n \mathbf{f}_n(\mathbf{r_t}) \tag{13.2}$$

where $\mathbf{r}_t = x\hat{x} + y\hat{y}$ denotes the two-dimensional space vector. Figure 13.3 shows subdomain triangular basis functions, but entire domain basis functions can be used as well. Due to the periodicity of the problem, the analysis is reduced to that of a single periodic cell by resorting to the Floquet theorem. Let us indicate by $\mathbf{k} = k_x\hat{x} + k_y\hat{y}$ the impressed vector wavenumber; by $k_{x\xi} = k_x + 2\pi\xi/d_x$, $k_{y\eta} = k_y + 2\pi\eta/d_y$ ($\xi, \eta = \pm1, \pm2, \ldots$) the FW wavenumbers in x and y, respectively; and by $\mathbf{k}_q = k_{x\xi}\hat{x} + k_{y\eta}\hat{y}$ the relevant vector form, where q denotes the two FW indices (ξ, η). By indicating with $\boldsymbol{\beta}_q$ the nodes of the reciprocal lattice, that is,

$$\boldsymbol{\beta}_q = \frac{2\pi\xi}{d_x}\hat{x} + \frac{2\pi\eta}{d_y}\hat{y},$$

we obtain $\mathbf{k}_q = \mathbf{k} + \boldsymbol{\beta}_q$, with $q = 0, 1, 2 \ldots$ and $\mathbf{k}_0 = \mathbf{k}$ by definition. It is also useful to introduce the normalized spectral vectors

$$\hat{\sigma}_q = \frac{\mathbf{k}_q}{\sqrt{\mathbf{k}_q \cdot \mathbf{k}_q}} \qquad \hat{\alpha}_q = \hat{z} \times \hat{\sigma}_q \tag{13.3}$$

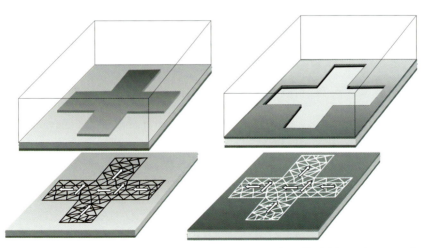

Figure 13.3 Application of equivalence principle to basic cell of (a) patch-type and (b) aperture-type FSS. Phase shift conditions are imposed on the vertical walls. A triangular mesh is shown, with subdomain basis functions used for expansion of (a) electric and (b) magnetic current distribution.

as a spectral basis to describe the TM and TE field components, respectively. By using a Galerkin spectral MoM approach, (13.1) is reduced to the matrix equation

$$\overline{\overline{Z}}_{\text{MoM}} \, \overline{I} = \overline{V} \tag{13.4}$$

where $\overline{V} = \{V_m\}_{m=1,N}^{\text{T}}$ is the known column vector of the complex amplitude of the impressed field on the $\mathbf{f}_n$ basis, and its components are given by

$$V_m = -\tilde{\mathbf{F}}_m^*(\mathbf{k}) \cdot \mathbf{E}_{\text{imp}}(\mathbf{k}) \tag{13.5}$$

In (13.5), the asterisk denotes the complex conjugate and $\mathbf{E}_{\text{imp}}(\mathbf{k})$ is the spectrum of the impressed field. The latter is given by two space harmonics representing the incident field and the field reflected by the grounded dielectric slab without patches. In (13.4), $\overline{I} = \{I_n\}_{n=1,N}^{\text{T}}$ is the column vector of the current expansion and $\overline{\overline{Z}}_{\text{MoM}} = \{Z_{nm}^{\text{MoM}}\}_{n,m=1,N}$ is the MoM impedance matrix, with entries given in an appropriate TE/TM form via

$$Z_{mn}^{\text{MoM}} = \sum_{q=0}^{M-1} \tilde{\mathbf{F}}_m^*(\mathbf{k}_q) \cdot \left[Z_{\text{GF}}^{\text{TM}}(\mathbf{k}_q) \hat{\sigma}_q \hat{\sigma}_q + Z_{\text{GF}}^{\text{TE}}(\mathbf{k}_q) \hat{\alpha}_q \hat{\alpha}_q \right] \cdot \tilde{\mathbf{F}}_n(\mathbf{k}_q) \tag{13.6}$$

In (13.6), $\tilde{\mathbf{F}}_n(\mathbf{k})[\tilde{\mathbf{F}}_m(\mathbf{k})]$ is the Fourier transform of the basis (test) function $\mathbf{f}_n(\mathbf{r}_t)[\mathbf{f}_m(\mathbf{r}_t)]$ sampled at the FW wavenumbers $\mathbf{k}_q$; $Z_{\text{GF}}^{\text{TM/TE}}(\mathbf{k})$ are the TM–TE components of the individual element spectral electric field GF, sampled in (13.6) at the vector FW wavenumber. In (13.6), the modal FW expansion is truncated at the integer $M - 1$ with M in general larger than N. This is an obvious consequence of the continuity of the FWs on the periodic cell, which implies the use of more FW modes than basis functions to describe the patch current. The GF impedances can be found by solving the pertinent transmission line problem representing the stratification for the TE and TM case; in the present case (single grounded substrate), one obtains

$$Z_{\text{GF}}^{\text{TM/TE}}(\mathbf{k}) = \left[Y_0^{\text{TM/TE}}(\mathbf{k}) - j Y_1^{\text{TM/TE}}(\mathbf{k}) \cot(k_{z1} h) \right]^{-1} \tag{13.7}$$

where

$$Y_0^{\text{TM}}(\mathbf{k}) = \frac{\omega \varepsilon_0}{k_z} \qquad Y_0^{\text{TE}}(\mathbf{k}) = \frac{k_z}{\omega \mu_0} \qquad Y_1^{\text{TM}}(\mathbf{k}) = \frac{\omega \varepsilon_r \varepsilon_0}{k_{z1}} \qquad Y_1^{\text{TE}}(\mathbf{k}) = \frac{k_{z1}}{\omega \mu_0}$$

are the modal z transmission line TM/TE characteristic admittances relevant to the free-space (subscript 0) and the dielectric (subscript 1) regions, respectively, $k_z = \sqrt{k^2 - k_x^2 - k_y^2}$, $k_{z1} = \sqrt{\varepsilon_r k^2 - k_x^2 - k_y^2}$, and k is the free-space wavenumber. The MoM matrix can be expressed in the compact form

$$\overline{\overline{Z}}_{\text{MoM}} = \overline{\overline{Q}}^H \overline{\overline{Z}}_{\text{GF}} \overline{\overline{Q}} \tag{13.8}$$

where $\overline{\overline{Z}}_{\text{GF}} = \text{diag}\{Z_{\text{GF}}^{\text{TM}}(\mathbf{k}_q), Z_{\text{GF}}^{\text{TE}}(\mathbf{k}_q)\}_{q=0,M-1}$ is a diagonal $2M \times 2M$ matrix, the superscript H denotes transpose conjugate, $\overline{\overline{Q}}^H = \{Q_{m,q}^{\text{TM*}}, Q_{m,q}^{\text{TE*}}\}_{m=1,N}^{\text{T}}{}_{q=0,M-1}$ is an $N \times 2M$ matrix, and $\overline{\overline{Q}} = \{Q_{q,n}^{\text{TM}}, Q_{q,n}^{\text{TE}}\}_{q=0,M-1}^{n=1,N}$ is a $2M \times N$ matrix. The entries of the Q matrices are given by $Q_{i,q}^{\text{TM}} = \tilde{\mathbf{F}}_i(\mathbf{k}_q) \cdot \hat{\sigma}_q$, $Q_{i,q}^{\text{TE}} = \tilde{\mathbf{F}}_i(\mathbf{k}_q) \cdot \hat{\alpha}_q$ $(i = n, m)$.

13.2.2   Aperture-Type FSS (Magnetic Current Approach)

For an inductive-type FSS (i.e., perforated screens), the FSS is substituted by a continuous, infinitely thin perfect electric conductor (PEC) screen with magnetic current distribution on both sides (Fig. 13.3); these currents have equal amplitude and opposite signs on the two different sides to ensure the continuity of the electric field through the aperture. The integral equation which imposes the continuity of the magnetic field follows from the relation $\mathbf{H}_s^+(\mathbf{M}) + \mathbf{H}_{\text{imp}} = \mathbf{H}_s^-(-\mathbf{M})$, where the superscript plus and minus refer to the GF of the upper and lower regions, respectively. The magnetic current is expanded in terms of basis functions $\mathbf{g}_n(\mathbf{r}_t)$ associated with the electric field

$$\mathbf{M}(\mathbf{r}_t) = \sum_{n=1}^{N} V_n \mathbf{g}_n(\mathbf{r}_t) \times \hat{z} \tag{13.9}$$

Imposing the continuity of the magnetic field through the aperture leads to the expression

$$\overline{\overline{Y}}_{\text{MoM}} \, \overline{V} = \overline{I} \tag{13.10}$$

where $\overline{V} = \{V_n\}_{n=1,N}^{\text{T}}$ is the unknown column vector, $\overline{I} = \{I_m\}_{m=1,N}^{\text{T}}$, $I_m = -\tilde{\mathbf{G}}_m^*(\mathbf{k}) \cdot \mathbf{H}_{\text{imp}}(\mathbf{k})$ is the known column vector of the impressed magnetic field on the MoM basis, and $\tilde{\mathbf{G}}_i(\mathbf{k})$ is the Fourier transform of $\mathbf{g}_i(\mathbf{r}_t)$. The MoM matrix may be expressed in the compact form

$$\overline{\overline{Y}}_{\text{MoM}} = \overline{\overline{P}}^H \, \overline{\overline{Y}}_{\text{GF}} \, \overline{\overline{P}} \tag{13.11}$$

where $\overline{\overline{Y}}_{\text{GF}} = \text{diag}\{Y_{\text{GF}}^{\text{TM}}(\mathbf{k}_q), \; Y_{\text{GF}}^{\text{TE}}(\mathbf{k}_q)\}_{q=1,M}$ [with $Y_{\text{GF}}^{\text{TM/TE}}(\mathbf{k}) = Y_0^{\text{TM/TE}}(\mathbf{k}) - jY_1^{\text{TM/TE}}(\mathbf{k}) \cot(k_{z1}h)$] is a diagonal $2M \times 2M$ matrix, $\overline{\overline{P}}^H = \{P_{m,q}^{\text{TM}*}, \; P_{m,q}^{\text{TE}*}\}_{\substack{m=1,N \\ q=0,M-1}}^{\text{T}}$ is an $N \times 2M$ matrix, and $\overline{\overline{P}} = \{P_{q,n}^{\text{TM}}, \; P_{q,n}^{\text{TE}}\}_{\substack{q=0,M-1 \\ n=1,N}}$ is a $2M \times N$ matrix whose components are given by $P_{i,q}^{\text{TM}} = \tilde{\mathbf{G}}_i(\mathbf{k}_q) \cdot \hat{\sigma}_q$, $Q_{i,q}^{\text{TE}} = \tilde{\mathbf{G}}_i(\mathbf{k}_q) \cdot \hat{\alpha}_q$ $(i = n, m)$

13.2.3   Dispersion Equation

The dispersion equation of the artificial surface can be obtained by assuming the existence of nontrivial solutions for zero impressed field, which implies

$$\det[\overline{\overline{Z}}_{\text{MoM}}(k_x, k_y, \omega)] = 0 \qquad \text{for patch-type FSS} \tag{13.12}$$

$$\det[\overline{\overline{Y}}_{\text{MoM}}(k_x, k_y, \omega)] = 0 \qquad \text{for aperture-type FSS} \tag{13.13}$$

where we have explicitly added the dependence of the MoM matrix on the angular frequency. The solution of the dispersion equation in the range $k_x^2 + k_y^2 > \omega^2/c^2$ leads to the wavenumbers of the SW supported by the artificial surface.

Equivalent expressions of the dispersion equations can be obtained by diagonalizing the MoM matrix and annulling the product of the eigenvalues,

$$\xi_1(k_x, k_y, \omega)\xi_2(k_x, k_y, \omega) \cdots \xi_N(k_x, k_y, \omega) = 0 \quad \text{for patch-type FSS} \tag{13.14}$$

$$\eta_1(k_x, k_y, \omega)\eta_2(k_x, k_y, \omega) \cdots \eta_N(k_x, k_y, \omega) = 0 \quad \text{for aperture-type FSS} \tag{13.15}$$

where $\xi_1, \xi_2, \ldots, \xi_N$ and $\eta_1, \eta_2, \ldots, \eta_N$ are the eigenvalues of $\overline{\overline{Z}}_{\text{MoM}}$ and $\overline{\overline{Y}}_{\text{MoM}}$, respectively. Equations (13.14) and (13.15) can be numerically convenient in comparison to (13.12) and (13.13) if a fast method to track the zeros of each eigenvalue is applied. To this regard, the covariance matrix approach in [18] can be used. The EBG of the artificial surface is defined as a frequency range where no solution of (13.14) or (13.15) exists for (k_x, k_y) real and below the light cone $(k_x^2 + k_y^2 > \omega^2/c^2)$. This means that no SW can propagate along the artificial surface for any frequency within the EBG. An alternative way to define the dispersion equation is to resort to the concept of accessible modes presented in the next section.

13.3   ACCESSIBLE MODE ADMITTANCE NETWORK

Let us assume that we are observing the field at a certain distance z from the artificial surface. In this case, the FW modes that are completely attenuated do not contribute to the field at z. In a multimode network description, this implies that the relevant modal ports can be considered as not "accessible" to the observer and therefore are neglected. This concept was introduced by Rozzi [19] for waveguide problems and is commonly used to calculate the coupling between FSSs located at different levels [16]. Denote by $2M_A$ the number of accessible TM/TE ports and consider the $2M_A$-port network in Figure 13.4, where each port is associated to an FW mode of TM or TE type. This network consists of a multiport "FSS network" loaded in parallel at each port by a modal TM or TE transmission line representing the unprinted grounded slab. The FSS network is conveniently characterized by $2M_A \times 2M_A$ admittance (impedance) matrices,

$$\overline{I}_{\text{FW}}^{\text{FSS}} = \overline{\overline{Y}}_{\text{FSS}} \, \overline{V}_{\text{FW}} \quad \text{for patch-type FSS} \tag{13.16a}$$

$$\overline{V}_{\text{FW}} = \overline{\overline{Z}}_{\text{FSS}} \, \overline{I}_{\text{FW}}^{\text{FSS}} \quad \text{for aperture-type FSS} \tag{13.16b}$$

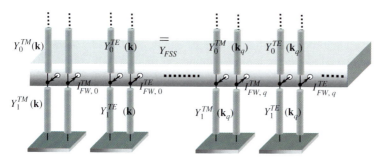

Figure 13.4   Multiport accessible FW-mode network and relevant transmission line parameters.

where $\overline{I}_{\mathrm{FW}}^{\mathrm{FSS}} = [I_{\mathrm{FW},q}^{\mathrm{TM}}, I_{\mathrm{FW},q}^{\mathrm{TE}}]_{q=0,M_A-1}^{\mathrm{T}}$ ($\overline{V}_{\mathrm{FW}} = [V_{\mathrm{FW},q}^{\mathrm{TM}}, V_{\mathrm{FW},q}^{\mathrm{TE}}]_{q=0,M_A-1}^{\mathrm{T}}$), is the vector of the FW amplitudes of the magnetic (electric) field expansion at the FSS level and denotes the electric current flowing into the FSS network (the FW mode voltage at the ports). In the following sections we will show how to construct the FSS admittance/impedance matrix for different FSSs.

13.3.1   Patch-Type FSS

For patch-type FSSs, $\overline{I}_{\mathrm{FW}}^{\mathrm{FSS}}$ is obtained by projecting the MoM vector currents $\overline{I}$ into M_A FW accessible modal currents $\overline{I}_{\mathrm{FW}}^{\mathrm{FSS}} = \overline{\overline{q}}\,\overline{I}$, where $\overline{\overline{q}} = \{Q_{q,n}^{\mathrm{TM}}, Q_{q,n}^{\mathrm{TE}}\}_{\substack{q=0,M_A-1 \\ n=1,N}}$ is the $2M_A \times N$ matrix obtained by the first $2M_A$ rows of $\overline{\overline{Q}}$. Analogously, the forcing term associated with the incident field can be projected onto the FW modal vector via TM/TE decomposition, thus leading to $\overline{V} = \overline{\overline{q}}^H \overline{V}_{\mathrm{FW}}^{\mathrm{imp}}$, where $\overline{\overline{q}}^H = \{Q_{m,q}^{\mathrm{TM}*}, Q_{m,q}^{\mathrm{TE}*}\}_{\substack{m=1,N \\ q=0,M_A-1}}$ is the $N \times 2M_A$ matrix consisting of the first $2M_A$ TE–TM columns of $\overline{\overline{Q}}^H$ and $\overline{V}_{\mathrm{FW}}^{\mathrm{imp}} = \{-\hat{\sigma}_0 \cdot \mathbf{E}_{\mathrm{imp}}(\mathbf{k}), -\hat{\alpha}_0 \cdot \mathbf{E}_{\mathrm{imp}}(\mathbf{k}), 0, \dots, 0\}^{\mathrm{T}}$ is a $2M_A$ element column vector. Using the above projections onto the MoM system $\overline{\overline{Z}}_{\mathrm{MoM}}\,\overline{I} = \overline{V}$ and assuming that this latter is invertible, one obtains

$$\overline{I}_{\mathrm{FW}}^{\mathrm{FSS}} = \overline{\overline{Y}}_{\mathrm{FW}}\,\overline{V}_{\mathrm{FW}}^{\mathrm{imp}} \qquad \overline{\overline{Y}}_{\mathrm{FW}} = \overline{\overline{q}}\,\overline{\overline{Z}}_{\mathrm{MoM}}^{-1}\,\overline{\overline{q}}^H \qquad (13.17)$$

where $\overline{\overline{Z}}_{\mathrm{MoM}} = \overline{\overline{Q}}^H \overline{\overline{Z}}_{\mathrm{GF}}\,\overline{\overline{Q}}$. Equation (13.17) establishes the relationship between the *impressed* modal voltages at $z = 0$ and the total modal FSS currents associated to $2M_A$ accessible FW modes. The *total* modal voltage $\overline{V}_{\mathrm{FW}}$ is obtained by adding the modal voltage $\overline{V}_{\mathrm{FW}}^{\mathrm{rad}} = -\overline{\overline{Z}}_{\mathrm{GF}}\,\overline{I}_{\mathrm{FW}}^{\mathrm{FSS}}$ induced by the *radiation* of the FSS current in the stratified medium to the *impressed* modal voltage field, thus obtaining

$$\overline{V}_{\mathrm{FW}} = \overline{V}_{\mathrm{FW}}^{\mathrm{imp}} - \overline{\overline{Z}}_{\mathrm{GF}}\,\overline{I}_{\mathrm{FW}}^{\mathrm{FSS}} = \overline{V}_{\mathrm{FW}}^{\mathrm{imp}} - \overline{\overline{Z}}_{\mathrm{GF}}\,\overline{\overline{Y}}_{\mathrm{FW}}\,\overline{V}_{\mathrm{FW}}^{\mathrm{imp}} = [\overline{\overline{I}} - \overline{\overline{Z}}_{\mathrm{GF}}\,\overline{\overline{Y}}_{\mathrm{FW}}]\overline{V}_{\mathrm{FW}}^{\mathrm{imp}}$$
$$(13.18)$$

From (13.18) we obtain $\overline{V}_{\mathrm{FW}}^{\mathrm{imp}} = [\overline{\overline{I}} - \overline{\overline{Z}}_{\mathrm{GF}}\,\overline{\overline{Y}}_{\mathrm{FW}}]^{-1}\overline{V}_{\mathrm{FW}}$, which, when inserted into (13.17), leads to

$$\overline{I}_{\mathrm{FW}}^{\mathrm{FSS}} = \overline{\overline{Y}}_{\mathrm{FSS}}\,\overline{V}_{\mathrm{FW}} \qquad \overline{\overline{Y}}_{\mathrm{FSS}} = \overline{\overline{Y}}_{\mathrm{FW}}[\overline{\overline{Y}}_{\mathrm{GF}} - \overline{\overline{Y}}_{\mathrm{FW}}]^{-1}\overline{\overline{Y}}_{\mathrm{GF}} \qquad (13.19)$$

13.3.2   Aperture-Type FSS

For aperture-type FSSs, the MoM vector $\overline{V}$ is written in terms of the FW modal voltage vector $\overline{V}_{\mathrm{FW}}$, that is, $\overline{V}_{\mathrm{FW}} = \overline{\overline{p}}^H \overline{V}$, where $\overline{\overline{p}}^H$ is the $N \times 2M_A$ matrix consisting of the first $2M_A$ TM/TE columns of $\overline{\overline{P}}^H$. The forcing term vector associated with the incident magnetic field is also projected onto the FW modal vector via TM/TE decomposition, which leads to $\overline{I} = \overline{\overline{p}}\,\overline{I}_{\mathrm{FW}}^{\mathrm{imp}}$, where

$\overline{I}_{\text{FW}}^{\text{imp}} = \{-2\hat{\sigma}_0 \cdot \mathbf{H}_{\text{inc}}(\mathbf{k}), -2\hat{\alpha}_0 \cdot \mathbf{H}_{\text{inc}}(\mathbf{k}), 0, \ldots .0\}^{\text{T}}$ and $\overline{\overline{p}}$ is the $2M_A \times N$ matrix obtained by the first $2M_A$ rows of $\overline{\overline{P}}$. By using the above projections in the MoM system $\overline{\overline{Y}}_{\text{MoM}} \, \overline{V} = \overline{I}$ and assuming that the latter is invertible, we obtain

$$\overline{V}_{\text{FW}} = \overline{\overline{Z}}_{\text{FW}} \, \overline{I}_{\text{FW}}^{\text{imp}} \qquad \overline{\overline{Z}}_{\text{FW}} = \overline{\overline{p}} \, \overline{\overline{Y}}_{\text{MoM}}^{-1} \, \overline{\overline{p}}^{H} \qquad (13.20)$$

where $\overline{\overline{Y}}_{\text{MoM}} = \overline{\overline{P}}^{H} \, \overline{\overline{Y}}_{\text{GF}} \, \overline{\overline{P}}$. Equation (13.20) establishes the relationship between the impressed modal current on the upper side of the continuous PEC screen and the total modal electric field. The total modal current $\overline{I}_{\text{FW}}$ is obtained as the sum of the current induced by the magnetic current radiation on the upper $(\overline{I}_{\text{FW}}^{+})$ and lower $(\overline{I}_{\text{FW}}^{-})$ sides of the continuous screen; that is,

$$\overline{I}_{\text{FW}} = (\overline{I}_{\text{FW}}^{+}) + \overline{I}_{\text{FW}}^{-} = (\overline{I}_{\text{FW}}^{\text{imp}} - \overline{\overline{Y}}_{\text{FW}}^{+} \, \overline{V}_{\text{FW}}) - \overline{\overline{Y}}_{\text{FW}}^{-} \, \overline{V}_{\text{FW}}$$

$$= \overline{I}_{\text{FW}}^{\text{imp}} - \overline{\overline{Y}}_{\text{GF}} \, \overline{V}_{\text{FW}} = [\overline{\overline{I}} - \overline{\overline{Y}}_{\text{GF}} \, \overline{\overline{Z}}_{\text{FW}}] \overline{I}_{\text{FW}}^{\text{imp}} \qquad (13.21)$$

where we used the identity $\overline{\overline{Y}}_{\text{FW}}^{+} + \overline{\overline{Y}}_{\text{FW}}^{-} = \overline{\overline{Y}}_{\text{GF}} = \text{diag}[Y_{\text{GF}}^{\text{TM}}(\mathbf{k}_q),$ $Y_{\text{GF}}^{\text{TE}}(\mathbf{k}_q)]_{q=0, M_A-1}$. Using (13.21) in (13.20) we obtain $\overline{\overline{Z}}_{\text{FSS}} \overline{I}_{\text{FW}}^{\text{FSS}} = \overline{V}_{\text{FSS}}$, where $\overline{\overline{Z}}_{\text{FSS}} = \overline{\overline{Z}}_{\text{FW}}[\overline{\overline{I}} - \overline{\overline{Y}}_{\text{GF}} \, \overline{\overline{Z}}_{\text{FW}}]^{-1}$. The latter can be given in terms of the impedance matrices as

$$\overline{V}_{\text{FW}} = \overline{\overline{Z}}_{\text{FSS}} \, \overline{I}_{\text{FW}}^{\text{FSS}} \qquad \overline{\overline{Z}}_{\text{FSS}} = \overline{\overline{Z}}_{\text{FW}} \left(\overline{\overline{Z}}_{\text{GF}} - \overline{\overline{Z}}_{\text{FW}} \right)^{-1} \overline{\overline{Z}}_{\text{GF}} \qquad (13.22)$$

13.3.3 Dispersion Equation in Terms of Accessible Modes

The dispersion equation of the structure can be obtained directly through the accessible mode description by looking for a nontrivial solution for vanishing impressed fields. Let us refer to Eq. (13.17) for the capacitive FSS and to Eq. (13.20) for the inductive FSS. To find a nontrivial solution for the zero impressed field, one must have $\det [\overline{\overline{Y}}_{\text{FW}}^{-1}] = 0$ for the patch-type FSS and $\det [\overline{\overline{Z}}_{\text{FW}}^{-1}] = 0$ for the aperture-type FSS; alternatively,

$$\{\delta_0(k_x, k_y, \omega)\delta_1(k_x, k_y, \omega) \cdots \delta_{M_A-1}(k_x, k_y, \omega)\}^{-1} = 0 \quad \text{for patch-type FSS}$$
$$(13.23)$$

$$\{\gamma_0(k_x, k_y, \omega)\gamma_1(k_x, k_y, \omega) \cdots \gamma_{M_A-1}(k_x, k_y, \omega)\}^{-1} = 0 \quad \text{for aperture-type FSS}$$
$$(13.24)$$

where $\delta_0, \delta_1, \ldots$ are the eigenvalues of $\overline{\overline{Y}}_{\text{FW}} = \overline{\overline{q}} \left(\overline{\overline{Q}}^{H} \, \overline{\overline{Z}}_{\text{GF}} \, \overline{\overline{Q}} \right)^{-1} \overline{\overline{q}}^{H}$ and $\gamma_0, \gamma_1, \ldots$ are the eigenvalues of $\overline{\overline{Z}}_{\text{FW}} = \overline{\overline{p}} \left(\overline{\overline{P}}^{H} \, \overline{\overline{Y}}_{\text{GF}} \, \overline{\overline{P}} \right)^{-1} \overline{\overline{p}}^{H}$, respectively. We note that the convenience in using (13.23) and (13.24) in place of the standard form (13.14) and (13.15) is implied by the fact that M_A can be much smaller than N. However, it is clear that the process of reconstructing the dispersion equation through (13.23) and (13.24) is hindered by numerical difficulties. Indeed, around the solution of the dispersion equation, the MoM matrix $\overline{\overline{Z}}_{\text{MoM}} = \overline{\overline{Q}}^H \overline{\overline{Z}}_{\text{GF}} \overline{\overline{Q}}$ $\left(\overline{\overline{Y}}_{\text{MoM}} = \overline{\overline{P}}^H \overline{\overline{Y}}_{\text{GF}} \overline{\overline{P}} \right)$ is nearly singular, and the inversion required to obtain $\overline{\overline{Y}}_{\text{FW}} (\overline{\overline{Z}}_{\text{FW}})$ is inaccurate. This results in an inaccuracy in detecting the position of the poles of the relevant eigenvalues. To overcome this difficulty, one may resort to an alternative process, based on an analytical approximation of the matrix $\overline{\overline{Y}}_{\text{FSS}}$ and followed by an analytical determination of the dispersion equation. This process is illustrated next, with reference to a single accessible mode ($M_A = 1$).

13.4 POLE–ZERO MATCHING METHOD FOR DISPERSION ANALYSIS

13.4.1 Dominant-Mode Two-Port Admittance Network

Let us assume that only one pair of TM/TE propagating FW modes are accessible for a given z level ($M_A = 1$). As a special case of (13.19) (capacitive FSS) and (13.20) (inductive FSS), the FSS is modeled by the two-port network shown in Figure 13.5a, which is characterized by the admittance matrices

$$\overline{\overline{Y}}_{\text{FSS}}(k_x, k_y; \omega) = \overline{\overline{q}} \left(\overline{\overline{Q}}^H \overline{\overline{Z}}_{\text{GF}} \overline{\overline{Q}} \right)^{-1} \overline{\overline{q}}^H$$
$$\times \left[\overline{\overline{Y}}_{\text{GF}} - \overline{\overline{q}} \left(\overline{\overline{Q}}^H \overline{\overline{Z}}_{\text{GF}} \overline{\overline{Q}} \right)^{-1} \overline{\overline{q}}^H \right]^{-1} \overline{\overline{Y}}_{\text{GF}} \qquad (13.25)$$

for the patch-type FSS and

$$\overline{\overline{Z}}_{\text{FSS}}(k_x, k_y; \omega) = \overline{\overline{p}} \left(\overline{\overline{P}}^H \overline{\overline{Y}}_{\text{GF}} \overline{\overline{P}} \right)^{-1} \overline{\overline{p}}^H$$
$$\times \left[\overline{\overline{Z}}_{\text{GF}} - \overline{\overline{p}} \left(\overline{\overline{P}}^H \overline{\overline{Y}}_{\text{GF}} \overline{\overline{P}} \right)^{-1} \overline{\overline{p}}^H \right]^{-1} \overline{\overline{Z}}_{\text{GF}} \qquad (13.26)$$

for the aperture-type FSS. In (13.25) and (13.26), the dependence on the frequency and on the impressed wave vector has been emphasized, and

$$\overline{\overline{Y}}_{\text{GF}} = \text{diag}[Y_{\text{GF}}^{\text{TM}}, Y_{\text{GF}}^{\text{TE}}] = \text{diag}[Y_0^{\text{TM}}(\mathbf{k}) - j Y_1^{\text{TM}}(\mathbf{k}) \cot(k_{z1}h), Y_0^{\text{TE}}(\mathbf{k})$$
$$- j Y_1^{\text{TE}}(\mathbf{k}) \cot(k_{z1}h)] \qquad (13.27)$$

$\overline{\overline{Z}}_{\text{GF}} = \overline{\overline{Y}}_{\text{GF}}^{-1}, \overline{\overline{q}}, \overline{\overline{p}}$ are matrices of size $2 \times N$, and $\overline{\overline{q}}^H, \overline{\overline{p}}^H$ of size $N \times 2$. Let us anticipate that the SW dispersion equation associated with the metamaterial surface is given by $\det[\overline{\overline{Y}}_{\text{GF}} + \overline{\overline{Y}}_{\text{FSS}}] = 0$ for the patch-type FSS and $\det[\overline{\overline{Z}}_{\text{GF}} + \overline{\overline{Z}}_{\text{FSS}}] = 0$ for the aperture-type FSS, that is, by the resonance of the circuit in the lower part of Figure 13.5a. We will solve the dispersion equation after approximating the FSS matrices (13.25) and (13.26).

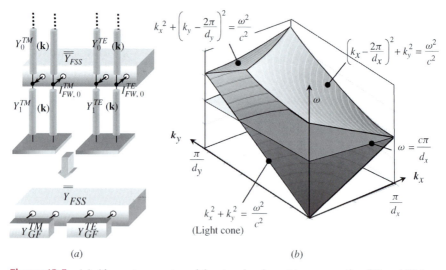

Figure 13.5 (a) Above: two-port modal network relevant to propagating TE and TM FW mode. Below: reduction at modal ports of transmission lines in absence of any excitation. (b) Brillouin diagram $(k_x, k_y) - \omega$ (the figures refer to a case $d_x > d_y$). Below the two portions of the upper conical surfaces, the higher order FW modes are cut off. This region identifies the validity of the FSS network in (a) with purely reactive entries of its admittance matrix $\overline{\overline{Y}}_{FSS}$. The free-space speed of light is denoted by c. The "light cone" is also depicted, and its surface identifies the cutoff of the dominant propagating mode.

The utilization of a two-port network is subject to the existence of an observation level z where the dominant TE and TM FW modes are the *only* accessible modes. This implies that all the higher order FW modes must be cut off. The cutoff condition of the higher order FW modes implies a limitation to the observable dispersion diagram. Figure 13.5b shows a dispersion diagram with angular frequency ω on the vertical axis and the wavenumbers k_x and k_y on the horizontal axes. Due to the periodicity of the FW spectrum, the observation may be restricted to the Brillouin region $(-\pi/d_x < k_x < \pi/d_x, -\pi/d_y < k_y < \pi/d_y)$, with a further (due to the symmetry of the structure) restriction to positive values of k_x and k_y. The cutoff region for higher order modes is imposed by the condition $k_{x\xi}^2 + k_{y\eta}^2 > \omega^2/c^2$ for $(\xi, \eta) \neq (0, 0)$. As a consequence, within the observed wavenumber plane, the cutoff region is delimited by portions of two cones whose vertices are at the FW wavenumbers closest to the origin (details are shown in Fig. 13.5b. A third cone is depicted in the same figure; its surface $k_x^2 + k_y^2 = \omega^2/c^2$ defines the cutoff of the dominant mode. Although this cone is not essential for the validity of the two-port model, it bounds the slow-wave region and is therefore important for representing the SW dispersion curves. Intersections of this cone with the vertical planes $\omega - k_x$ and $\omega - k_y$ identify the well-known "light lines" in these two planes. Figure 13.5b also shows the horizontal plane $\omega = \omega_M = c\pi/\max(d_x, d_y)$, which is the minimum frequency at which the higher order FW modes are attenuated for *any* wavenumber. In all cases of practical interest, the EBG of the artificial surface is located within the range $(0, \omega_M)$.

As far as the FSS admittance network is concerned, we would like to emphasize the following points:

(i) The $\overline{\overline{Y}}_{\mathrm{FSS}}$ entries are associated with the ratio at the FSS level between the transverse discontinuity of the H field and the transverse E field pertaining to the dominant FW mode. These quantities may be considered as relevant to an average field concept [12]. Consequently, the derivation of the FSS admittance matrix from the MoM matrix, expressed by (13.25) and (13.26), can be seen as a *homogenization process* of the periodic impedance.

(ii) The homogenization process does not require the hypothesis of small periodicity compared with the wavelength; it has only been assumed that all the higher order FW modes are cut off in the Brillouin region. The two-port network in Figure 13.5a is able to recover the transverse total field only above the level at which the only accessible mode is the dominant one.

(iii) Through the inverse of the MoM matrix, $\overline{\overline{Y}}_{\mathrm{FSS}}$ incorporates the information from all the higher order TE and TM FWs associated with the FSS discontinuity. This implies that the information on the overall dispersion equation in the cutoff region of higher order modes (Fig. 13.5b) can be obtained directly from $\overline{\overline{Y}}_{\mathrm{FSS}}$. Outside of this region, a model with more ports is needed to recover the dispersion properties.

13.4.2 Diagonalization of FSS Admittance Matrix

From here on, we will deal with the case of patches by treating the $\overline{\overline{Y}}_{\mathrm{FSS}}$ matrix, but analogous steps may be performed for aperture-type FSSs by using the $\overline{\overline{Z}}_{\mathrm{FSS}}$ matrix. The FSS admittance matrix in (13.25) can be diagonalized by a rotation matrix $\overline{\overline{R}}(\alpha)$,

$$\overline{\overline{Y}}_{\mathrm{FSS}}(k_x, k_y; \omega) = \overline{\overline{R}}(\alpha(k_x, k_y; \omega)) \,\mathrm{diag}\,[Y^{(1)}_{\mathrm{FSS}}(k_x, k_y; \omega), Y^{(2)}_{\mathrm{FSS}}(k_x, k_y; \omega)]$$
$$\times \overline{\overline{R}}(-\alpha(k_x, k_y; \omega)) \tag{13.28}$$

where

$$\overline{\overline{R}}(\alpha) = \begin{bmatrix} \cos\alpha & -\sin\alpha \\ \sin\alpha & \cos\alpha \end{bmatrix} \tag{13.29}$$

and $\overline{\overline{R}}^{-1}(\alpha) = \overline{\overline{R}}(-\alpha)$. The rotation angle α and the eigenvalues $Y^{(1)}_{\mathrm{FSS}}, Y^{(2)}_{\mathrm{FSS}}$ are dependent on k_x, k_y, and ω, as stressed in the notation. The angle α can be calculated from the entries $Y^{(i,j)}_{\mathrm{FSS}}$ of $\overline{\overline{Y}}_{\mathrm{FSS}}$ as $\alpha = \frac{1}{2} \tan^{-1}(2Y^{(1,2)}_{\mathrm{FSS}}/(Y^{(2,2)}_{\mathrm{FSS}} - Y^{(1,1)}_{\mathrm{FSS}}))$, where the tangent function is inverted in the range $(-\pi/2, \pi/2)$. The eigenvalues of $\overline{\overline{Y}}_{\mathrm{FSS}}$ can be found by inversion of (13.27) as

$$\mathrm{diag}\,[Y^{(1)}_{\mathrm{FSS}}, Y^{(2)}_{\mathrm{FSS}}] = \overline{\overline{R}}(-\alpha) \,\overline{\overline{Y}}_{\mathrm{FSS}}\, \overline{\overline{R}}(\alpha) \tag{13.30}$$

The following properties can be easily demonstrated:

(i) Since, in the absence of losses, the $Y_{FSS}^{(i,j)}$ are purely reactive in the cutoff region of the higher order modes (Fig. 13.5b), this implies a real value of α for any wavenumber. This also occurs when $\omega^2/c^2 < k_x^2 + k_y^2$.

(ii) The limit $\omega \to 0$ and $k_x, k_y \to 0$ implies that $\alpha \to 0$ and $\overline{\overline{R}} \to \overline{\overline{I}}$. Thus, the TE and TM transmission line networks are decoupled in such a limit.

(iii) Decoupling between the TE and TM ports also occurs when the direction of wave propagation is along any plane of geometric symmetry of the FSS. The planes of symmetry define the contour of the "irreducible" Brillouin region (IBR). As a consequence, the dominant TE and TM modes are decoupled along the segments of the IBR contour which converge at the origin.

To approximate and store the values of α, it is convenient to define the unwrapped angle $\alpha_u = n(\pi/4) + \alpha$, with $n = 1, 2, \ldots$ chosen so as to have a continuous value of α_u. As an illustration of this, we show the case of printed crossed dipoles in Figure 13.6 (see the geometry in the inset). The unwrapped angle $\alpha(\pi/d_x, k_y; \omega)$ is presented there as a function of the frequency and wavenumber along the segment $X \to M$ of the IBR triangular boundary $\Gamma \to X \to M \to \Gamma \equiv (0,0) \to (\pi/d_x, 0) \to (\pi/d_x, \pi/d_x) \to (0,0)$. We note that, along $\Gamma \to X$

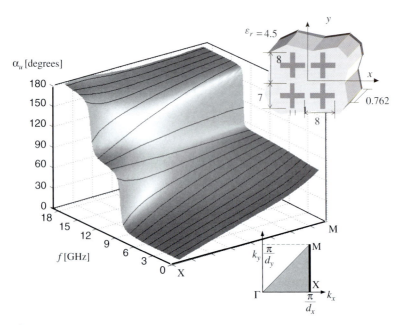

Figure 13.6 Unwrapped angle α_u versus frequency and wavenumber along path $X-M$ of Brillouin zone for crossed-dipole FSSs (see inset, dimensions in millimeters). The constant-frequency lines are shown.

and $M \rightarrow \Gamma$, α approaches 0 and α_u approaches a stairlike variation with steps of $\pi/2$. As apparent from the constant-frequency line of the surface $\alpha(\pi/d_x, k_y; \omega)$, α_u exhibits a smooth variation against the wavenumber. This regular behavior allows us to approximate $\alpha_u(\pi/d_x, k_y; \omega^{(j)})$ for a dense sampling at $\omega^{(j)}$ starting from $\alpha_u(\pi/d_x, k_y^{(i)}; \omega^{(j)})$ for a few values of $k_y^{(i)}$.

13.4.3 Foster's Reactance Theorem and Rational Approximation of Eigenvalues

In the absence of losses, the equivalent FSS admittance is purely reactive for every (k_x, k_y). It can be demonstrated that the imaginary parts of the eigenvalues $Y_{FSS}^{(i)}(k_x, k_y; \omega)$, seen as a function of frequency, respect the Foster's reactance theorem for every real value of (k_x, k_y). The demonstration can be performed as in [20] by referring to an equivalent waveguide with periodic boundary conditions. Note that this is valid within the cutoff region of the higher order FW modes described in Figure 13.5b; indeed, for frequencies where another pair of TM/TE modes is propagating, the two-port FSS matrix loses its purely reactive property. The main implication of Foster's reactance theorem is that the function $Y_{FSS}^{(i)}(k_x, k_y; \omega)$ possesses the same pole–zero analytical properties as a passive "*driving point*" *LC* function of the frequency. These properties are as follows:

(i) The poles and zeros lie on the real ω axis and are *simple* and *alternate*.

(ii) A pole or a zero must be at $\omega = 0$.

(iii) Poles and zeros are symmetrically displaced with respect to the origin.

An important consequence of these properties is that the admittance frequency function can be well approximated by a rational function after the positions of the poles and zeros along the real ω axis have been determined. Since, in the quasi-static limit, the patch-FSS behaves like a shunt capacitance (see Fig. 13.1), the FSS admittance has a simple zero at the origin. On the basis of (i) to (iii), the eigenvalues for the patch-type FSS can be approximated as

$$
\begin{aligned}
&Y_{FSS}^{(i)}(k_x, k_y, \omega) \\
&= \frac{j\omega C_0^{(i)}(k_x, k_y)(1 - [\omega/\omega_{z1}^{(i)}(k_x, k_y)]^2)(1 - [\omega/\omega_{z2}^{(i)}(k_x, k_y)]^2) \cdots}{(1 - [\omega/\omega_{p1}^{(i)}(k_x, k_y)]^2)(1 - [\omega/\omega_{p2}^{(i)}(k_x, k_y)]^2) \cdots}
\end{aligned} \tag{13.31}
$$

where $\omega_{p1}^{(i)} < \omega_{z1}^{(i)} < \omega_{p2}^{(i)} < \omega_{z2}^{(i)}$. For the aperture-type FSS, we have the following eigenvalues for the impedance matrix, which exhibit a simple zero at the origin:

$$
\begin{aligned}
&Z_{FSS}^{(i)}(k_x, k_y, \omega) \\
&= \frac{j\omega L_0^{(i)}(k_x, k_y)(1 - [\omega/\omega_{z1}^{(i)}(k_x, k_y)]^2)(1 - [\omega/\omega_{z2}^{(i)}(k_x, k_y)]^2) \cdots}{(1 - [\omega/\omega_{p1}^{(i)}(k_x, k_y)]^2)(1 - [\omega/\omega_{p2}^{(i)}(k_x, k_y)]^2) \cdots}
\end{aligned} \tag{13.32}
$$

where $\omega_{p1}^{(i)} < \omega_{z1}^{(i)} < \omega_{p2}^{(i)} < \omega_{z2}^{(i)}$. As mentioned in Section 4.2 [point (ii)], in the limit for $\omega \rightarrow 0$, the angle α tends to zero, and the TE and TM networks are decoupled. Consequently the FSS is capacitive (inductive) for the patch-type (aperture-type) FSS, thus recovering the expected network shown in Figure 13.1. It is important to anticipate the following points, which will be further investigated in the next sections:

(a) The quantities $C_0^{(i)}(k_x, k_y)[L_0^{(i)}(k_x, k_y)]$ are independent on ω and represent the quasi-static capacitance (inductance) of the FSS. Their dependence on (k_x, k_y) is found to be very weak and thus very easy to approximate.

(b) Equations (13.30) and (13.31) allow an analytical definition of the admittance, over a broad frequency range, on the basis of the determination of the wavenumber-dependent poles and zeros. As will be discussed below, $\omega_{z_j}^{(i)}(k_x, k_y)$ and $\omega_{p_j}^{(i)}(k_x, k_y)$ can be calculated directly for a few values of the wavenumber and can then be approximated straightforwardly.

(c) The numerical calculation of $\overline{\overline{Y}}_{\text{FSS}}$ in (13.25) and [or $\overline{\overline{Z}}_{\text{FSS}}$ in (13.26)] is accurate at those wavenumbers where $Y_{\text{FSS}}^{(i)}$ exhibits poles or zeros because the MoM matrix is well conditioned. Indeed, the solution of the dispersion equation (which implies a noninvertible MoM matrix) given by $\det [\overline{\overline{Y}}_{\text{FW}} + \overline{\overline{Y}}_{\text{GF}}] = 0$ is incompatible with the conditions obtained at the zeros and poles of $Y_{\text{FSS}}^{(i)}$.

(d) The sequence of poles and zeros in a certain frequency range is sometimes not sufficient to describe the eigenvalues in *the same* range. As discussed in [17], to investigate the range $\Omega = (0, \omega_{\max})$, the pole−zero pairs internal to Ω and one more pole−zero pair, the closest to $\omega_{\max}$, should be included. This practical rule is subject to the limitation that the last zero−pole pair should be included inside the cutoff region of the higher order FW modes (see Fig. 13.5b). To extend the frequency range of approximation, the process can be applied to a four-port accessible-mode matrix. However, it should be noted that the present approach applied to a two-port matrix model is sufficient to describe the first bandgap of the artificial surface, which is typically the most important one.

13.4.4   Poles and Zeros of FSS and Metamaterial Admittance

From the approximation (13.30) and (13.31), the analytical representation of the FSS in a broad frequency range can be derived from the following functions:

$$\omega_{z_j}^{(i)}(k_x, k_y), \omega_{p_j}^{(i)}(k_x, k_y), \alpha(\omega, k_x, k_y), C_0(k_x, k_y), (L_0(k_x, k_y)) \qquad (13.33)$$

All these functions show a very weak variation with respect to the wavenumber and are easy to approximate from the data related to only a few wavenumbers. In particular, the functions $\omega = \omega_{z_j}^{(i)}(k_x, k_y)$ and $\omega = \omega_{p_j}^{(i)}(k_x, k_y)$, respectively, are denoted as the *dispersion curves* of the poles and zeros of $Y_{\text{FSS}}^{(i)}$. These curves are very regular and can be approximated by a simple second-order polynomial form.

For a triangular IBR, the dispersion curves of poles and zeros are approximated along the segments $\Gamma \rightarrow X \rightarrow M \rightarrow \Gamma$ by

$$
\left.\begin{aligned}
\omega_{p_j}^{(i)}(k_x, 0) &= A_{p_j}^{(i)} + B_{p_j}^{(i)} k_x + C_{p_j}^{(i)} k_x^2 \\
\omega_{z_j}^{(i)}(k_x, 0) &= A_{z_j}^{(i)} + B_{z_j}^{(i)} k_x + C_{z_j}^{(i)} k_x^2
\end{aligned}\right\} \quad (\Gamma - X)
$$

$$
\left.\begin{aligned}
\omega_{p_j}^{(i)}\left(\frac{\pi}{d_x}, k_y\right) &= D_{p_j}^{(i)} + E_{p_j}^{(i)} k_y + F_{p_j}^{(i)} k_y^2 \\
\omega_{z_j}^{(i)}\left(\frac{\pi}{d_x}, k_y\right) &= D_{z_j}^{(i)} + E_{z_j}^{(i)} k_y + F_{z_j}^{(i)} k_y^2
\end{aligned}\right\} \quad (X - M) \qquad (13.34)
$$

$$
\left.\begin{aligned}
\omega_{p_j}^{(i)}(\kappa, \kappa) &= G_{p_j}^{(i)} + H_{p_j}^{(i)} \kappa + L_{p_j}^{(i)} \kappa^2 \\
\omega_{z_j}^{(i)}(\kappa, \kappa) &= G_{p_j}^{(i)} + H_{p_j}^{(i)} \kappa + L_{p_j}^{(i)} \kappa^2
\end{aligned}\right\} \quad (M - \Gamma)
$$

The coefficients $A, \ldots, H, L$ in (13.34) are independent of the frequency and wavenumber and are calculated by matching the values obtained from (13.25), (13.26), and (13.30). Figure 13.7a shows the dispersion curves of poles and zeros for a crossed-dipole FSS on the triangular contour of the IBR. The arms of the crosses have length 7 mm and width 1 mm. The periodicity is 8 mm along both directions. The dielectric substrate has relative permittivity $\varepsilon_r = 4.5$ and thickness $h = 0.762$ mm. We stress that the TE and TM ports are uncoupled along the two segments $\Gamma \rightarrow X$ and $M \rightarrow \Gamma$, whereas they are coupled (hybrid-mode region) along $X \rightarrow M$. However, at the first pair of TM or TE poles, the decoupling also persists approximately in the segment $X \rightarrow M$, thus justifying the notation of "quasi TM" and "quasi TE" poles.

Assume now that the FSS admittance network is placed in parallel with the admittance from the short-circuited transmission line, that is,

$$
\overline{\overline{Y}}_{\mathrm{cc}} = \mathrm{diag}\left[-j\frac{\omega \varepsilon_r \varepsilon_0}{k_{z1}}\cot(k_{z1}h), -j\frac{k_{z1}}{\omega \mu_0}\cot(k_{z1}h)\right]
$$

so that

$$
\overline{\overline{Y}}_{\mathrm{MM}}(k_x, k_y; \omega) = \overline{\overline{Y}}_{\mathrm{FSS}} + \overline{\overline{Y}}_{\mathrm{cc}} \qquad (13.35)
$$

where $\overline{\overline{Y}}_{\mathrm{MM}}$ can be defined as the equivalent admittance matrix of the *metamaterial,* to intend that this admittance incorporates the effects of both the FSS and the dielectric material slab. The metamaterial matrix $\overline{\overline{Y}}_{\mathrm{MM}}$ has the same Foster reactance theorem properties as the matrix $\overline{\overline{Y}}_{\mathrm{FSS}}$, since it is the sum of two reactance-type functions. However, the poles and zeros of $\overline{\overline{Y}}_{\mathrm{MM}}$ are more numerous than those of $\overline{\overline{Y}}_{\mathrm{FSS}}$. This general fact motivates the choice to synthesize, via pole–zero matching, the *FSS admittance* instead of the *metamaterial admittance.* The poles and zeros of $\overline{\overline{Y}}_{\mathrm{MM}}$ are shown in Figure 13.7b with reference to the same case as that in Figure 13.7a. The poles of $\overline{\overline{Y}}_{\mathrm{MM}}$ denote the locus of the Brillouin space where the surface behaves like an equivalent PEC. Thus, the poles of $\overline{\overline{Y}}_{\mathrm{FSS}}$ are a subset of the poles of $\overline{\overline{Y}}_{\mathrm{MM}}$. The zeros of $\overline{\overline{Y}}_{\mathrm{MM}}$ mark a perfect magnetic conductor surface in the Brillouin space. Note that this way of introducing the

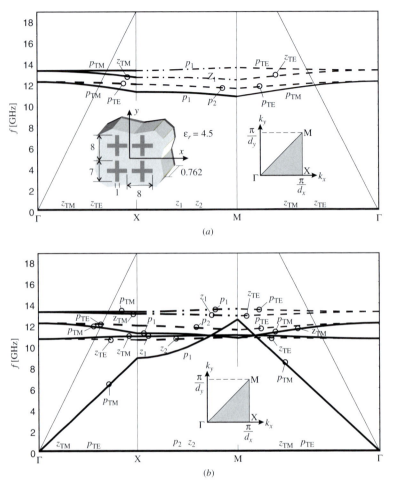

Figure 13.7 Pole and zero dispersion diagrams for crossed-dipole FSS surface along IBR. Curves relevant to (*a*) FSS admittance matrix $\overline{\overline{Y}}_{FSS}$ (*b*) metamaterial admittance matrix $\overline{\overline{Y}}_{MM}$ for structure shown in the inset of (*a*) (dimensions in millimeters). The continuous and dashed lines denote the TM (quasi-TM) and TE (quasi-TE) curves, respectively, using the abbreviation *p* for poles and *z* for zeros. Dot–dashed lines denote the "hybrid" poles and zeros, where the TE and TM modes are coupled. Pole (zero) curves relevant to the eigenvalues numbered by $i = 1, 2$ use the abbreviations p_i, z_i, respectively. Note that the pole and zero curves of $\overline{\overline{Y}}_{FSS}$ are fewer and never intersect with each other.

magnetic conductor is independent of whether or not the wavenumber is below the light cone. As is apparent from Figure 13.7*b*, the dispersion curves of poles and zeros of $\overline{\overline{Y}}_{MM}$ can intersect while preserving, *for each wavenumber*, the alternating values of the frequencies of the poles and zeros. Instead, no intersection occurs for the FSS poles and zeros (Fig. 13.7*a*).

We would like to remark that analogous considerations hold for the aperture-type FSS, except that the metamaterial impedance matrix is given by

$$\overline{\overline{Z}}_{MM}(k_x, k_y; \omega) = \overline{\overline{Z}}_{FSS}\left[\overline{\overline{Z}}_{FSS} + \overline{\overline{Z}}_{cc}\right]^{-1}\overline{\overline{Z}}_{cc} \qquad (13.36)$$

where $\overline{\overline{Z}}_{cc} = \overline{\overline{Y}}_{cc}^{-1}$.

13.4.5 Analytical Form of Dispersion Equation

The construction of the FSS admittance $\overline{\overline{Y}}_{FSS}$ proceeds through the following steps: (a) diagonalization of $\overline{\overline{Y}}_{FSS}$ through (13.30) and interpolation of the angle α; (b) identification of the pole and zero frequencies of the eigenvalues of $\overline{\overline{Y}}_{FSS}$; (c) second-order polynomial approximation of the pole and zero dispersion diagrams; and (d) approximation of the eigenvalues via a rational function of frequency on the segments of the contour of the IBR.

The final output is an analytical form for $\overline{\overline{Y}}_{FSS}$ for any frequency and wavenumber. This allows us to deal with an analytical form of the SW dispersion equations $\det\left[\overline{\overline{Y}}_{FSS} + \overline{\overline{Y}}_{GF}\right] = 0$ (det $\left[\overline{\overline{Z}}_{FSS} + \overline{\overline{Z}}_{GF}\right] = 0$ for the aperture-type FSS), as derived from the equivalent networks in Figure 13.5a. After some straightforward algebraic manipulations, the dispersion equation can be explicitly rewritten in terms of α and of $Y_{FSS}^{(i)}$ as

$$(Y_{FSS}^{(1)} + Y_{GF}^{TM})(Y_{FSS}^{(2)} + Y_{GF}^{TE}) + \sin^2 \alpha(Y_{FSS}^{(1)} - Y_{FSS}^{(2)})(Y_{GF}^{TM} - Y_{GF}^{TE}) = 0 \quad (13.37)$$

(or analogous relation for the aperture-type FSS, with impedance quantities). Within the slow-wave region $k_x^2 + k_y^2 > \omega^2/c^2$, this equation identifies the wavenumbers of the SWs supported by the artificial surface. The solutions to this equation can be easily found by using conventional numerical procedures. Analysis of the signs of the various terms in (13.37) reveals that for those frequencies at which we found poles and zeros of $Y_{FSS}^{(i)}$, Eq. (13.37) cannot be satisfied, as stated in Section 13.4.3, point (c).

13.4.6 Examples

We illustrate in this section the dispersion diagrams obtained for different types of artificial surfaces consisting of thin dipole and crossed-dipole FSSs and of crossed-aperture FSSs. The numerical results relevant to the dispersion diagrams have been successfully validated through a comparison with those obtained with both the full-wave MoM procedure presented in Section 13.2.3 and the commercial software CST Microwave Studio. Except for the case of Figure 13.10 below, these comparative results are not explicitly shown to avoid an overcrowding of the diagrams.

The SW dispersion diagram for crossed dipoles of Figure 13.7 is plotted in Figure 13.8a, along the triangular path $\Gamma - X - M - \Gamma$ relevant to the IBR

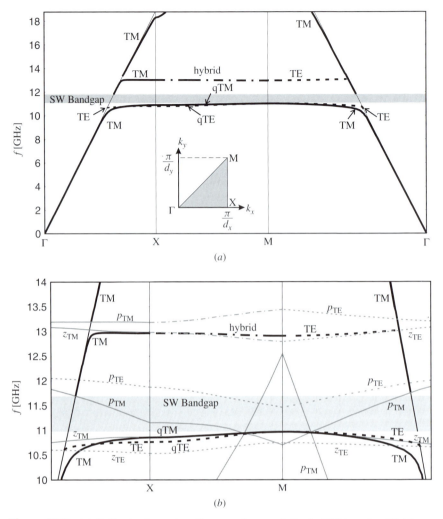

Figure 13.8 (*a*) Brillouin dispersion diagram of crossed-dipole FSS-based artificial surface and (*b*) zoom image around bandgap region. Continuous and dashed lines denote the TM (quasi-TM) and TE (quasi-TE) modes, respectively, using the abbreviation p for poles and z for zeros.

boundary (see the inset). The artificial surface is the one shown in the inset of Figure 13.7. In the zoom image in Figure 13.8*a*, the poles and zeros of $\overline{\overline{Y}}_{MM}$ are drawn for convenience. It is found that TM SWs change from improper to proper at the frequencies where the curves of the poles of the artificial surface admittance intersect the light line, that is, where the surface acts as a PEC for grazing propagation. Conversely, TE SWs experience the same phenomenon at the frequencies where the surface behaves as a perfect magnetic conductor for grazing propagation. Improper SWs are not shown in the figure. We would like to stress that the paths $\Gamma-X$ and $M-\Gamma$ are associated with the pure TM or TE SW modes, while $X-M$ is related to hybrid SW modes with a TM or TE predominance. It is found that the dominant SW mode in the x direction has a TM nature and possesses a zero cutoff frequency; for propagation in oblique directions, the SW mode becomes hybrid (quasi-TM) and again becomes a pure TM SW mode in the symmetry plane $M-\Gamma$. We found a second mode which is TE for propagation in the symmetry planes and quasi-TE elsewhere. This mode exhibits a very flat dispersion diagram for any wavenumber (similar to the response of a notch filter); it almost degenerates with the quasi-TM modes in the segment $X-M$. We found an EBG between 11 and 11.7 GHz. As the dependence on substrate thickness is concerned, we found that varying the thickness from 0.762 to 0.128 mm reduces the bandgap and moves it at lower frequencies (from 10.6 to 11 GHz).

The complementary aperture-type FSS results are shown in Figure 13.9. The crossed-aperture FSS, with arm length 7 mm, arm width 1 mm, periodicity 8 mm, dielectric substrate with $\varepsilon_r = 4.5$, and thickness 0.127 mm, is shown in the inset. The dominant mode is a TM wave that is a slow wave also in the

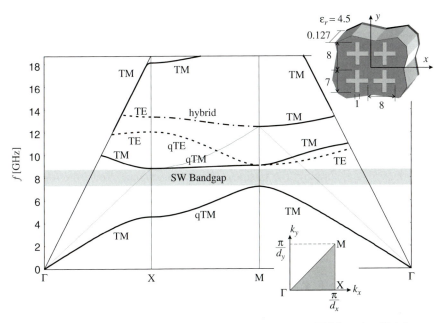

Figure 13.9 Brillouin dispersion diagram of crossed-aperture FSS-based artificial surface. Dimensions are given in the inset. Continuous and dashed lines denote TM (or quasi-TM) and TE (or quasi-TE) modes, respectively, whereas the dash–dotted line denotes a hybrid mode. Thin solid line denotes the light line in the dielectric, corresponding to the nondispersive curve of the TEM mode in the dielectric-filled parallel-plate waveguide.

dielectric region. The higher frequency of this mode (at point M) individuates the lower limit of the EBG. The upper limit of the EBG is obtained at the frequency where the TM mode becomes propagating along x (point X). A very large EBG for SWs is obtained for this structure in the frequency range 7.2 to 8.84 GHz. Note that only proper SWs are shown in the dispersion diagram. When the substrate thickness increases, the bandgap tends to decrease until it disappears, due to a shift of the dispersion curve of the dominant mode at the M point toward higher frequencies. A similar behavior is found for the Jerusalem-cross aperture-type structure in [21]. It is worth noting that in comparison with the complementary dipole structure with the same thickness (EBG from 10.6 to 11 GHz), the aperture-type structure exhibits a bandgap four times larger and located at much lower frequencies. It is apparent (cf. Figs. 13.8 and 13.9) that a very different situation occurs for complementary capacitive-type and inductive-type FSS structure as concerned with the bandgap location. This behavior can be explained as follows. For the aperture-type structure, the first dominant TM mode is an FSS-tied SW bilaterally attenuated at opposite sides of the FSS; indeed, its phase velocity is less than the speed of light in both the dielectric and the free space medium. This wave can exist also for the FSS floating in free space. The grounded dielectric slab behaves as a *capacitive reactance loading* at the FSS level. The upper medium wave admittance is also capacitive being the quasi-TM (qTM) mode. The resonance condition is implied by the balance of the total capacitive admittance with the FSS inductance.

The behavior is indeed very different for the patch-type FSS. For such a structure, the dominant mode is still qTM; however, the exponential attenuation is only on the free-space side, while a propagation regime occurs within the slab, at least in the major part of the Brillouin region. This implies an inductive loading of the grounded slab at the FSS level which resonates with the sum of the capacitive admittances of the FSS and free space. A qTE mode can also exist, with a similar configuration of attenuation toward free space and propagation within the slab. The wave impedance is inductive on both sides of the FSS, and the resonance is ensured by the intrinsic dipole capacitance. Since the resonance condition is always found with a propagating wave regime inside the slab, the lower bound of the bandgap is always higher with respect to the one of the complementary aperture-type structure.

Obtaining a full bandgap for every azimuth direction of SW propagation requires elements as symmetric as possible. It is interesting for some applications (e.g., qTEM waveguides, soft wall horns) to use asymmetrical elements to stop or to pass particular directions of propagation while enhancing or eliminating the EBG for some directions. Printed dipoles, printed slots, or "gangbuster" FSSs [15] may be used for this purpose. The obtained anisotropy also creates different responses to the two polarizations, which leads to "soft" (high impedance for TM polarization, low impedance for TE polarization) or "hard" (high impedance for TE polarization, low impedance for TM polarization)

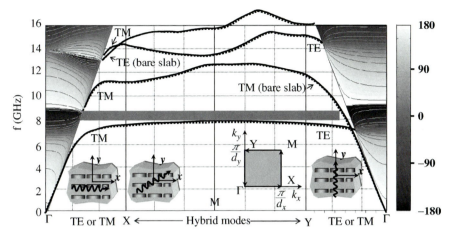

Figure 13.10 Brillouin dispersion diagram along boundary of IBR (path $\Gamma-X-M-Y-\Gamma$ shown in inset). The dispersion curves of the SWs are calculated by the pole–zero matching method (dotted line) and by a conventional dispersion full-wave analysis (continuous line). The regions beyond the light line show the phase of the reflection coefficient. Width and length of dipoles, 1 and 6 mm, respectively; dielectric substrate relative permittivity $\varepsilon_r = 10.2$ and thickness 0.762 mm; periodicity in both directions 8 mm.

properties [22]. Figure 13.10 shows the SW dispersion diagram associated with printed x-oriented dipoles. The dipoles have length 6 mm, width 1 mm, and periodicity 8 mm, while the dielectric substrate has relative permittivity 10.2 and thickness 0.762 mm. In this figure, results obtained with the equivalent network (solid line) are compared with those from the full-wave approach presented in section 13.2.3 (dotted line). In this case, the IBR is bounded by the square $\Gamma-X-M-Y-\Gamma$. Consider first the path $\Gamma-X$. There, only a TM wave can provide an interaction with the FSS, since the E field is aligned with the electric dipoles. Since the dipoles are thin, the TE SW coincides with those of a bare (unprinted) grounded slab. Along the $Y-M$ direction, the TE wave interacts with the dipoles, while the TM wave is that of the bare slab. When the direction of propagation is oblique, the mode is hybrid. We note that the EBG is not extended to any wavenumber, because the bare slab TM SW (with zero cutoff frequency) exists for propagation along y. As shown in Figure 13.11a, the partial bandgap decreases to the point of disappearing as the dipole width increases. Square patches do not have any EBG, not even a partial EBG. It is interesting to compare the previous results with those obtained with the insertion of shorting pins (Fig. 13.11b), which leads to a mushroom structure [1]. It is worth noting that the introduction of the shorting pins brings in additional dominant TM modes at a lower frequency and modifies the cutoff of the preexisting TM mode. This leads to a low-frequency bandgap which is maximized for square patches.

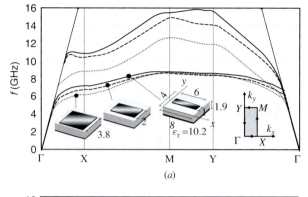

(a)

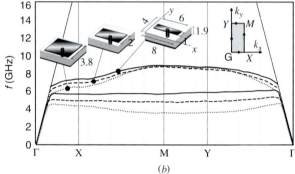

(b)

Figure 13.11 Comparison between dispersion diagrams of (a) dipole-type artificial surfaces and (b) mushroomlike artificial surfaces for different sizes of width dipoles (dimensions in millimeters).

13.5 CONCLUSIONS

In this chapter, we have presented a method for the dispersion analysis of the class of artificial EBG surfaces realized by FSSs printed on a grounded slab. This method has been illustrated here with reference to a patch FSS or aperture FSS on a single layer but has general applicability to a multilayer structure. On the basis of the MoM solution, a two-port admittance matrix is defined with the ports corresponding to the dominant TE and TM FW of the exact Floquet expansion. The admittance matrix is then characterized by the poles and zeros of its eigenvalues for a few values of the wavenumber. The identification of a set of dispersion curves associated with the poles and zeros of the FSS and their regularity allows the interpolations of these curves by low-order polynomials. Furthermore, the eigenvalues of the FSS admittance matrix respect Foster's reactance theorem and thus the properties of the lumped-element LC driving-point functions of the frequency. These properties allow the approximation of the eigenvalues in terms of rational functions. The consequent closed-form expression is applied to formulate the dispersion equation for SWs. Dispersion of leaky-wave modes, not treated here, is addressed in [17] and is still under investigation.

We would like to remark that the full-wave analysis for each k_x and k_y is very efficient, since it implies the inversion of a moderate-size MoM matrix; however, obtaining accurate information on the continuous spectrum (k_x, k_y) requires a large amount of computational time. The main peculiarity of the method presented in this chapter is the possibility of reconstructing the dispersion diagram in the continuous (k_x, k_y) wavenumber domain over a large frequency range, starting from the response of the structure in a few points of the (k_x, k_y) spectral domain. This procedure also opens very interesting possibilities for Green's function analysis, which requires an integration of the continuous spectrum.

ACKNOWLEDGMENTS

The authors wish to thank Marco Caiazzo and Massimo Nannetti of the University of Siena for help in preparing the numerical results.

REFERENCES

1. D. Sievenpiper, L. Zhang, R. F. Jimenez Broas, N. G. Alexopolous, and E. Yablonovitch, "High-impedance electromagnetic surfaces with a forbidden frequency band," *IEEE Trans. Microwave Theory Tech.*, vol. 47, pp. 2059–2074, Nov. 1999.
2. F.-R. Yang, K.-P. Ma, Y. Qian, and T. Itoh, "A novel TEM waveguide using uniplanar compact photonic-bandgap (UC-PBG) structure," *IEEE Trans. Microwave Theory Tech.*, vol. 47, pp. 2092–2098, Nov. 1999.
3. P.-S. Kildal, "Artificially soft and hard surfaces in electromagnetics," *IEEE Trans. Antennas Propag.*, vol. 38, pp. 1537–1544, Oct. 1990.
4. F. Yang and Y. Rahmat-Samii, "Microstrip antennas integrated with electromagnetic band-gap (EBG) structures: A low mutual coupling design for array applications," *IEEE Trans. Antennas Propag.*, vol. 51, pp. 2936–2946, Oct. 2003.
5. F. Yang and Y. Rahmat-Samii, "Reflection phase characterizations of the EBG ground plane for low profile wire antenna applications," *IEEE Trans. Antennas Propag.*, Special Issue on Metamaterials, vol. 51, pp. 2691–2703, Oct. 2003.
6. M. Caiazzo, S. Maci, and N. Engheta, "A metamaterial slab for compact cavity resonators," *IEEE Antennas Wireless Propag. Lett.*, vol. 3, pp. 261–264, 2004.
7. Y. Zhang, J. von Hagen, M. Younis, C. Fischer, and W. Wiesbeck, "Planar artificial magnetic conductors and patch antennas," *IEEE Trans. Antennas Propag.*, Special Issue on Metamaterials, vol. 51, pp. 2704–2712, Oct. 2003.
8. R. Leone and H. Y. D. Yang, "Design of surface-wave band-gaps for planar integrated circuits using multiple periodic metallic patch arrays," in *2001 IEEE MTT International Microwave Symposium Digest*, Phoenix, AZ, vol. 2, pp. 1213–1216, May 20–25, 2001.
9. S. Clavijo, R. E. Díaz, and W. E. McKinzie III, "Design methodology for Sievenpiper high-impedance surfaces: An artificial magnetic conductor for positive gain electrically small antennas," *IEEE Trans. Antennas Propag.*, Special Issue on Metamaterials, vol. 51, pp. 2678–2690, Oct. 2003.
10. D. S. Lockyer, J. C. Vardaxoglou, and R. A. Simpkin, "Complementary frequency selective surfaces," *IEE Proc. Microwaves Antennas Propag.*, vol. 147, pp. 501–507, Dec. 2000.
11. S. B. Savia and E. A. Parker, "Equivalent circuit model for superdense linear dipole arrays," *IEE Proc. Microwaves Antennas Propag.*, vol. 150, pp. 37–42, Feb. 2003.
12. S. Tretyakov, *Analytical Modelling in Applied Electromagnetics*, Artech House, New York, 2003, pp. 217–224.
13. A. Sanada, C. Caloz, and T. Itoh, "Planar distributed structures with negative refractive index," *IEEE Trans. Microwave Theory Tech.*, vol. 52, pp. 1252–1263, Apr. 2004.

14. A. Grbic and G. V. Eleftheriades, "Periodic analysis of a 2-D negative refractive index transmission line structure," *IEEE Trans. Microwave Theory Tech.*, vol. 51, pp. 2604–2611, Oct. 2003.

15. B. A. Munk, *Frequency Selective Surfaces: Theory and Design*, Wiley, New York, 2000.

16. R. Orto and R. Tascone, "Planar periodic structures," in *Frequency Selective Surfaces*, (J. C. Vardaxoglou Ed.), Research Studies Press, Taunton, England, 1997, Chapter 7, pp. 221–275.

17. S. Maci, M. Caiazzo, A. Cucini, and M. Casaletti, "A pole-zero matching method for EBG surfaces composed of a dipole FSS printed on a grounded dielectric slab," *IEEE Trans. Antennas Propag.*, Special Issue on Artificial Magnetic Conductors, Soft/Hard Surfaces, and Other Complex Surfaces, vol. 53, pp. 70–81, Jan. 2005.

18. M. Bozzi, S. Germani, L. Minelli, L. Perregrini, and P. deMaagt, "Efficient calculation of the dispersion diagram of planar electromagnetic band-gap structures by the MoM/BI-RME method," *IEEE Trans. Antennas Propag.*, Special Issue on Artificial Magnetic Conductors, Soft/Hard Surfaces, and Other Complex Surfaces, vol. 53, pp. 29–35, Jan. 2005.

19. T. E. Rozzi, "Network analysis of strongly coupled transverse apertures in waveguide," *Int. J. Circuit Theory Appl.*, vol. 1, pp. 161–178, 1973.

20. R. E. Collin, *Foundations for Microwave Engineering*, McGraw-Hill, New York, 1992.

21. R. Coccioli, F.-R. Yang, K.-P. Ma, and T. Itoh, "Aperture-coupled patch antennas on UC-PBG substrate," *IEEE Trans. Microwave Theory Tech.*, vol. 47, pp. 2123–2130, Nov. 1999.

22. S. Maci and P.-S. Kildal, "Hard and soft gangbuster surfaces," in *URSI International Symposium on Electromagnetic Theory Digest*, Pisa, Italy, vol. 2, pp. 290–292, May 23–27, 2004.

CHAPTER 14

SPACE-FILLING CURVE HIGH-IMPEDANCE GROUND PLANES

John McVay, Nader Engheta, and Ahmad Hoorfar

Ever-increasing demands for high-performance, low-profile, conformal and flush-mounted antennas with improved radiation characteristics for various communications and radar applications have resulted in considerable interest by the electromagnetic research community in high-impedance surfaces, also known as artificial magnetic conductors [1–5]. These surfaces have a reflection coefficient $\Gamma \cong +1$ when illuminated with a plane wave, instead of the typical $\Gamma \cong -1$ for a conventional perfectly electric conducting (PEC) surface. These structures can obviously offer interesting applications for antenna designs [1–10] and for thin absorbing screens [11]. For example, a horizontal dipole antenna placed above such a metamaterial surface will have an image current with the same phase as the current on the dipole, resulting in enhanced radiation performance [6–10]. Several different types of high-impedance ground planes have been studied by various research groups (see, e.g., [1–5]).

Since magnetic conducting surfaces do not exist naturally, it is necessary to artificially create a surface with magnetic conduction properties in a certain band of frequencies. This can be achieved by utilizing resonant inclusions on a nonconducting host substrate layer in parallel with a conducting ground plane. Near the resonance of the inclusion, strong currents are induced on the surface, and together with the conducting ground plane, this structure may provide an equivalent magnetic conductor for a frequency range corresponding to the frequency range in the vicinity of a resonance of the surface. One possibility to form inclusions that are resonant but have an electrically small footprint at their resonant frequency is the use of the space-filling curves.

Space-filling curves are, in general, a continuous mapping of the normalized interval [0, 1] onto the normalized square [0, 1] × [0, 1]. In 1890, Giuseppe Peano suggested the first space-filling curve, now called the Peano curve [12]. In 1891, David Hilbert introduced his version of a space-filling curve [12]. These curves are both expressed in terms of iteration numbers, and in both cases, as this iteration number (also called the order number) approaches infinity, the representative curve occupies the entire square; that is, the curve itself passes through

Metamaterials: Physics and Engineering Explorations, Edited by N. Engheta and R. W. Ziolkowski
Copyright © 2006 the Institute of Electrical and Electronics Engineers, Inc.

every point within the square. While both curves pass through every point in the square for infinite iteration orders, they have a different means to this same end, as can be seen in the comparison of the first three orders of the Peano and Hilbert curves shown in Figure 14.1.

These curves offer certain attractive properties; that is, a structure of this shape can be made from an electrically long metallic wire compacted within a very small area footprint. Moreover, these space-filling geometries can be planar structures, thus allowing for ease of fabrication. The total length of these Peano and Hilbert curves as a function of the iteration order number N is shown in Table 14.1. As can be seen, the Peano curve has a higher compression rate (e.g., a longer total length) than the Hilbert curve for a fixed order of N.

As the iteration order of the curve increases, a space-filling curve may maintain its footprint size while its length increases. This property is what allows the antennas represented by these curves to possess a relatively low resonant frequency, that is, a long resonant wavelength with respect to the linear dimension of its footprint. The Hilbert and Peano curves have been used for the design of small antennas [13–18] and frequency-selective surfaces (e.g., [19]).

In this chapter, we review some of our work on exploring the role of space-filling curves in constructing metamaterial surfaces in which many inclusions in the shape of the Hilbert curve or Peano curve are placed, in a two-dimensional (2D) periodic arrangement, on a host surface whose distance above a conducting ground plane is small [20, 21]. We also briefly review some of our numerical

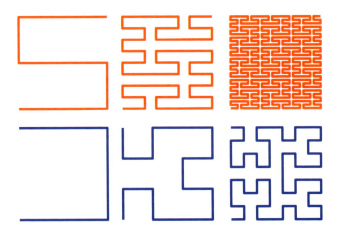

Figure 14.1   Peano (red) and Hilbert (blue) curves of orders 1 through 3.

TABLE 14.1   Total Length S for Peano and Hilbert Curves with Respect to Iteration Order Number N

Peano	$S = (3^{2N} - 1)d$	$d = L/(3^N - 1)$
Hilbert	$S = (2^{2N} - 1)d$	$d = L/(2^N - 1)$

Note: L is the linear side dimension of the curve.

results on the radiation performances of short dipoles placed above such a space-filling curve high-impedance surface (HIS) [6–10]. In addition, we briefly discuss some of the results of our ongoing research on the use of space-filling curve inclusions in forming thin absorbing layers [22] and double-negative (DNG) bulk media [23–25].

14.1 RESONANCES OF SPACE-FILLING CURVE ELEMENTS

Plane-wave scattering from single Peano and Hilbert curve elements of varying iteration orders were simulated in free space using a method-of-moments (MoM) code [26]. These simulations were used to determine the resonant frequencies of each of the orders of the space-filling curve structures shown in Figure 14.1 when they were contained within a 30-mm × 30-mm footprint. A normally incident plane wave with two separate polarizations, that is, E_x and E_y, illuminated the space-filling curve elements. A frequency sweep was applied and the maximum current induced within the structure was obtained as a function of the frequency. Linear interpolation was then utilized to find the frequencies for which this maximum current falls to the -3-dB point with respect to its maximum. The difference between these two frequencies is the frequency width, which we denote by Δf. The fractional bandwidth of each element was then defined as $BW = \Delta f / f_{max}$.

The current distributions on the Hilbert elements, orders 1 through 3, are shown in Figures 14.2 and 14.3 for the x- and y-polarized excitations,

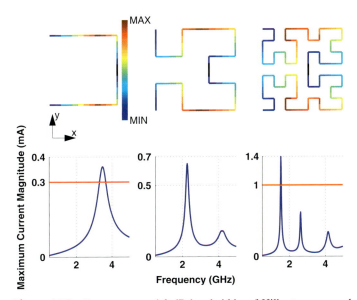

Figure 14.2 Resonances and 3-dB bandwidths of Hilbert curves, orders 1 through 3, for induced current (shown in color) due to normally incident plane-wave polarized in x direction.

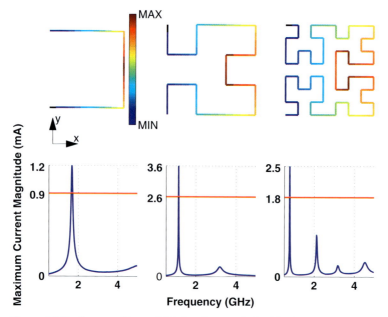

Figure 14.3   Same as Figure 14.2, but for y-polarized incident wave.

respectively. As can be seen, the currents on each element go to zero at the ends, as expected; they also go to zero at the symmetric center of the curve for the x-polarized case. This is due to the symmetry of the Hilbert elements with respect to the x axis. For the case of the y-directed polarization, the induced current goes to zero only at the wire ends and its maximum value occurs at the symmetric center. Due to this polarization-dependent nature of the Hilbert curve, the resonant frequency for the y-directed polarization is found to be about half of that for the x-directed polarization. This is due to the fact that the wire appears to the current to be electrically twice as long for the x-directed polarization as compared to the y-directed case.

The maximum induced currents on each space-filling curve as a function of the frequency are also shown in these figures. As can be seen, the resonant frequency is reduced as the length of the curve increases, that is, as the iteration order is increased. A comparison between the maximum induced currents in Figures 14.2 and 14.3 show the effects of the polarization on the resonant frequency.

The corresponding results for the single Peano curve elements under the influence of the normal incidence plane wave are shown in Figures 14.4 and 14.5. In Figure 14.6, the 30-mm side dimension of each Peano curve when it is normalized with respect to the resonant wavelength λ_{RES} and the corresponding bandwidth are plotted as a function of the iteration order of the curve. The bandwidths are defined as was mentioned at the beginning of this section. [It should be noted that for the x-polarized cases, only the "dominant" resonance, which is the second resonance shown in Figure 14.4, is utilized to evaluate the

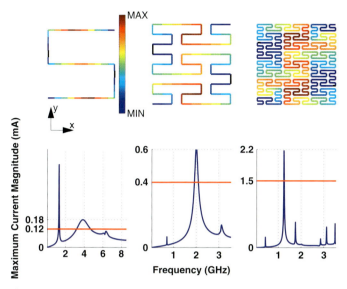

Figure 14.4 Same as Figure 14.2 for the *x*-polarized incident wave, but for the Peano curve. From [21]. Reproduced/modified by permission of American Geophysical Union.

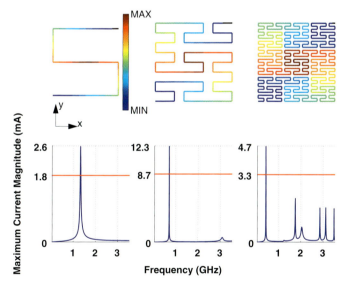

Figure 14.5 Same Peano curves as in Figure 14.4, but for the *y*-polarized incident wave. From [21]. Reproduced/modified by permission of American Geophysical Union.

data in Figure 14.6. This resonance is considered as dominant since it will be shown (see Fig. 14.9 below) that it corresponds to the Peano high-impedance surface resonance for the *x*-polarized cases.] It can be seen that as the order of the curve is increased, the electrical footprint of the curve decreases since the resonant frequency decreases, as expected and is evident from Figures 14.4

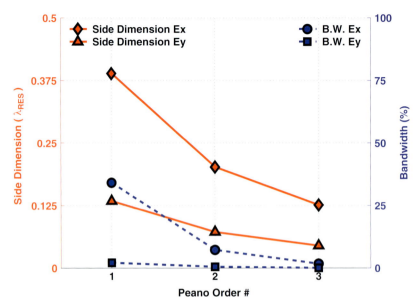

Figure 14.6 Normalized side dimensions (with respect to resonant wavelength λ_{RES}) and relative 3-dB bandwidths of maximum value of induced current on Peano curve versus Peano curve iteration order. From [21]. Reproduced/modified by permission of American Geophysical Union.

and 14.5. As in the case of the Hilbert inclusion, the dependence of the resonant frequency on the polarization of the incident wave for the Peano curve can also be seen from these figures. For the cases where the incident electric field is polarized in the y direction, the resonant frequency is approximately a third of that of the x-polarized case. Further insight into this polarization dependence can be obtained from the current distributions along the Peano curve, shown in Figures 14.4 and 14.5.

From the results reviewed above, it can be seen that the space-filling curve elements can resonate at frequencies where the footprint of the curve can be considered electrically very small. The higher the order of the curve, the lower the resonant frequency and, thus, the smaller the footprint of the curve with respect to the resonant wavelength. The cost of achieving such a compact resonant structure is clearly seen in the effect on the bandwidth. This effect on the bandwidth is in general expected, as a resonant structure becomes effectively smaller with respect to the resonant wavelength. Such an effect has also been observed in the design of electrically small antennas patterned after Peano or Hilbert curve elements [13–15]. As compared to a Hilbert curve element, however, a Peano curve element of identical footprint and iteration order resonates at a much lower frequency, albeit at the expense of a much smaller bandwidth, due to the higher compression rate of the Peano curve algorithm. The interested reader is referred to [13–21] for further details.

14.2 HIGH-IMPEDANCE SURFACES MADE OF SPACE-FILLING CURVE INCLUSIONS

14.2.1 Peano Surface

To construct a surface of Peano curve inclusions, the Peano curve elements can be placed in a planar, 2D array as shown in Figure 14.7. To evaluate the scattering properties of this array, which has an infinite extent in its plane, a periodic MoM code was utilized [27]. In this case, each element was modeled as a thin metallic strip with a strip width of 0.5 mm. The footprint dimensions remain identical to the previous cases (30 × 30 mm). The Peano array was placed a distance (15 mm) above a conducting ground plane of infinite extent. The supporting dielectric substrate is considered air here, although any other dielectric can be considered in our analysis. Again, a time-harmonic, normally incident plane wave was utilized to excite this structure, and the reflection coefficient from the surface was numerically evaluated as a function of the frequency. Both polarizations were again studied.

Figure 14.8 shows the magnitude and phase of the reflection coefficient Γ versus frequency for the Peano surface comprised of an array of Peano curves of order 2 located at a height of 15 mm above the conducting ground plane and for a separation distance of 3.75 mm between each Peano curve inclusion within the array. This distance was chosen to be equal to the length of a single section of the curve itself (parameter d from Table 14.1). The structure was illuminated with a normally incident plane wave polarized in the x and y directions separately. Since a ground plane of infinite extent is present under the Peano surface, the magnitude

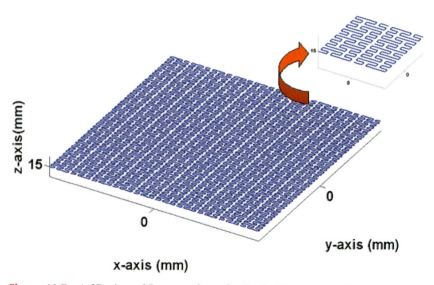

Figure 14.7 A 3D view of Peano surface of order 2 above a conducting ground plane. From [21]. Reproduced/modified by permission of American Geophysical Union.

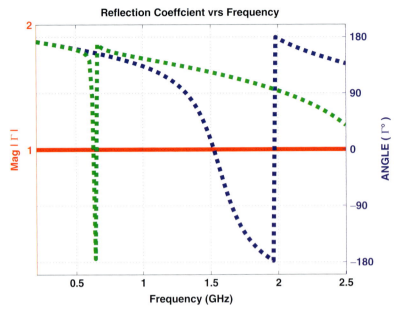

Figure 14.8 Magnitude and phase of reflection coefficient from Peano surface of order 2 above conducting ground plane for normally incident wave with polarizations in x (blue) and y (green) directions. From [21]. Reproduced/modified by permission of American Geophysical Union.

of the reflection coefficient is always unity since all of the incident energy is always reflected. It should be noted here that all metals present in our numerical simulations are considered lossless; therefore conductor losses are not taken into account. Also, since the substrate between the Peano surface and the conducting ground plane is assumed to be air, no dielectric losses are present in this structure.

In Figure 14.8, the phase of the reflection coefficient at 0.5 GHz is shown to be approximately $180°$. As the frequency increases, this phase passes through $0°$ degrees and goes toward $-180°$. At the frequency where the phase is $0°$ (1.53 GHz for the case of the x-polarized incident wave) the Peano surface above the ground plane achieves an overall reflection coefficient of $+1$ and, therefore, acts as a HIS (i.e., an artificial magnetic conductor). Far from this resonance denoted by F_{HIS}, this surface has an overall reflection coefficient of -1 and therefore behaves effectively as a traditional electric-conducting ground plane. It can be noted here that the footprints of the inclusions are approximately $0.063\lambda_{HIS}$ and $0.153\lambda_{HIS}$ at the corresponding frequency F_{HIS} in Figure 14.8. Moreover, the heights above the ground plane are approximately $0.031\lambda_{HIS}$ and $0.076\lambda_{HIS}$, respectively. Thus, both the inclusions and the heights above the substrate are considered to be electrically small at the resonant frequency for both polarizations [21].

We have also performed similar analyses for the surfaces made from the Peano curve inclusions of orders 1, 2, and 3 [21]. The corresponding resonant

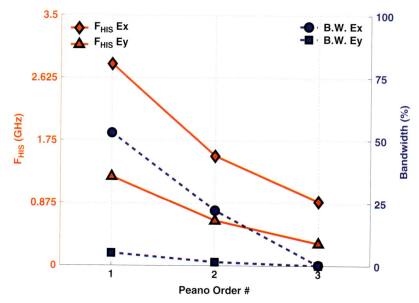

Figure 14.9 F_{HIS} and bandwidths of Peano surface above conducting ground plane for Peano curve inclusions of orders 1 to 3. From [21]. Reproduced/modified by permission of American Geophysical Union.

frequencies and relevant bandwidths are shown in Figure 14.9. The bandwidths here are defined by the frequency values where the reflection coefficient phase falls between ±90°.

14.2.1.1 Effects of Substrate Height and Interelement Spacing   In our recent work, we have numerically analyzed the roles of the substrate heights and the interelement spacings for the Peano high-impedance surface [21]. Here we briefly review these results. To investigate these effects, the height of the Peano surface composed of an array of Peano curves of order 2 with an interelement separation distance of 3.75 mm was varied from 5 to 15 mm in steps of 1 mm. The heights were chosen to ensure that the surface could still be considered to be electrically close to the ground plane for the smallest operating wavelength. Figure 14.10 shows the predicted resonant frequency F_{HIS} as well as the ±90° bandwidth as a function of the height of the surface above the conducting ground plane. It can be seen that for the x-polarized cases the resonant frequency decreases and the bandwidth increases as the height above the ground plane increases. Very little change is found for the y-polarized resonances. This result is due to the fact that since these resonances occur at lower frequencies, the relative change in the height with respect to the resonant wavelength is less pronounced.

A parametric study was also performed with respect to the interelement spacing between the Peano curve inclusions within the infinite 2D array. The separation distances were varied from 1 to 15 mm in steps of 2 mm [21]. The results of this study are shown in Figure 14.11, which shows the F_{HIS} and the

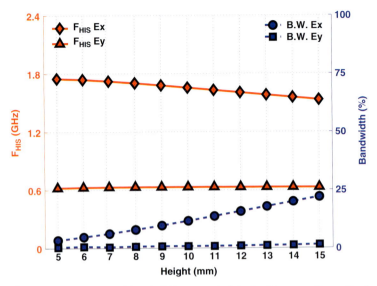

Figure 14.10 F_{HIS} and bandwidth versus height for Peano surface of order 2. From [21]. Reproduced/modified by permission of American Geophysical Union.

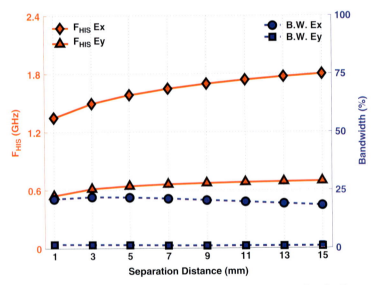

Figure 14.11 F_{HIS} and bandwidths versus interelement spacing for Peano surface of order 2. From [21]. Reproduced/modified by permission of American Geophysical Union.

$\pm 90°$ bandwidth as functions of this interelement distance. We can see that for the x-polarized cases the resonant frequency increases whereas the bandwidth decreases as the separation distance increases. This trend is also present for the y-polarized resonances, albeit less pronounced, due to the fact that these resonances occur at lower frequencies, and again the relative change in the separation distance with respect to the corresponding resonant wavelengths is less noticeable.

14.2.2 Hilbert Surface

We have also studied the case of Hilbert curve–based high-impedance surfaces [20], as shown in Figure 14.12, and have obtained analogous results. In this case, however, we have also studied the effect of the angle of incidence of the incoming wave on the frequency at which the surface behaves as a high-impedance surface, that is, as an artificial magnetic conductor. Figure 14.13 shows the phase of the reflection coefficient of a surface composed of an array of Hilbert curve elements of order 3 on a substrate over a conducting ground plane versus frequency for different angles of incidence. As in the Peano case, each curve is contained within a 30-mm × 30-mm footprint, and for this case the separation distance between adjacent elements within the array is 4.285 mm. The surface is again 15 mm above an infinitely conducting ground plane. The vertically polarized plane wave is arriving from the $\phi = 0°$ direction and θ is varied from $0°$ to $60°$ in $20°$ increments. Figure 14.13 clearly shows that as the angle of incidence becomes more oblique, the resonance frequency F_{HIS} increases while the $\pm 90°$ bandwidth decreases.

For comparison purposes, plots are provided for the resonant frequency F_{HIS} and the order number, height variation, and separation distance parametric studies for the surfaces comprised of Hilbert curve inclusions. These respective plots are shown in Figures 14.14 to 14.16.

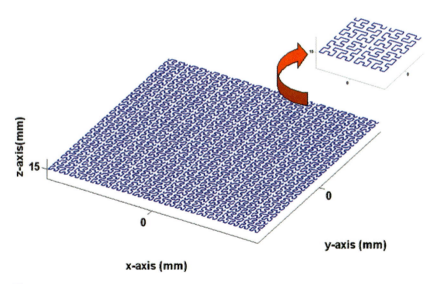

Figure 14.12   A 3D view of the Hilbert surface of order 3 above a conducting ground plane.

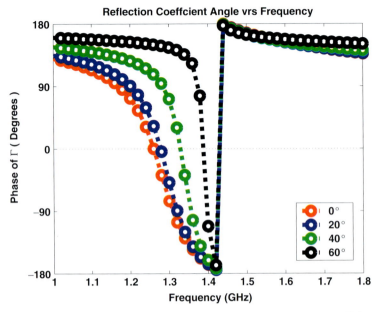

Figure 14.13 Zero-degree phase transition of phase of reflection coefficient for various angles of incidence of incoming plane wave.

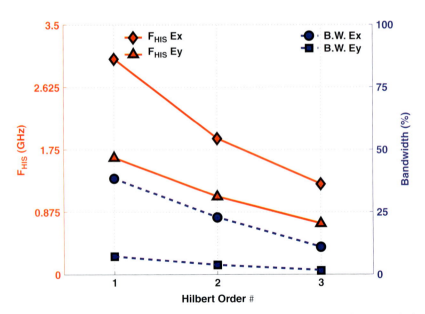

Figure 14.14 F_{HIS} and bandwidths of Hilbert surface above conducting ground plane for Hilbert curve inclusions of orders 1 to 3.

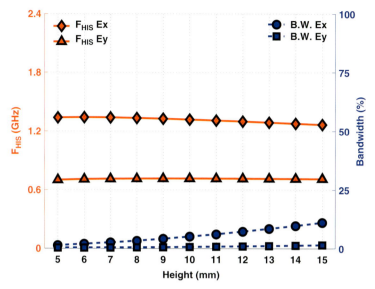

Figure 14.15 F_{HIS} and bandwidth versus height for Hilbert surface of order 3.

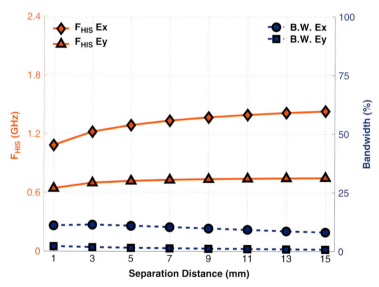

Figure 14.16 F_{HIS} and bandwidth versus interelement spacing for Hilbert surface of order 3.

14.2.2.1   *Hilbert Surface of Order 3: Experimental Results*   To experimentally measure and verify the high-impedance performance of the Hilbert surface of order 3, the geometry needed to be scaled such that the resonant frequency would fall within the frequency range of the rectangular waveguide which would be utilized for the measurement process. Also, substrate effects would additionally

need to be taken into account since the fabricated surface would require a nonair substrate. Based on the fabrication options available, the Hilbert surface was fabricated on a 1.575-mm FR-4 substrate with dielectric constant and loss tangent equal to 4.4 and 0.02, respectively. The Hilbert elements were scaled such that, in the presence of the substrate, the resonant frequency would fall within the range of a WR-430 waveguide (1.7 to 2.6 GHz). The final dimensions for the Hilbert curve elements was 12 mm × 12 mm and the surface, shown in Figure 14.17, was formed by 19 × 23 elements. The measurement setup, shown in Figure 14.18, consisted of the WR-430 waveguide placed directly above and in direct contact with a ground plane (shorted at the aperture). The Hilbert surface was cut to a 7 × 3 element array to fit within the waveguide and was placed inside the waveguide on a 5-mm-thick foam spacer above the ground plane. The distance of the Hilbert surface above the ground plane was therefore equal to 6.575 mm, that is, the thickness of the FR-4 substrate, plus the foam spacer thickness. We then collected the S_{11} values as a function of the frequency. The Agilent ENA5071B vector network analyzer was first calibrated up to the waveguide transition using a standard cable calibration package. A measurement was then performed with the waveguide "shorted" to the ground plane in order to provide a calibration to the waveguide aperture. Figure 14.19 shows the measured S_{11} (i.e., reflection coefficient) values as a function of the frequency. The losses of the substrate are evident in the decrease in the magnitude of S_{11} near the resonance of the surface. Figure 14.19 also shows the calculated S_{11} values as predicted by a finite-element method (FEM) simulation [28], which includes the finite waveguide, the finite Hilbert element array, and the substrate and conductivity effects. The slight shift in the frequency at which the phase goes through zero in the simulated results

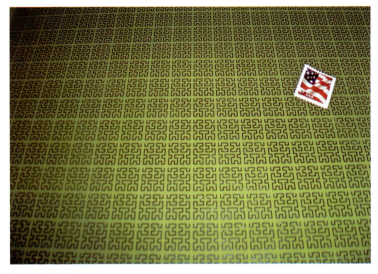

Figure 14.17 Constructed Hilbert surface of order 3, 19 × 23 elements on FR-4 substrate.

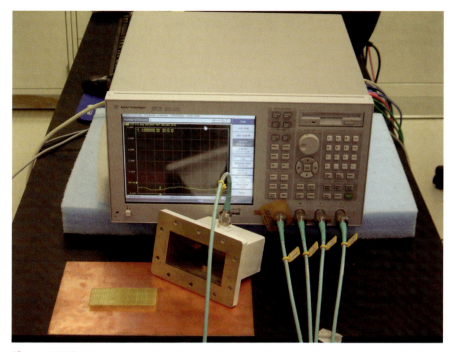

Figure 14.18   Vector network analyzer and waveguide setup for measurement of reflection coefficient from Hilbert surface of order 3.

may be due to the uncertainty in the value of the dielectric constant of the FR-4 substrate in the FEM modeling.

14.2.2.2   Use of Space-Filling Curve High-Impedance Surfaces for Thin Absorbing Screens   In the measurement process presented above, an interesting effect is noted involving the magnitude of the reflection coefficient. For a substrate with losses, the magnitude of the reflection coefficient is reduced around the surface resonance. To exploit this property and to explore the possibility of using space-filling curve surfaces in the design of thin absorbers [22], we analyzed, using the MoM-based IE3D simulation code [27], both the Peano surface of order 2 and the Hilbert surface of order 3 on a substrate with a thickness of 1.575 mm and various loss tangents ranging from 0.02 to 0.10. The surface was excited with a normally incident, x-polarized plane wave and the magnitude of the reflection coefficient was obtained versus frequency for each loss tangent considered. The results of this study are shown in Figure 14.20.

As can be seen, the reflection coefficient magnitude drops to a low value for a loss tangent of 0.08, approximately 4 times the loss tangent of the FR-4 substrate. This could offer some interesting applications in the area of absorbing materials and low observables [11]. In a conventional Salisbury screen [29], it is known that the resistive sheet is placed at a distance of $\lambda/4$ above the ground plane, where the electric field is maximum, thus maximizing losses. For the

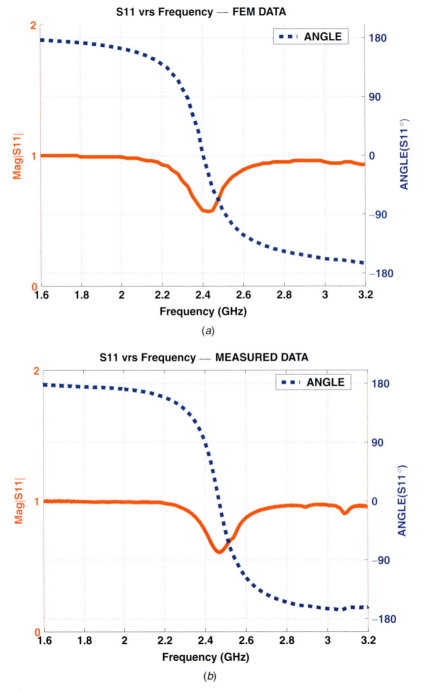

Figure 14.19 (*a*) Simulated and (*b*) measured reflection coefficient versus frequency for Hilbert surface of order 3, shown in Figure 14.17.

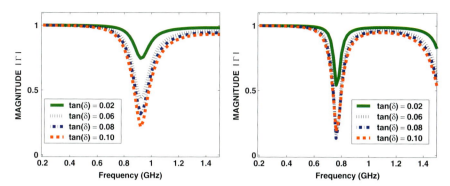

Figure 14.20 MoM-simulated magnitudes of reflection coefficients for Peano of order 2 (left) and Hilbert of order 3 (right) surfaces above substrate with various loss tangents. The design parameters are given in the text.

high-impedance ground plane, the electric field may be maximum at the surface and thus when this surface is close to the conductive ground and in the presence of a loss mechanism (i.e., a substrate with losses); this surface can offer high losses with a much smaller total thickness than that of the conventional Salisbury screen.

14.3 USE OF SPACE-FILLING CURVE HIGH-IMPEDANCE SURFACES IN ANTENNA APPLICATIONS

One important application of high-impedance surfaces is in the enhancement of the performance of low-profile antennas. To demonstrate the potential usefulness of the space-filling curve high-impedance surfaces, we studied the performance of an electrically small dipole antenna above the Hilbert surface of order 3 [6–9]. As considered in the previous section, a height of 15 mm was chosen as the height of the Hilbert surface from the ground plane. A center-fed small dipole of $\lambda/20$ length was then placed at an additional 15 mm above the Hilbert surface. The surface was modeled by a finite array of 11×11 Hilbert curves of order 3. The dipole, the Hilbert inclusions, and the ground plane were assumed to be made of copper with a conductivity of 5.813×10^7 S/m.

Figure 14.21 shows a comparison between the input impedance (real and imaginary) of the short dipole antenna above the ground plane both with and without the Hilbert surface. The results in the top row of Figure 14.21 were obtained using the MoM software package IE3D [27]. The multiresonant behavior of the radiation resistance near the resonance of the surface (i.e., in the region 1.4 to 1.7 GHz) was also independently confirmed using the NEC-4 based code, GNEC [26]. Those results are shown in the bottom row of Figure 14.21. By using four times the number of frequency points in the GNEC simulations than in the IE3D runs, other peaks, in addition to (and in between) those obtained by IE3D, were observed. Figure 14.22 shows the zoomed-in section of the GNEC

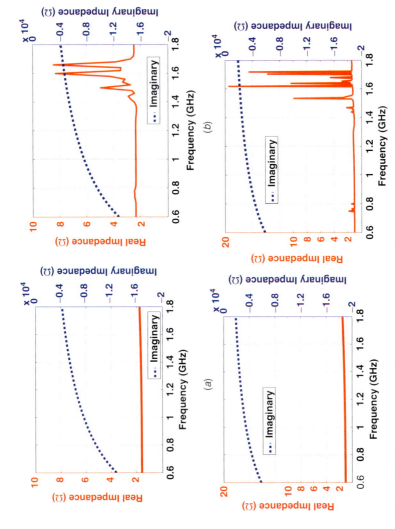

Figure 14.21 Comparison of input impedance of x-directed short dipole: (*a*) antenna alone, above ground plane; (*b*) antenna above Hilbert surface with ground plane using two independent simulation packages, IE3D (top row) and GNEC (bottom row).

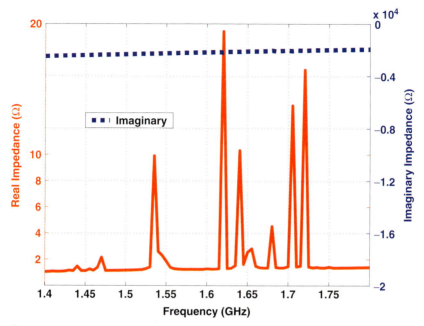

Figure 14.22 Zoomed-in section of NEC-4 simulation of input impedance of x-directed short dipole above Hilbert surface with ground plane.

simulated results in the region 1.4 to 1.8 GHz. The differences in the numerical values obtained using these two separate simulation codes can be attributed to the fact that the two codes, although both MoM, use different approaches in their implementations and also to the fact that, unlike the flat-strip assumption for the inclusions in the IE3D simulation, in the GNEC modeling of the Hilbert surface an equivalent wire radius $a = 0.25w$, where w is the strip width of the inclusions, was used. The nature of these resonances, which are mostly due to the various length scales of the Hilbert elements and finite-sized Hilbert surface, will be studied in more detail in the future. Similar behaviors have also been noted for the short dipole element above different orders of Hilbert curves for different polarizations with respect to the short dipole element and for short dipoles above Peano surfaces [10]. Figure 14.23 shows the corresponding results for the radiation efficiency and the directivity of the dipole obtained from the IE3D simulations. As expected, the radiation efficiency improves near the resonant frequency of the Hilbert surface. This can be a useful result when one attempts to improve the efficiency of electrically small radiating elements. Figure 14.23 also shows the maximum directivity. It becomes clear that the surface itself, consisting of many small resonant inclusions, is radiating and thus affecting the pattern of the nearby dipole antenna. The relatively drastic changes in the efficiency and directivity as a function of the frequency illustrate the potential challenges in the design and application of these high-impedance surfaces in an antenna design. When a dipole is oriented along the y axis, similar characteristics are shown at

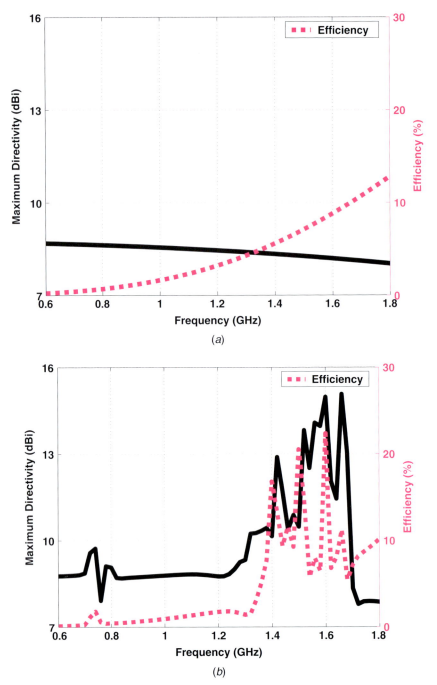

Figure 14.23 Comparison of directivity and efficiency of x-directed short dipole using IE3D numerical simulations: (a) antenna alone, above a ground plane; (b) antenna above Hilbert surface with ground plane.

lower frequencies which correspond to the E_y resonance of the surface [7]. We have also studied the performance of other antennas in the presence of Hilbert surfaces [6–9].

14.4 SPACE-FILLING CURVE ELEMENTS AS INCLUSIONS IN DNG BULK MEDIA

With recent interest in metamaterials with negative permittivity and permeability [30–40], we have explored, using numerical tools, the electromagnetic wave interactions with metallic inclusions formed using space-filling curve algorithms. When they are embedded in a host medium, these space-filling curves may lead to composite media with negative material parameters [23–25]. We have analyzed the electric and magnetic polarizability tensors for these inclusions, and we have then applied the mixing formulation (e.g., Maxwell–Garnett formula) to obtain approximate values for the effective permittivity and permeability of the bulk media. In this section, we give a quick review of this topic.

The electric and magnetic dipole moments of the space-filling curve inclusions were numerically calculated using the current distributions given by a MoM-based code. The space-filling curve geometry was modeled as a thin-wire structure of radius 0.125 mm. Figure 14.24 shows the electric and magnetic dipole moments versus frequency for the Hilbert curve of order 3 for the y-polarized electric field (E_y). It can be seen that in the presence of a y-directed electric field an electric dipole moment is present and at that same frequency a magnetic dipole moment is also present due to the looping nature of the curve. It is interesting to note that for the Peano curve these resonances occur at different frequencies due to the particular symmetries of the Peano curve.

A bulk medium can be conceptually constructed by embedding many identical space-filling curve inclusions within a host medium. The Maxwell–Garnett mixing formula, as one of the commonly used mixing rules, evaluates the effective relative permittivity and permeability. Figure 14.25 shows the effective permittivity and permeability for the Hilbert curve of order 3 and air as the host medium as evaluated using the Maxwell–Garnett mixing formula on the polarizability tensors of the elements. Figure 14.25 reveals that for a specific frequency range the bulk media formed with the Hilbert curve inclusions may have simultaneously both negative effective permittivity and permeability, and thus it can be a DNG medium. From Figure 14.25, it also appears that the bandwidth for the negative effective permittivity is different from that for the negative effective permeability. It is interesting to note that a bulk medium made of Peano curve inclusions may also display the negative permittivity and negative permeability properties. However, as we have found recently, these effective negative properties occurred at different frequencies and thus do not exhibit DNG properties. Rather, they exhibit single-negative characteristics, which can offer potential applications as multifunction media [41]. These properties are currently under study.

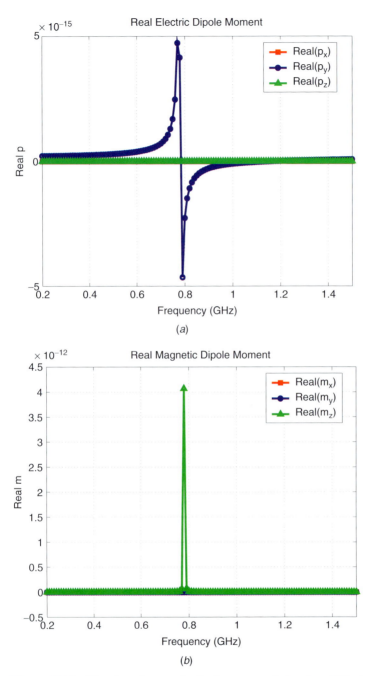

Figure 14.24 Electric and magnetic dipole moments induced on Hilbert curve of order 3 for y-polarized electric field evaluated using MoM-based numerical code.

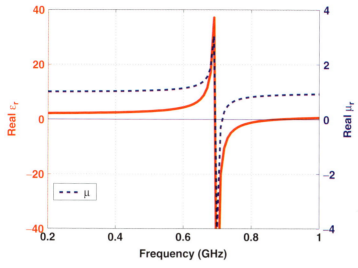

Figure 14.25 Effective permittivity and permeability of bulk media formed by Hilbert curve inclusions of order 3.

14.5 CONCLUSIONS

Peano and Hilbert space-filling curve inclusions can offer many salient characteristics for electromagnetic applications. One of the most interesting of these characteristics is the ability for these curves to resonate at frequencies where the electrical footprint, that is, the area the curve fills, is electrically very small. This allows space-filling curves to be excellent candidates in the formation of surfaces and bulk media where electrically small inclusions, which can be engineered for specific electromagnetic properties, are desired. Since the footprint of these inclusions can be considered electrically small while its total wire length with respect to the operational wavelength is not small, these metamaterials can be studied with respect to their equivalent bulk properties (effective medium), thus simplifying otherwise very complicated interactions. This effect was highlighted in the presented study of an electrically small dipole antenna above such surfaces. Since it is desirable to have small, low-profile radiating systems, it becomes apparent that any surface near the radiating element, which contains such inclusions, must also exhibit electrically small characteristics. Moreover, as presented above, it is the relatively small size of these curves which allows one to use the standard mixing formulas to evaluate the effective parameters of bulk media comprised of them. The cost paid for this "electrical compression" is evident in the relatively narrow bandwidths associated with these inclusions. We are currently looking for and studying potential techniques to address this particular bandwidth issue as well as the issues involving the dependence of the response of these curves on the polarization [42].

REFERENCES

1. D. Sievenpiper, L. Zhang, R. F. Jimenez Broas, N. G. Alexopolous, and E. Yablonovitch, "High-impedance electromagnetic surfaces with a forbidden frequency band," *IEEE Trans. Microwave Theory Tech.*, vol. 47, no. 11, pp. 2059–2074, Nov. 1999.

2. F. Yang and Y. Rahmat-Samii, "Reflection phase characterization of an electromagnetic band-gap (EBG) surface," in *2002 IEEE Antennas and Propagation Society International Symposium Digest*, San Antonio, TX, vol. 3, pp. 744–747, June 16–21, 2002.

3. Y. Zhang, J. von Hagen, and W. Wiesbeck, "Patch array as artificial magnetic conductors for antenna gain improvement," *Microwave Opt. Technol. Lett.*, vol. 35, no. 3, pp. 172–175, Nov. 2002.

4. R. Gonzalo, P. de Maagt, and M. Sorolla, "Enhanced patch-antenna performance by suppressing surface waves using photonic-bandgap substrates," *IEEE Trans. Microwave Theory Tech.*, vol. 47, pp. 2131–2138, Nov. 1999.

5. R. C. Hansen, "Effects of a high-impedance screen on a dipole antenna," *IEEE Antennas Wireless Propag. Lett.*, vol. 1, pp. 46–49, 2002.

6. J. McVay, N. Engheta, and A. Hoorfar, "Dipole radiation near a metamaterial surface made of Hilbert-curve inclusions," in *Bianisotropics 2002 Digest of Abstracts*, Marrakech, Morocco, p. 38, May 8–11, 2002.

7. J. McVay, A. Hoorfar, and N. Engheta, "Applications of Hilbert artificial magnetic conductor ground planes to performance enhancement of electrically small antennas," in *2003 IEEE Sarnoff Symposium Digest*, pp. 297–300, March 11–12, 2003.

8. J. McVay, A. Hoorfar, and N. Engheta, "Radiation characteristics of microstrip dipole antennas over a high-impedance metamaterial surface made of Hilbert inclusions," in *2003 IEEE MTT Int. Microwave Symp. Digest*, Philadelphia, PA, pp. 587–590, June 8–13. 2003.

9. J. McVay, A. Hoorfar, and N. Engheta, "Patch antennas above Hilbert high-impedance surfaces," in *2003 IEEE Antennas and Propagation Society International Symposium and USNC-CNC-URSI North American Radio Science Meeting URSI Digest of Abstracts*, Columbus OH, p. 164, June 22–27, 2003.

10. J. McVay, A. Hoorfar, and N. Engheta, "Small dipole antenna near Peano high-impedance surfaces," in *2004 IEEE Antennas Prop. Soc. Int. Symp. Digest*, Monterey, CA, vol. 1, pp. 305–308, June 20–26, 2004.

11. N. Engheta, "Thin absorbing screens using metamaterial surfaces," in *2002 IEEE AP-S Int. Symp. Digest*, San Antonio, TX, vol. 2, pp. 392–395, June 16–21, 2002.

12. H. Sagan, *Space-Filling Curves*, Springer-Verlag, New York, 1994.

13. J. Zhu, A. Hoorfar, and N. Engheta, "Peano antennas," *IEEE Antennas Wireless Propag. Lett.*, vol. 3, pp. 71–74, 2004.

14. J. Zhu, A. Hoorfar, and N. Engheta, "Bandwidth, cross polarization, and feed-point characteristics of matched Hilbert antennas," *IEEE Antennas Wireless Propag. Lett.*, vol. 2, pp. 2–5, 2003.

15. J. Zhu, A. Hoorfar, and N. Engheta, "Feed point effects in Hilbert-curve antennas," in *IEEE Antennas and Propagation Society International Symposium and USNC/URSI National Radio Science Meeting URSI Digest of Abstracts*, San Antonio, TX, p. 373, June 16–21, 2002.

16. D. H. Werner and S. Ganguly, "An overview of fractal antenna engineering research," *IEEE Antennas Propag. Mag.*, vol. 45, no. 1, pp. 38–56, Feb. 2003.

17. K. J. Vinoy, K. A. Jose, V. K. Varadan, and V. V. Varadan, "Hilbert curve fractal antenna: A small resonant antenna for VHF/UHF applications," *Microwave Opt. Tech. Lett.*, vol. 29, no. 4, pp. 215–219, May 2001.

18. S. R. Best, "A comparison of the performance properties of the Hilbert curve fractal and meander line monopole antennas," *Microwave Opt. Tech. Lett.*, vol. 35, no. 4, pp. 258–262, Nov. 2002.

19. E. A. Parker and A. N. A. El Sheikh, "Convoluted array elements and reduced size unit cells for frequency-selective surfaces," *Proc. IEE Part H: Microwaves Antennas Propag.*, vol. 138, pp. 19–22, Feb. 1991.

20. J. McVay, N. Engheta, and A. Hoorfar, "High impedance metamaterial surfaces using Hilbert-curve inclusions," *IEEE Microwave Wireless Components Lett.*, vol. 14, no. 3, Mar. 2004.

21. J. McVay, A. Hoorfar, and N. Engheta, "Peano high impedance surfaces," *Radio Science*, vol. 40, RS6S03, 2005.

22. J. McVay, A. Hoorfar, and N. Engheta, "Thin absorbers using space-filling curve high impedance surfaces," in *2005 IEEE Antennas and Propagation Society International Symposium Digest*, Washington, DC, July 3–8, 2005, vol. 2A, pp. 22–25.

23. J. McVay, N. Engheta, and A. Hoorfar, "Space-filling-curve elements as possible inclusions for double-negative metamaterials," in *2004 IEEE Antennas and Propagation Society (AP-S) International Symposium and USNC-URSI National Radio Science Meeting URSI Digest of Abstracts*, Monterey, CA, p. 136, June 20–26, 2004.

24. J. McVay, N. Engheta, and A. Hoorfar, "Numerical study and parameter estimation for double-negative metamaterials with Hilbert-curve inclusions," in *2005 IEEE Antennas and Propagation Society (AP-S) International Symposium and USNC-URSI National Radio Science Meeting Digest*, Washington, DC, vol. 2B, pp. 328–331, July 3–8, 2005.

25. J. McVay, N. Engheta, and A. Hoorfar, "Numerical study of phase variation through double-negative and single-negative media formed by space-filling curve inclusions," paper presented at the *International Conference in Electromagnetics and Advance Applications (ICEAA'05)*, Torino, Italy, Sept. 12–15, 2005.

26. GNEC, version 1.4, Nittany Scientific, Riverton, UT.

27. IE3D, version 9.3, Zeland Software, Fremont, CA.

28. High Frequency Structure Simulator, version 9.2, Ansoft Corporation, Pittsburgh, PA.

29. R. L. Fante and M. T. McCormack, "Reflection properties of the Salisbury screen," *IEEE Trans. Antennas Propag.*, vol. AP-36, pp. 1443–1454, Oct. 1988.

30. D. R. Smith, W. J. Padilla, D. C. Vier, S. C. Nemat-Nasser, and S. Schultz, "Composite medium with simultaneously negative permeability and permittivity," *Phys. Rev. Lett.*, vol. 84, no. 18, pp. 4184–4187, May 2000.

31. D. R. Smith and N. Kroll, "Negative refractive index in left-handed materials," *Phys. Rev. Lett.*, vol. 85, no. 14, pp. 2933–2936, Oct. 2000.

32. J. B. Pendry, "Negative refraction makes a perfect lens," *Phys. Rev. Lett.*, vol. 85, no. 18, pp. 3966–3969, Oct. 2000.

33. R. A. Shelby, D. R. Smith, S. C. Nemat-Nasser, and S. Schultz, "Microwave transmission through a two-dimensional, isotropic, left-handed metamaterial," *Appl. Phys. Lett.*, vol. 78, no. 4, pp. 489–491, Jan. 2001.

34. R. A. Shelby, D. R. Smith, and S. Schultz, "Experimental verification of a negative index of refraction," *Science*, vol. 292, no. 5514, pp. 77–79, Apr. 2001.

35. V. G. Veselago, "The electrodynamics of substances with simultaneously negative values of ε and μ," *Sov. Phys. Usp.*, vol. 10, no. 4, pp. 509–514, 1968.

36. M. M. I. Saadoun and N. Engheta, "Theoretical study of electromagnetic properties of non-local omega media," in *Progress in Electromagnetic Research (PIER) Monograph series*, vol. 9, EMW Publishing, Chapter 15, Editor Alain Priou, pp. 351–397, 1994.

37. I. V. Lindell, S. A. Tretyakov, K. I. Nikoskinen, and S. Ilvonen, "BW media—Media with negative parameters, capable of supporting backward waves," Report No. 366, *Electromagnetic Laboratory report series*, Helsinki University of Technology, Apr. 2001.

38. R. W. Ziolkowski, "Superluminal transmission of information through an electromagnetic metamaterials," *Phys. Rev. E*, vol. 63, 046604, Apr. 2001.

39. R. W. Ziolkowski and E. Heyman, "Wave propagation in media having negative permittivity and permeability," *Phys. Rev. E*, vol. 64, 056625, Oct. 2001.

40. N. Engheta, "An idea for thin subwavelength cavity resonators using metamaterials with negative permittivity and permeability," *IEEE Antennas Wireless Propag. Lett.*, vol. 1, no. 1, pp. 10–13, 2002.

41. A. Alù and N. Engheta, "Pairing an epsilon-negative slab with a mu-negative slab: Resonance, anomalous tunneling and transparency," *IEEE Trans. Antennas Prop.*, Special Issue on Metamaterials, vol. 51, no. 10, pp. 2558–2571, Oct. 2003.

42. J. McVay, A. Hoorfar, and N. Engheta, "Bandwidth enhancement and polarization-dependence reduction for space-filling curve artificial magnetic conductors," presented at the *USNC-URSI National Radio Science Meeting*, Washington, DC, July 3–8, 2005.

INDEX

Note: Page numbers followed by t and f indicate tables and figures, respectively.

Metamaterials: Physics and Engineering Explorations, Edited by N. Engheta and R. W. Ziolkowski
Copyright © 2006 the Institute of Electrical and Electronics Engineers, Inc.